T0314091

Engineering Project Management

Engineering Project Management

Neil G. Siegel
The IBM Professor of Engineering Management
University of Southern California
Los Angeles, US

This edition first published 2019

Registered Offices
John Wiley & Sons, Inc., 111 River Street, Hoboken, NJ 07030, USA
John Wiley & Sons Ltd, The Atrium, Southern Gate, Chichester, West Sussex, PO19 8SQ, UK

Editorial Office
The Atrium, Southern Gate, Chichester, West Sussex, PO19 8SQ, UK

For details of our global editorial offices, customer services, and more information about Wiley products visit us at www.wiley.com.

Wiley also publishes its books in a variety of electronic formats and by print-on-demand. Some content that appears in standard print versions of this book may not be available in other formats.

Library of Congress Cataloging-in-Publication Data

Names: Siegel, Neil G., author.
Title: Engineering Project Management / professor Neil G. Siegel, Ph.D., the IBM Professor of Engineering Management, University of Southern California, LosAngeles, US.
Description: Hoboken, NJ, USA: John Wiley & Sons, Inc., [2019] | Includes bibliographical references and index. |
Identifiers: LCCN 2019016035 (print) | LCCN 2019018487 (ebook) | ISBN 9781119525783 (Adobe PDF) | ISBN 9781119525790 (ePub) | ISBN 9781119525769 (hardback)
Subjects: LCSH: Engineering–Management. | Project management.
Classification: LCC TA190 (ebook) | LCC TA190 .S586 2019 (print) | DDC 620.0068/8–dc23
LC record available at https://lccn.loc.gov/2019016035

Cover Design: Wiley
Cover Image: © Westend61/Getty Images

Set in 10/12pt WarnockPro by SPi Global, Chennai, India

Printed in Great Britain by TJ International Ltd, Padstow, Cornwall

10 9 8 7 6 5 4 3 2 1

To my wife, Robyn.

In paraphrase of Maulana:

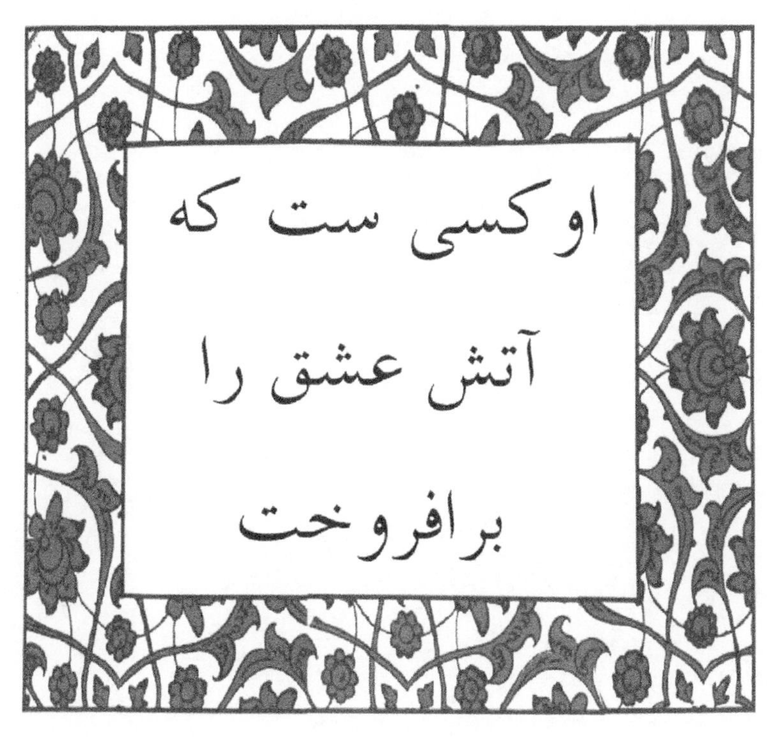

Contents

About the Author

Neil Siegel spent many years successfully managing engineering projects, big and small. He managed successful projects in aerospace and defense, civil government (at the federal, state, and city level), health-care, the steel industry, the energy industry, higher education, the entertainment industry, and others. His inventions are also widely used in the consumer electronics industry.

He has won many honors and awards for his work as a manager of engineering projects, including election to the U.S. National Academy of Engineering, the U.S. National Academy of Inventors, the IEEE Simon Ramo Medal for Systems Engineering, and many others. In addition to these personal awards, projects that he led have received honors and awards, such as the inaugural Crosstalk Award for the best-managed software project across the entire U.S. Government.

He earned a Ph.D. in systems engineering at the University of Southern California, where his advisor was the noted computer scientist and systems engineering Barry Boehm.

He retired after 18 years as a vice-president of the Northrop Grumman Corporation at the end of 2015, and now teaches systems engineering and engineering project management at the University of Southern California.

More information is available at https://neilsiegel.usc.edu.

Acknowledgments

I started working on actual engineering projects in 1976, and did so full time until the end of 2015, building systems for customers in aerospace and defense, but also civil government (at the federal, state, and city level), health-care, the steel industry, the energy industry, higher education, entertainment, and others. My inventions are widely used in the consumer electronics industry (I sometimes say "used in a billion devices worldwide," but a rigorous count is beyond my means). I also had the opportunity to build such systems for customers in several countries outside of the United States, to travel extensively in those countries, and even to live overseas for a couple of years on such an assignment. I draw upon those experiences in writing this book.

Over that period of time when I worked on these engineering projects, I benefited from many people, a few of whom I will name herein.

First, I want to recognize the many people who taught me how to be the manager of an engineering project. I will let Peter Karacsony, Dr. Joe Mason, and Jack Distaso represent the large number of people who helped me along this wonderful (but demanding) life path.

Next, my many customers, most of whom truly believed in being effective partners in the difficult enterprise of building complex engineered systems. LTG (ret) William Campbell will represent this group of great people.

My childhood friend Dr. Mitch Allen is an archeologist who spent much of his professional career as an academic editor and publisher. When I conceived the idea of writing this book, Mitch (despite being retired from the publishing business) taught me everything that I needed to know in order to write a book proposal, and actually found me acquisition editors by name to whom I could submit my proposal.

I also wish to thank my collaborators at Wiley, my publisher: Eric Willner and his staff.

I wish to thank my former company, TRW/Northrop Grumman (TRW was acquired by Northrop Grumman in 2002). In addition to offering me an amazing career – with the opportunity truly to save lives, improve the defense of the United States and our allies, aid humanity, and enjoy continuous intellectual stimulation – they kindly allowed me to create a set of teaching and research materials that drew upon data and lessons learned from real projects, and allowed me to release that information about those real engineering project experiences to my students and the public. This book could not exist in this form without my ability to tell those stories.

Two real engineering projects were key learning experiences during my career, and are the source of some of the lessons learned and stories described herein:

- *The Forward-Area Air Defense Command-Control-and-Intelligence System.* Peter Karacsony was the manager of this project; I was the chief engineer.
- *Force XXI Battle Command Brigade-and-Below* (also known as the Blue-Force Tracker, the Appliqué, or the Digitized Battlefield). I was the project manager; Jack Distaso was my direct supervisor during this time. LTC (ret) William Campbell was the senior customer (called in Army nomenclature the *program executive officer*) for this project, and for many more projects that I had the opportunity to build for the US Army, as well.

I was always blessed with an amazing team of engineers and other professionals when I set out to manage an engineering project; several are named at appropriate places within the stories told in this book. I owe a great deal to those colleagues at TRW and Northrop Grumman, and to those colleagues at various other companies who worked with me as subcontractors on these great projects.

I used an early draft of the book with my undergraduate engineering students at the University of Southern California, and am grateful for the feedback that they provided. Some of them were willing to be recognized by name, and so I would like to acknowledge (in alphabetical order) Terry Lam, Aaron Lew, Seema Snitkovsky, Sara Stevens, Kathleen Sullivan, and Tal Volk.

Lastly (but always first) my wife, Dr. Robyn Friend, who – in addition to everything else – was always the first to read every chapter, and provided useful and thoughtful feedback.

About the Companion Website

This book is accompanied by a companion website:

www.wiley.com/go/siegel/engineering_project_management

The website includes:

- Briefing charts for lectures
- Reviews in advance of examinations
- Solution manuals

Scan this QR code to visit the companion website.

Introduction

I spent many years as a practicing engineer, including many assignments as the manager of engineering projects. The projects that I managed ranged from very small to very large; as I got older and more experienced, the projects that I led tended to get larger and more complex. Our teams were in general successful in delivering systems and products that our customers found useful, and at times constituted revolutionary improvements over previous capabilities. I have been credited with saving lives, money, and time, all on a large scale.

As I progressed from project to project, I drew certain conclusions about managing such engineering projects, and developed my own techniques and methods. I took courses offered by my company in project management, and read books on the subject. I found a significant difference between what I experienced as a project manager, and what the books had to say. What I did as a project manager, what I spent my time doing and worrying about, seemed very different from what the books said.

I also, learned through my reading and research, that the overall track record of success in engineering projects is not very good. A shockingly large portion of the engineering projects that are started turn into failures.

Recently, I elected to retire from full-time work as a practicing engineer and engineering project manager, and took an appointment at a university as a full-time professor of engineering in a department of systems engineering. Systems engineering is my love and my passion, and I wanted to create courses that taught systems engineering my way, and to continue my researches into how to do systems engineering better than we do it now.

After I had a good start in creating my systems engineering courses, the university asked me to teach a course in engineering project management. It had never occurred to me to want to teach that; I was completely focused on systems engineering. Of course, systems engineering and engineering project management are, in my mind, very closely related. When I was the manager of an engineering project, I employed what I characterized as the "systems engineering mindset" in order to plan and manage the project.

I discovered that I had quite a lot to say about engineering project management, and enjoyed teaching the course quite a bit. The students seemed to find my approach – grounded in actual experience as an engineering project manager, and full of examples from actual projects – both informative and enjoyable.

And, of course, I had to select a textbook for my engineering project management course. I purchased and read several of them. They were all as I remembered them: they talked a lot about stuff that I didn't actually do, and said nothing about many things that I had found were vital. After I had taught the course a couple of times, I realized that perhaps there was room in the world for a textbook that would describe my approach to engineering project management, one that would teach the activities that I found myself actually spending time on and worrying about when I was the manager of large, complex engineering projects. This book is the result.

Concept of the Book

The book is "sized" for a one-semester course, but could easily be adapted to either one or two quarters of instruction.

The book is intended to serve both upper-division undergraduates (e.g. juniors and seniors) and students who are just starting graduate studies. I use essentially the same course materials for both of these student audiences; the graduate students will get additional readings and a lot more homework (in ways which I will describe within the text).

Engineering is all about achieving practical results, and in that spirit of hands-on, practical results, my class does *not* consist entirely of lectures. For most of the weeks within the course, I use at least one class session for what I call a "facilitated lab session." During these facilitated lab sessions, I take only a short amount of time to explain a technique, and then provide the students with a problem which they work on (in teams) during the class session. They are allowed to consult with each other, to look at their class notes and books, to ask me questions, to show me their in-progress work, and get feedback on the spot. What I find is that they seldom actually finish the work during this class session (and therefore they still have to do that work as homework), but by the end of each facilitated lab session, they understand the method properly, and are able to do the problem correctly.

So the course – and therefore this book – is laid out as two parallel, week-by-week tracks: one a progressive set of lectures, and one a progressive set of practical techniques and team exercises. As the students' knowledge and practical techniques are built up and mastered, this leads to a set of gradable course materials, and an integrated set of learning (via a combination of reading, lectures, hands-on facilitated sessions, individual homework assignments, and team homework assignments) for each student. I supplement these gradable materials with a mid-term and a final examination, in order to develop grades for each student.

At the ends of the chapters that correspond to the weeks which have these facilitated lab sessions, therefore, there will be a subsection that addresses the topic, technique, and assignment for that week's lab session.

Many of the artifacts created by the students could be viewed as sections of a project plan. The graduate students get assigned to create more elements of a project plan than the undergraduates, commensurate with their extra year or two of previous instruction, and most especially the fact that a large portion of engineering graduate students who are taking a course in engineering project management have a few years of actual work experience. That experience is likely to have been focused in just a couple of small teams buried within large projects at their companies, but they are at least aware of the larger world of project management, and are motivated to want to learn quickly about that larger world.

The organization of the book is described in the following table:

Chapter/ week	Lecture	In-class facilitated workshop
1	**The role and the challenge**. What is engineering project management? Why do we teach engineering project management? Do engineering projects matter to society? Do projects matter to business? What is a "project?" What is an "engineering project?" What is a "project manager?" In this chapter, we discuss all of these questions and also provide you basic information about the role of engineering project manager and the opportunity that this role represents for you.	Team exercise: the value of engineering projects to society
2	**Performing engineering on projects (part I)**. How do we do engineering on projects? Engineering projects are different from other projects, so learning to be an effective manager of an engineering project starts by understanding how we do engineering on projects. We accomplish this engineering through the engineering life-cycle. In this chapter, I summarize key aspects of how we do the initial stages of the engineering life-cycle, which are called "requirements analysis" and "design."	This week is all lectures
3	**Performing engineering on projects (part II)**. In this chapter, I continue our summary of the key aspects of how we do engineering on projects, covering the remaining stages of the engineering life-cycle, from "implementation" all the way through to "phase-out and disposal."	This week is all lectures
4	**Understanding your users and your other stakeholders**. We have two coordinate systems of value and engineering the user experience. Engineering projects often create products and/or services that never existed before. Under these circumstances, it is easy to lose sight of what aspects of the new item are essential, and which are less so. We solve this dilemma by rigorous and continuous focus on our eventual users and customers. What are they trying to accomplish? How do they do it now? What are the shortfalls? What are their needs and desires? At the same time, our degrees of engineering freedom are usually entirely within the technical domain: choices about materials, parts, algorithms, mechanical structures, and so forth. In this chapter, you will learn how to understand your users, how to relate that understanding of your user to the engineering choices that are your degrees of design freedom. We then extend this focus on our users to all the "stakeholders" of our project. We end the chapter with a discussion of how to use good engineering and good management to achieve a compelling and effective experience for your users and your customers when they operate your system, through what we call the user experience.	Team exercise: the customer's coordinate system of value, the engineer's coordinate system of value, relating them, use of operational performance measures (OPMs) and technical performance measures (TPMs)

(Continued)

Chapter/ week	Lecture	In-class facilitated workshop
5	**How do engineering projects get created**. Creating winning proposals. When we get our first job, we are likely to be assigned to work on an existing engineering project; we are not troubled by the question of how this engineering project came into existence. Who created it? Why? How is it being paid for? How did it come to pass that it is our company that is doing the work? But as we progress in our careers, we come to realize that these aspects matter a lot. In fact, understanding them, so that you can help your company win new projects, is an important path for you to achieve attractive assignments and career success. In this chapter, I will therefore teach you the basics about winning engineering projects for your company, which centers around something called the "proposal."	Team exercise: proposals, the Heilmeier questions, win themes
6	**Organizing and planning**. Congratulations! You have been named the manager of our new engineering project. What do you do next? You decompose the work entailed in performing the project into smaller pieces, using a hierarchy. When this is done in a particular fashion, it is called a work-breakdown structure. Projects all over the world are managed using a work-breakdown structure. In this chapter, I both teach you the basics of creating and using a work-breakdown structure and show you how to do it effectively within the specific context of engineering projects. Then, we move on to discuss the organizational structure of your project, and finally, I show you how to use your work-breakdown structure as the basis to create a complete project plan for your engineering project.	Team exercises: the work-breakdown structure and its essential components
7	**Creating credible predictions for schedule and cost**. In Chapters 2 and 3, I provided you with insight about some of the key factors regarding how we do engineering on projects. We will now use that knowledge as I start discussing the processes that we use for performing actual project management on our engineering project. In this chapter, I focus on the activity network, which allows us to make credible predictions regarding the schedule (that is, how long it will take us to do all of the work entailed in our project). We will also see how this same activity network is an essential first step toward estimating how much our project will cost. Predicting schedule and cost in a credible fashion are among the basic expectations for a good engineering project manager.	Team exercise: the *activity network* as the primary schedule management artifact; exercises in creating an activity network, including three-point (optimistic, nominal, and pessimistic) durations for tasks in an activity network. Characterization of the magnitude of the impact the use of such three-point estimates has on the outcome, in contrast to the outcome using single-point estimates

Chapter/ week	Lecture	In-class facilitated workshop
8	**Drawing valid conclusions from numbers**. Invalid data and poor statistical methods can lead to bad decisions! There are many ways for an engineering project manager to make mistakes, but one of the most common and most insidious is through making logical and procedural mistakes that cause us to draw erroneous and invalid conclusions from quantifiable data, and as a result make poor data-based decisions. As engineers, we measure things, and then we often make decisions based on those numbers. For example, we predict when our project will be done, how much it will cost when it is done, and what the technical capabilities of our product will be (e.g. how far will our new airplane be able to fly safely without refueling), and use those data to make decisions for our project. Whenever we use numbers, however, there is a chance or error: our measurements always involve uncertainties, a particular assumption is only true under certain circumstances, we may not have collected appropriate samples, and so forth. In this chapter, I show you the most common ways that we undermine our own credibility through poor data collection, errors in logic, procedural mistakes, weak statistics, and other errors, and how you can instead use valid methods and strong statistics to create credible predictions for all of our project management roles and measures.	No facilitated lab session this week; this is often the time for a mid-term examination Optional team exercise: building and using a control chart; the five tests that separate signal from noise
9	**Risk and opportunity management**. Some things can (and probably will) go wrong on your project; how can you still get the job done? Every human endeavor entails uncertainty; things may not go as planned. In light of that uncertainty, how do you get the project done within the promised parameters, such as schedule, cost, technical capability, and so forth? In this chapter, I show you how to combine good engineering and good statistics in a manner that allows you to cope with these uncertainties.	Team exercise: risk management
10	**Monitoring the progress of your project (part I)**. Once you have created a great project plan, you can start work on your engineering project. You now need a set of tools and mechanisms to allow you to monitor what is going on, and to determine if progress is as expected (or not!). In this chapter, I show you how to assess progress on schedule and cost. I also introduce you to the principal financial measures that your company will use to measure the business performance of your project.	Team exercise: earned value

(Continued)

Chapter/ week	Lecture	In-class facilitated workshop
11	**Monitoring the progress of your project (part II)**. There is much more to monitor on your engineering project than just schedule and cost (although sometimes it will seem like those are the only things your boss cares about!). In this chapter, I teach you a comprehensive method for monitoring schedule, cost, technical capability, and risk ... and to make sure that all of these parts fit together correctly. Since on most projects your contract will require you to assess progress at least once a month, I used to call this method the monthly management rhythm, but have (reluctantly) switched to the more general phrase "periodic management rhythm."	Team exercise: the monthly management cycle
12	**Four special topics** (a) Congratulations! You have won a competition for a new engineering project. How do you go about getting this new project started? Starting a new project turns out to be a special problem that requires a special set of skills, which I will teach you in this chapter. (b) Most systems today have large amounts of software, which provides particular benefits, but creates many particular liabilities and risks too. I show you how to deal with projects that involve large amounts of software development. (c) People will come and ask if you are going to use agile software development methods on your project. I show you some of the key differences between agile and conventional development methodologies, show you how to decide if your project is suitable for agile methods, and show you how to cope with some of the most common risks and weaknesses of the agile methods. (d) Projects are defined as temporary activities, so every project ends. I tell you what you need to do to end a project.	Team exercise: project start-up
13	**The social aspects of engineering project management**. It is a cute cliché that engineers lack social and interpersonal skills. In actual fact, being a good engineering project manager is a highly social activity, and you will find that exercising effective interpersonal skills is in actual fact an important part of your success as an engineering project manager. In this chapter, I teach you what you need to know: aligning and building an effective team, motivating and inspiring people, managing conflict, and other topics. The good news is that we engineers can actually learn to do this aspect of the job just fine ... despite the clichés. I will then discuss how you can get ahead in your career. I conclude with a couple of special topics: how to deal with the special problems presented by those projects whose work is geographically distributed across more than one work site, and those projects that include teams located in multiple countries.	Team exercise: the social aspects of engineering project management

Chapter/ week	Lecture	In-class facilitated workshop
14	**Achieving quality**. Many engineering textbooks teach that what you need to focus your management efforts on are schedule, cost, and technical capability. In the real world, that is insufficient. Factors such as reliability, safety, low latent defect rates, and "environmental friendliness" increasingly play a role in product and company success. In this chapter, I teach you the basics of this aspect of your role as an engineering project manager, which we group together under the title of "quality."	The facilitated lab sessions in this week are used for student team presentations
15	**Applying our ideas in the real world and ethics in engineering**. In many aspects of engineering and engineering project management, there are factors in tension with each other. In fact, we have seen many examples of this throughout this book: making the airplane lighter achieves better fuel efficiency, but makes the plane more expensive; hiring more people for your project might get it finished sooner, but probably increases the total cost at complete. Every engineering management process that we have discussed is subject to similar tension when you go out into the real world and try to apply it. In this chapter, I help you understand how to use these techniques within the limitations of actual people, companies, and customers. I also discuss the sort of knowledge that you will need to acquire on a continuing basis as you move through your engineering project management career. I conclude with a discussion on the important subject of ethics in engineering.	The facilitated lab sessions in this week are used for student team presentations

1

The Role and the Challenge

What is engineering project management? Why do we teach engineering project management? Do engineering projects matter to society? Do engineering projects matter to business? What is a "project?" What is an "engineering project?" What is a "project manager?" In this chapter, we discuss all of these questions, and also provide you basic information about the role of engineering project manager and the opportunity that this role represents for you.

1.1 Introduction

In this book, we study the subject of *engineering project management.* Let's look carefully at the first two of those three words:

- "Project." By this word, we mean a deliberately undertaken endeavor to create something that we believe will be of value. This desired result might be a tangible artifact ("product"), or it might be a service. We might plan to use it ourselves, or we might be planning to offer it for sale.
- "Engineering project." By adding the modifier "engineering," we now indicate that the methods we will use to create this result involve "engineering." By this we mean the methods that we will use depend significantly on the application of technology and technological concepts in order to achieve a practical effect. Other disciplines – such as finance, art, and so forth – may also be made use of in order to create our desired product or service, but by identifying this as an "engineering project," we are indicating that we consider the role played by the application technology and technological concepts to be *central to the success of the project.*

The judgment about whether the role of technology and technological concepts is central to a project is sometimes fairly obvious (e.g. designing a new computer processor microchip), but in other cases it may be more subtle. For example, painting in oils on canvas involves quite a lot of technology; someone has to know how to create the paints, someone has to figure out what methods of preparing the canvas are likely to create a surface to which the paint will adhere well, and so forth. These are highly technological subjects! But since those technologies have *already* been worked out, and can today be made available in a form that allows the paint and the canvas to be used in a

Engineering Project Management, First Edition. Neil G. Siegel.

Companion website: www.wiley.com/go/siegel/engineering_project_management

manner that requires only minimal knowledge of the underlying technology, we probably would no longer consider creating an oil painting an engineering project. This is because, while painters must learn the technique for preparing a canvas, and for applying the paint, they generally no longer have to learn how to do the underlying technological work, such as creating the paint, and no longer have to understand how certain materials in the paint result in certain properties for the paint. There was certainly a time (a few hundred years ago), before these technologies had been worked out, that one could properly have considered painting in oils on canvas an engineering project. But by now, these technologies have matured to a point where a non-technologist can *apply* them by learning a *technique*; the artist need not know the underlying technologies. Given the maturity of these technologies, and their availability in a form that non-technologists can learn to use, we would probably consider that the creation of technology is no longer central to the success of our oil-painting project, and therefore, today we would likely not consider painting in oils on canvas an engineering project.

This is akin to the difference between *operating* a car or a computer and *designing* a car or a computer. Operating the car or the computer requires technique, but does not require detailed knowledge of the underlying technology.

Now, let's add our third word:

- "Engineering project management." By adding this third word, we have moved from focusing on the desired end result (that is, the product or service that we intend to create) to focusing on the *process and method by which we will create it*. By *management*, herein we mean the notion of planning and organizing the activities that will create the desired end product or service – e.g. things that are done *before* we start the actual project. It has the additional connotation of providing leadership to that activity while the effort is underway – e.g. things that are done *during* the conduct of the actual project.

Engineering project management is therefore a discipline that provides the method to get to our desired result: that product or service that we want to use or offer for sale, where the application of technology or technological concepts is central to the eventual success of the project.

1.1.1 Why Do We Care About Engineering Project Management?

I will start this part of the story with two assertions:

1. Engineering projects are vital for society.
2. Engineering projects are "of the essence" to business – they are the *only* source of revenue for many companies – and the creator of new methods and new products for almost every organization (business, government, non-profit, etc.).

There is a third assertion, too:

3. Being the manager of an engineering project is a *great* job!

Let's discuss each of these three assertions.

In your opinion, what is the most important human accomplishment of the last 2000 years? Think about it for a moment, and write down a phrase describing your answer.

Your answer:

When I ask this question on the first day of my engineering project management courses, I get lots of really good and interesting answers. Obviously, this is a matter of opinion, and we can all have an opinion. But here is my answer:

> The most important human accomplishment of the last 2000 years is the doubling of the average length of human life.

What the archeologists and other scientists who study these matters tell us is that for hundreds of thousands of years, the human lifespan averaged around *35 years* … until around 125 years ago, when the average human lifespan started increasing. Recently, the average human lifespan has reached more than *70 years*.

To depict this improvement, I created the graph in Figure 1.1 from data made public by the World Health Organization.

In fact, in many parts of the world, the typical human lifespan is now more than *80 years*.

Obviously, living to 70 rather than to 35 is viewed by most people as a very good thing indeed! But what caused this doubling of human life expectancy? This question has been studied by the US National Academies.[1] Apparently, *engineering projects* deserve most of the credit, due to the following types of large-scale societal systems that have been created by such projects:

- Water treatment and delivery
- Sewage treatment and transport

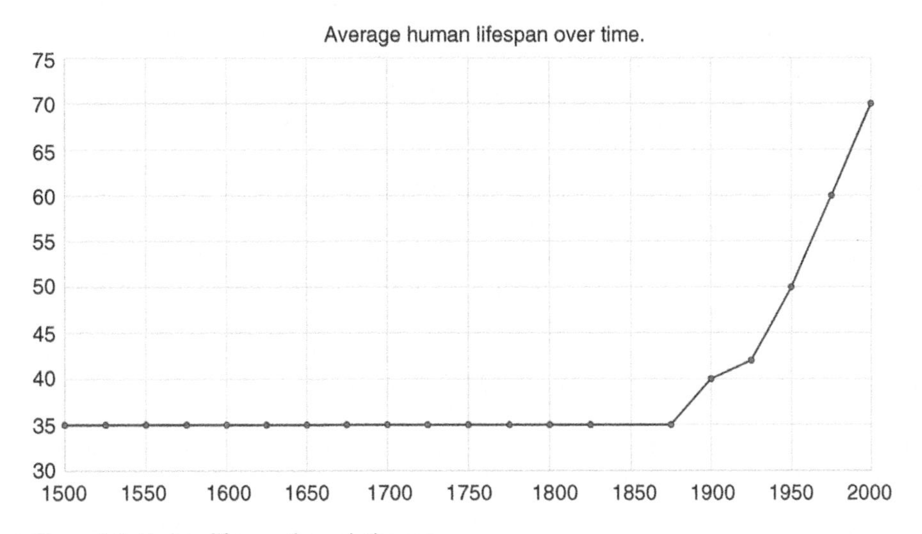

Figure 1.1 Human lifespan through the ages.

1 There are three US National Academies: Science, Engineering, and Medicine.

- Motor-powered tractors
- Motorized transport and delivery
- Large-scale electricity generation and delivery
- Affordable, mass-scale refrigeration
- Canning and other mass-scale food storage/preservation techniques
- ... and so forth.

To be a little more precise, the National Academies estimate that 80% of the increase in human life expectancy is due to such engineered systems; most of the rest is due to improved healthcare. And *all* of these important societal systems were created by engineering projects!

Ergo: If you want to change the world for the better, become an *engineering project manager*! (We get paid pretty well too.)

My second assertion was that engineering projects are "of the essence" to business – they are the *only* source of revenue for many companies – and the creator of new methods and new products for almost every organization (business, government, non-profit, etc.).

Consider an aerospace contractor. They obtain revenue by signing contracts with *external customers* (which might be their national government) who want to build something (e.g. a new military airplane, a government data processing system, an air traffic control radar, or whatever). The process of building this product is in fact an engineering project. It is the successful performance and completion of this engineering project that provides the company with its revenue.

Now, consider a consumer-oriented company like Apple or Microsoft. They don't have customers who give them contracts to build something; instead, they make their own decisions about what they want to build, use their own funds to build it, and then offer it for sale as a product or service in the market. But they would have nothing to sell if they did not undertake (in this case, at their own expense) an engineering project to create that product or service. The result of this *internally funded* engineering project might be a new version of a software product, or it might be a device (e.g. a new type of mobile phone), or it might be a service (e.g. it might be software that allows them to do your accounting for you, using that new accounting software package). But whatever the result of this internally funded engineering project, it is the *project* that created the artifact that the company is able to offer for sale. Without the result of that engineering project, the company would have no revenue.

So, whether the "customer" who is paying for the conduct of the engineering project is *external* (as is usually the case for an aerospace contractor), or is *internal* (as is usually the case for companies that build products for the general public), their revenue results from being able to offer for sale the results created by engineering projects. In most companies, in fact, this accounts for nearly 100% of their revenue. Hence, my assertion that engineering projects are "of the essence" to business.

My third assertion was that being the manager of an engineering project is a great job. It certainly was for me.

Why is it a great job?

- You have a lot of freedom
- You have a lot of responsibilities
- You have the ability to make a difference

- You have the ability to try out your ideas
- You have the ability to hone your people skills
- ... and you are in charge!

We will talk much more about this aspect in Chapter 13.

Make no mistake, being the manager of an engineering project is usually difficult, and at times can be downright stressful. But undertaking something that is hard, yet at the same time worth doing (engineering projects are important for society!), and doing your best, usually leads in the end to significant job satisfaction. I like to say that being the manager of an engineering project has two great attributes: It is both *interesting* and *important*.

- *Interesting*, because every project is different and difficult, presenting new problems to be solved.
- *Important*, because engineering projects, as described above, benefit both society and business.

1.1.2 The Opportunity For You

Becoming the manager of an engineering project might sound like something to which only a few people can aspire. That is not the case; in fact, there is *lots* of opportunity.

Think about it – many companies have lots of engineering projects, and every one of those engineering projects needs a project manager. And if the project is of any size, each engineering project needs many *subordinate managers* (e.g. people who are in charge of *segments* of the project) too. As we will learn in a later chapter, projects are usually organized as a *hierarchy*: someone is in charge of the entire project, but reporting to that person are people who each run a segment of the project, and so forth. So, in fact, engineering projects usually need lots of managers.

Being the manager of a large engineering project may sound a little scary. At times, it is. But there are lots of ways to prepare yourself for the role of being an engineering project manager (in addition to, of course, reading this book!). You might start by working on a big project ... then become the manager of a small piece of that same big project ... and then run a small project ... and then run a mid-sized project ... and thereby work your way up to running a big project. Each step along the way teaches you part of what you need to know, provides you with the necessary experiences, and allows you to improve your skills. By the time you are offered some sort of management role on an engineering project, you will likely be ready for that responsibility.

1.2 The Project

At the beginning of the chapter, we defined a project as a deliberately undertaken endeavor to create something that we believe will be of value. Let's now go into this in a little more detail, so as to elicit a better understanding of the characteristics of projects, and why it is important to distinguish them from other activities (e.g. activities that are *not* projects).

Definition: A *project* is a *temporary* activity intended to create a product or a service.

By "temporary," we mean that a project has a deliberate commencement and termination; the project must proceed forward toward some *well-defined conclusion*, an *ending*. Here are some examples, to help you understand the concept that a project must have a planned *ending*:

- *Building* an oil refinery is a project, but *operating* the refinery after it is complete is *not a project.*
- *Inventing and designing* the next iPhone is a project, but *building* 100 000 000 of them and then selling/supporting them is *not.*
- *Building* a new web page is a project, but *using it* for sales and advertising once it is complete is *not.*
- *Making* a flower bouquet is a project, but *tending your garden year after year* is *not.*

Notice that the result of the project may be offered for sale, or may just be used internally by the creating organization. Both would qualify as projects; the result need not be intended for sale.

If the actions of operating the oil refinery, continuing to build and sell iPhones until people stop buying them, using the web page after it is complete, and tending your garden year after year are not projects, what are they? I term these *continuous business operations*, rather than *projects*. These types of activities are characterized by the fact that they proceed forward on an *open-ended* basis, without a well-defined conclusion or ending.

The table in Figure 1.2 captures these differences (and a few more that we will come back to in a later chapter) in a comparative format.

Why do we make this distinction? After all, the oil refinery will be shut down someday. There are two reasons:

- The activities and challenges faced by a project and by a continuous business operation are quite different, and likely call for people with different experiences and expertise to act as their managers.
- The mechanisms through which projects and continuous business operations create revenue and repay their investments are quite different. This creates a need to use different techniques to manage the two different types of endeavor.

Think about that oil refinery. Each month, it costs us money to operate it (we have to buy the crude oil, pay the people who work at the refinery, maintain the machinery, pay property taxes, and so forth), but we also receive revenue every day, as we sell the refined

Projects	Continuous business operations
Temporary	On-going
Unique	Repetitive
Specific time frame	***Open-ended*** time frame
Situational	Standardized
Project-specific measures	Company-specific measures
Subtle performance indicators	Clear performance indicators
Focus is on creating something new	Focus is on reducing unjustified variation
Changes the status quo	***Is*** the status quo

Figure 1.2 Comparison of a project with a continuous business operation.

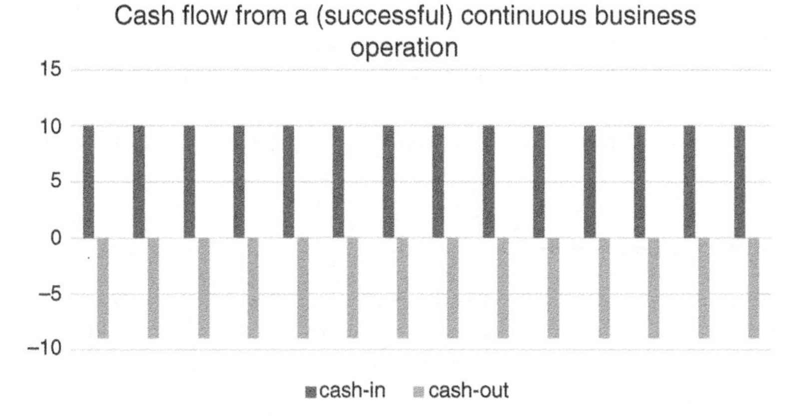

Figure 1.3 Cash-in and cash-out from a continuous business operation.

petroleum products that the refinery produces. Our monthly cash-out and monthly cash-in are continuous, and nearly equal (ideally, with cash-in being slightly more than cash-out, of course!) (see Figure 1.3).

This is completely different from the cash-out/cash-in profile (the finance people call this the "cash flow") of a *project* (see Figure 1.4).

As you can see, *projects spend money long before they return any benefits*. Because of this, managing the project to meet schedule and cost is *vital*. The most common ways to lose money – or otherwise run into difficulties – on a project are quite different from the most common ways to lose money (or run into difficulties) during a continuous business operation.

Note that the right-hand side of Figure 1.4 is actually likely a continuous business operation; projects generally make their money after the project per se is completed, and the resulting product or service is placed into service as a continuous business operation.

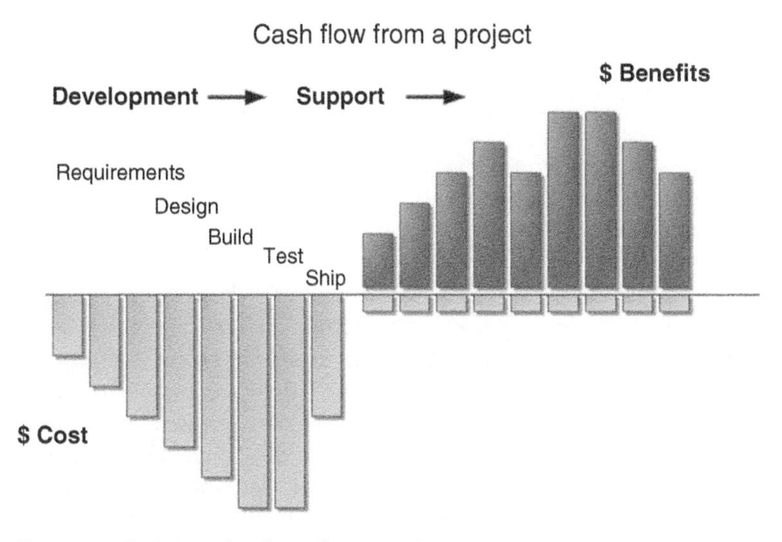

Figure 1.4 Cash-in and cash-out from a project.

Furthermore, since through your project you are creating a product or service anew, the types of things that could go wrong (in a later chapter, we will call these "risks") are usually quite different from the things that can go wrong during a continuous business operation.

For all of these reasons, we therefore distinguish between a project and a continuous business operation.

1.2.1 Where Do Projects Come From?

A project is created when an organization needs to *create* (or *replace*) a product or service. They therefore need a *temporary activity* to create that new (or revised) product or service. The project is *complete* when the new product or service has been created, and is ready to start being operated.

Someone pays for it. This person or organization is usually called the "customer" or the "client". The actual customer may hire someone to oversee the project acquisition process for them; this person or organization is called the "buyer".

1.2.2 Customers

The customer can be *external* (e.g. outside the organization that is building the project; an example would be if the US Federal Government hired Lockheed Martin to design and build a new type of military airplane; another example would be your local city hiring a company to repave a particular road).

The customer can also be *internal* (e.g. Apple funding the design of the next iPhone). In this case, the customer is one set of people inside the company; the project will be performed by a different set of people inside the company.

In fact, there are usually *multiple* customers for a project: those who are *paying*, those who are *buying*, and those who will be *using* the resultant product or service.

1.2.3 Attributes of Projects

Projects are *important*.

- Therefore, we aim to be *rigorous*.
  - We achieve this rigor via appropriate levels of planning, via the use of credible analytical methods, and via formal procedures to verify the correctness of our project's products and services

Projects are *difficult*.

- Therefore, we aim to *learn from past experiences*, and from *experts*.
  - We accomplish this learning via *engineering and management processes*. A *process* is written guidance about how to accomplish a task. The process is written in a fashion to incorporate the lessons our organization has learned from performing on previous projects: both those projects that were successful, and those projects that were not. We use such processes to guide engineering activities, management activities, personnel activities (e.g. recruiting and hiring), and many other things that we must do in order to complete our project.

Projects involve *more than one person.*

- Therefore, *communication among the team members* is "of the essence."
  - We accomplish this via *written artifacts, face-to-face meetings, and many other methods* so as to ensure that "we are all on the same page."

1.2.4 The Project Life-Cycle

It has been found useful to group the steps typically undertaken to start, perform, and complete a project into a standardized set of categories; these categories (I will call them *stages*), taken together, form a conceptual *life-cycle* for a project. Figure 1.5 shows one version of such a depiction of the project life-cycle.

Actually, "fielding and use" and "production" are likely to be "continuous business operations," as I defined that term above, rather than a "project," because they usually will not have a well-defined completion date. But almost everyone includes them in such depictions of an engineering project life-cycle; I will do that too.

In Chapter 2, I introduce my own particular version of the project life-cycle, which we will use for the remainder of the book.

You will see that as we discuss various steps and methods of engineering project management, they generally can be related to one or more of these six life-cycle categories.

1.2.5 Goals of the Project/Factors in Tension With Each Other

An engineering project will usually have more than one goal. For example:

- We want to complete the project, having built a product or service that has all the required capabilities.
- We want to complete the project on the end date that we promised.
- We want to complete the project within the amount of money we promised.

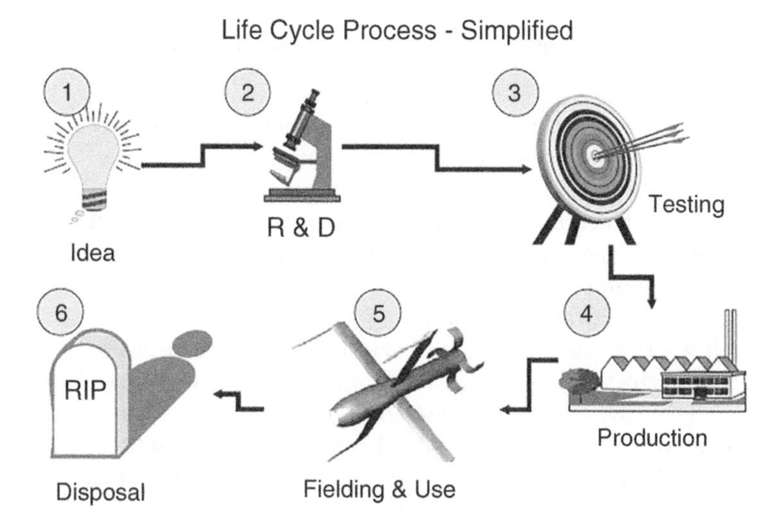

Figure 1.5 An example of a project life-cycle. Source: From http://www.almc.army/mil/hsv/2001-DL.PDF.

The goals usually tend to compete with, or be in tension with, each other. For example, one could usually increase the probability of completing the project on time (and within the allocated budget) if one omits some of the capabilities.

The essence of engineering project management, however, is achieving a *balance* between these competing goals. One may have to work very hard, for example, to create a design that provides the required capabilities while also staying within the constraints of time and money. It is often the case, in my experience, that it is relatively easy to find a design that seems to work, but much harder to find a design that works within the constraints of time and money. As the manager of your engineering project, you must find a design that *accomplishes a large portion of* ***all*** *the goals*. This is an example of resolving factors that are in tension (in this case, capability, schedule, cost) by achieving a *balance*. Maybe you decide that there are a few capabilities that are not actually needed, or not worth the cost of implementing them; you can go back and show your customer how much time and money you could save by omitting them, and obtain their written concurrence to do so. This is a very typical action for the manager of an engineering project.

You may also have to reject a design that would in fact work, but one that you determine would take too much time and cost too much; you must create a different design that will achieve a better balance. This, too, is a very typical action for the manager of an engineering project.

But you must do all of this without taking inordinate risks. For example, someone on your team may promise you that they can develop a revolutionary new algorithm that will perform the necessary computations 10 times faster than current algorithms, and such an improvement is needed to meet all of your requirements. But this revolutionary new algorithm is still just a promise, and many things can go wrong between the promise and completion of a useable algorithm. Later in the book, we will call the probability that something can go wrong a *risk*, and show you how to achieve balance between promises and failure.

Therefore, I (along with some others) place a fourth item on the list of key goals for your engineering project:

- Managing the risks that will be encountered by the project, so as to avoid catastrophic failure.

In addition to the risk that the project will not work, there is another category of risk: that in the conduct of the project, we might inadvertently harm people or the environment.

Risk is one key area where engineering projects are different from other sorts of projects; since we are often inventing new things via our engineering projects, we often face far more risks – and more profound risks – than do other types of projects. Engineering project managers must *always* think about risks.

Because of the centrality of risk to engineering projects, I always like to focus on these four constraints as the key factors in tension that I must manage over the course of my engineering project. I call this the *quadruple constraint* (Figure 1.6). Not surprisingly, those who only use the first three items on the list call their version of this list the "triple constraint." That shorter list may be fine for other types of projects, but for engineering projects we will usually need my version, the quadruple constraint.

If your project is for an external customer, the elements of the quadruple constraint are usually expressed in a legally binding document called a *contract*. One portion of

- We must complete the project having built a product or service that has all of the required capabilities
- We must complete the project on the end date that we promised
- We must complete the project within the amount of money that we promised
- We must manage the risks that will be encountered by the project, so as to avoid catastrophic failure

Figure 1.6 The quadruple constraint.

this contract (called *specifications*) will define the *capabilities* ("what" your project's product or service is supposed to do) and *quantifiable measures* ("how fast, how far, how long, how reliable"). The contract will also specify a *date* by which the product or service must be delivered, and the agreed-upon *price* that you will be paid when the work is complete to the satisfaction of the contractual terms.

Often, there are financial incentives and penalties; perhaps you don't get paid until you prove that the project satisfies all of these requirements; or there are penalties for being late; or there are bonuses for completing the product on time; or some combination of all these.

On complex projects, it is common for the contract also to mandate certain standards for *how* the work is accomplished, in addition to the specifications for the end product. For example, the contract may require certain engineering and/or business methods to be employed; these are one of the ways in which the customers show that they expect *you* to manage the risks that will be encountered by the project.

If your project is for an internal customer, the written documents that embody the quadruple constraint will still exist; they may be less formal, and probably are not legally binding. But your management still expects you to adhere to them! Your promotions, and perhaps even your job, will be on the line.

So, what constitutes success for an engineering project?

- Meet the quadruple constraint:
  - Within allocated time
  - Within budgeted cost
  - At proper capability and performance level
  - And with risks kept under control.
- But also:
  - Accepted and liked by the buying customer
  - Accepted and liked by the users
  - Without disturbing the main workflow of your company
  - Without exhausting, hurting, or demoralizing your people
  - And accomplish all of the above while complying with all applicable laws, regulations, and company policies.

The quadruple constraint defines a major portion of *all* engineering projects, whether external or internal. The specifics of your engineering project determine the relative importance of each dimension of the quadruple constraint. Sometimes, capability is the most important. Often, meeting the promised delivery time is the most important. Every customer cares about cost and risk too.

How do you know the right balance of emphasis for your project? Through frequent and forthright discussions with your customers (remember, however, there are usually *multiple* customers for your project, and they likely will not agree with each other on all matters), your technical personnel, your corporate management, and other stakeholders and experts. You must do this not just at the beginning of the project, but on a continuous basis.

1.3 The Project Manager

1.3.1 The Role

Whether the customer is internal or external, the project is performed by some set of people; this set of people are called the "project team" or the "development team." That project development team is led by someone; that person is the *project manager.*

Sometimes the same term is also used for the person in charge of the *buying organization*; but in this book, we will reserve the term "project manager" to signify the person in charge of the *project development team*: the team actually building the new product or service.

Being a project manager is about "doing," not "consulting." You are responsible for achieving a good outcome, creating the capability of the delivered product or service, completing the project on time, completing the project for the amount of money agreed to, and many other things. You are provided with the authority necessary to accomplish those ends.

You lead a large number of activities: planning the activities; recruiting, motivating, and aligning the people involved; managing, monitoring, and controlling the project's activities; ensuring that the work is actually accomplished according to the plan and according to the requirements in your contract (more about that later); verification of completeness and quality; and many other things too.

Projects are almost always accomplished by *teams*, not by single individuals. The image in the movies of one person sitting at home and building something over the weekend that changes the world seldom happens. It almost always takes *teams* – in fact, often large teams – to accomplish something important.

As the project manager, therefore, you will break the project into small pieces, and then assign responsibility for each piece to a named person.

Since the work is being performed by a team, you must ensure that all of the people on that team have:

- A shared vision of the desired outcome of the project
- A shared understanding of the means and methods that will be employed to create the desired outcomes
- A shared knowledge of the constraints (including schedule and cost).

This is the *alignment* that we will depict in Figure 1.8 below, and discuss in more detail in Chapter 13.

Furthermore, each member of the team must learn to derive *his or her* satisfaction from the accomplishments of the *team*, rather than solely from *their own* accomplishments.

You must create the conditions whereby this can take place. Note that this is quite different from what we are taught to do in school; in school, although we may occasionally participate in a team project, the general emphasis is that we do *our own* work, and derive our satisfaction and sense of accomplishment almost exclusively from our own work. This is *not* at all how real engineering projects work; what matters on an engineering project is the accomplishment of the *team*.

A good team can do so much more than any single person! Fashioning the team in such a manner that it can actually realize this potential is part of your responsibility as project manager.

As the manager of an engineering project, you have both *authority* and *responsibility*.

- *Authority* means that you have been designated by your company or organization as being the person allowed to make certain types of decisions regarding the project. The types of decisions that you are authorized to make are usually spelled out in writing, and likely include decisions about selecting personnel for your project (usually you also get a role in determining their compensation), technical decisions about the design, what companies and vendors to use to provide parts and services, and so forth.
- *Responsibility* means that you have accepted that you have the obligation to accomplish all the goals of the project within the constraints and limits that have been specified to you. These constraints and limits are also spelled out in writing, and likely include items such as how much time it will take to complete the project, how much money the project will cost, what you will deliver, that you will not do anything illegal or against your organization's rules and policies (or knowingly allow someone else on the project to do anything illegal or against your organization's rules and policies), and so forth.

Since the project you will be leading is an *engineering* project, and we are engineers, it is natural to think that most of our time, and most of our decisions, are about engineering. This is *not* correct; as the manager of an engineering project, we have many responsibilities other than engineering. In fact, we lead *every* aspect of the project, which includes social factors and matters of business, contracts, and finance (just to name a few), in addition to technical and engineering matters. Being the manager of an engineering project is a very multi-faceted role. We need to learn a little bit about a lot of subjects. We do not need to be an expert in all of them, but we need to know enough to be able to talk effectively to the experts in each discipline.

My own opinion, however, is that it is vitally important that the manager of a complex engineering project be an engineer. You will not be the lead designer; you will have someone working for you who fills that role. But in many of your interactions, you must judge the credibility of the people to whom you are talking, and on an engineering project the majority of those people are fellow engineers. And engineering considerations are central, as we said at the beginning of the chapter, to the success of the project (else we would not consider it an engineering project).

But, even though you are an engineer, you lead and integrate *all* of the aspects of the project, not just the engineering. As we will see in a later chapter, there are many different aspects to a project, and many different specialists with whom we must interact.

You also lead the interaction with all of the people involved in the project. This includes your team, the people who will actually do the work of the project. But you will interact with many more people than just your team; we use the term *stakeholders* to indicate all of the various people who have some sort of vital interest in the outcome and conduct of your project. Examples of such stakeholders include:

- Your management
- Corporate specialist staff functions (e.g. human resources, finance, law, contracts, quality, etc.)
- The people and organization who are paying for your project
- The people and organization who are buying your project (these are usually hired by the organization that is paying for the project)
- Your users – the people who will be using the product or service, once you complete it
- Other interested and affected parties, sometimes including the general public.

Let's examine some of this nomenclature; in particular, let's discuss the distinction between *paying* for the system, *buying* the system, and *using* the system.

Here is an example. In the US Federal Government, it is Congress that pays for a project (the US Constitution says that *only* Congress has the power to authorize the spending of public funds), but Congress always designates some executive agency to act as the *buyer*. For example, Congress might decide to acquire a new type of military aircraft, authorize the spending of a certain amount of money to pay for an engineering project to design and build this new aircraft, and might also designate the US Air Force as the agency to act as the buyer for this project. By the term *buyer*, we mean someone acting on behalf of those who are paying for the project to select someone to build it for them, and to oversee the builder's progress. Congress has neither the time nor the skills to select a company to design and build this aircraft, so they designate someone from an organization with the appropriate expertise to do it on their behalf. In fact, there will be a specific person within the Air Force named as the buyer for your project. That buyer will have an entire team of their own too. But neither Congress nor that team of people who act as the buyer are the people who are going to fly that new military aircraft when it is made; the actual *users* of the product of our engineering project will be Air Force pilots, and the senior officers within the Air Force who command them. So, we have funders, buyers, and users who in this example are all distinct from each other. In some simpler projects, the buyer might be the user, but most of the time, these are distinct roles.

Of course, there might be many other people who are interested and concerned about your project. Our new military aircraft probably has to fly over areas where there are US Army and US Marine Corps personnel, so those branches of the US military will also have ideas about how our airplane should be designed. Our airplane will be based at air bases within the United States (at least, during peacetime), and will make noise, cause pollution, and so forth, even during routine training missions over the United States, so local city and state governments, environmental groups, and even the general public are to some extent stakeholders too.

And since in our example it is the US Congress that is funding the project, we ought to point out that many different people and organizations have opinions about the purposes for which Congress ought to authorize the spending of public money; in some real sense, all of those people and organizations are influencers on your project.

As the manager of your engineering project, you have to interact with *all* of these people. You synchronize and align all of them. You work with them so that they understand and agree with your goals and limitations. You help them understand your approach to the project (e.g. in what order and sequence we will do the various steps, and so forth), and to some extent you must even help them understand your design. You must keep them working in alignment, keep them motivated, and keep them believing in the value of your project. You must collaborate with them to keep the project sold; we will learn that most engineering projects that are started are not completed successfully; most encounter problems, and many are stopped before completion. It is part of your job to deal with the problems as they arise (and ideally, to anticipate them in advance!), and do so in a fashion that allows all of the stakeholders to agree that the project should be allowed to continue until it is complete.

Let me say a few words about "synchronizing and aligning" people. We have all had experiences of a situation where it seemed like everyone in the room had a different opinion about how something ought to get done, even something simple like what toppings we want on tonight's pizza. In a complex engineering project, we must resolve these differences of opinion and reach consensus. This process of reaching consensus is not just a matter of getting people to agree with you; instead, it is a chance for many people to contribute ideas and create a plan that is better. One cannot make progress on a complicated endeavor like an engineering project unless one reaches such a consensus. There are literally hundreds, if not thousands, of such matters to be addressed. What programming language will we use? Will we use a licensed or unlicensed radio frequency spectrum for communications? How long does the battery have to last between recharging cycles? Some are design issues. Some are what we will call *process* issues (e.g. what are the methodologies and tools – that is, the process – that we will use to accomplish our goals). Some are personnel issues; for example, two team leaders might both want to hire the same engineer. Some are legal issues; for example, does our contract allow us to buy the jeweled bearings that we need from a vendor outside of the United States, or not?

We need to *align* our team. What do I mean by that expression? Here's an example. I like cats, but when we had four cats all at once it was often chaos at home; they each had their own ideas about priorities, objectives, methods, and who ought to be in charge! In that case, it was cute and often a little silly. But you cannot be an effective manager of an engineering project without herding those cats! When we start, everyone will have a different opinion, on each of hundreds of matters. Our team looks like Figure 1.7. That is, all heading in different "directions"; each has their own ideas about priorities, objectives, sequence, tools, methods, design concepts, and so forth.

What we need to do is to get the team to look like Figure 1.8 instead. That is, we need to reach a state where we have a shared set of goals; a shared plan for methods, tools, and techniques; an agreed-upon sequence of steps; shared priorities; agreed-upon concepts for the design; and so forth. We do not reach this state all at once; indeed, even if you have a great plan with which to start, every single day you will discover new areas and additional details requiring new discussions, new realignment, and additional decisions. That is part of what makes the job so interesting!

1.3.2 You as the Manager of an Engineering Project

According to the dictionary,[2] "management" is (i) [verb] the judicious use of means to accomplish an end (ii) [noun] the group of those who manage or direct an enterprise. So, by "management" we can mean either the process and activity of overseeing an

2 Merriam-Webster, © 2000 Zane Publishing, Inc.

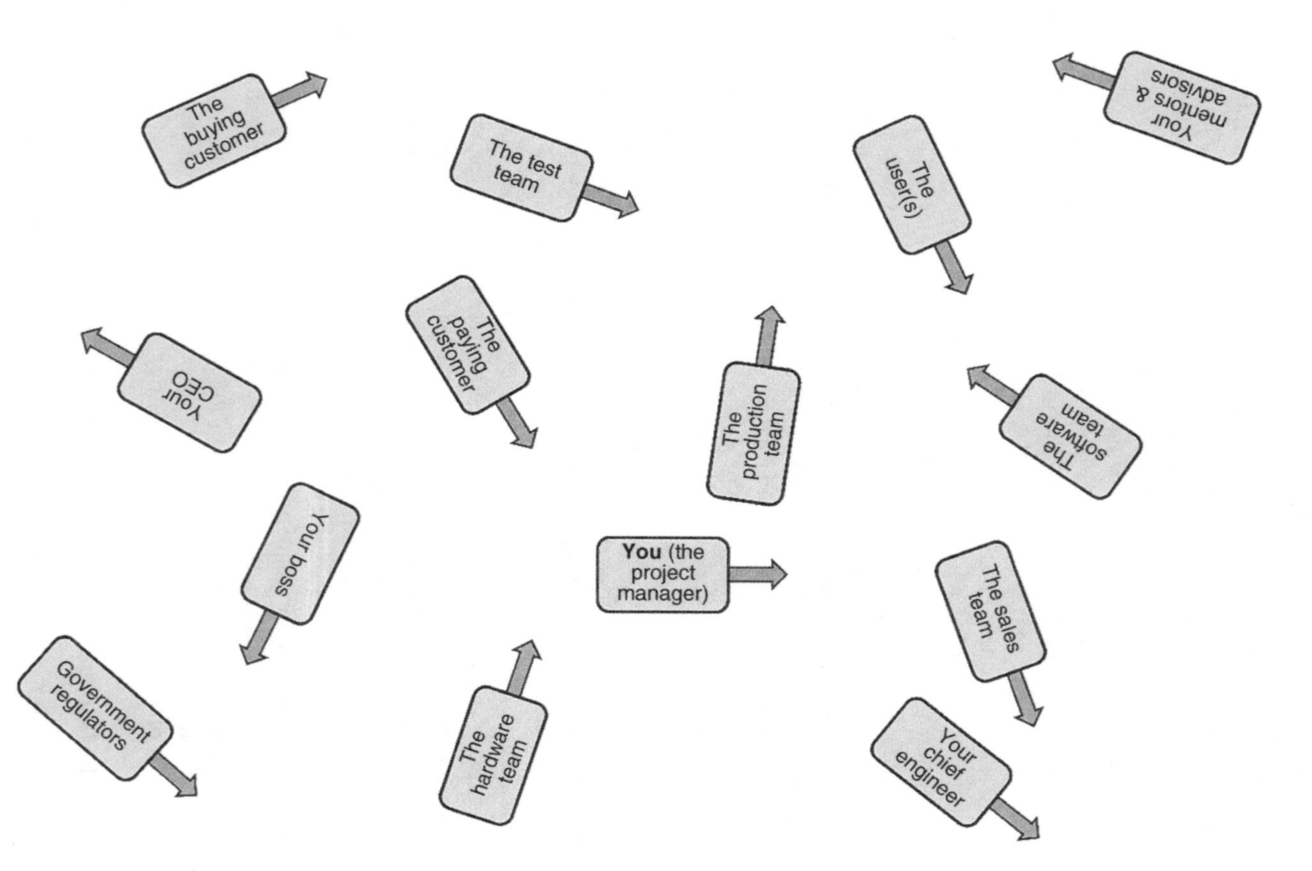

Figure 1.7 An unaligned team.

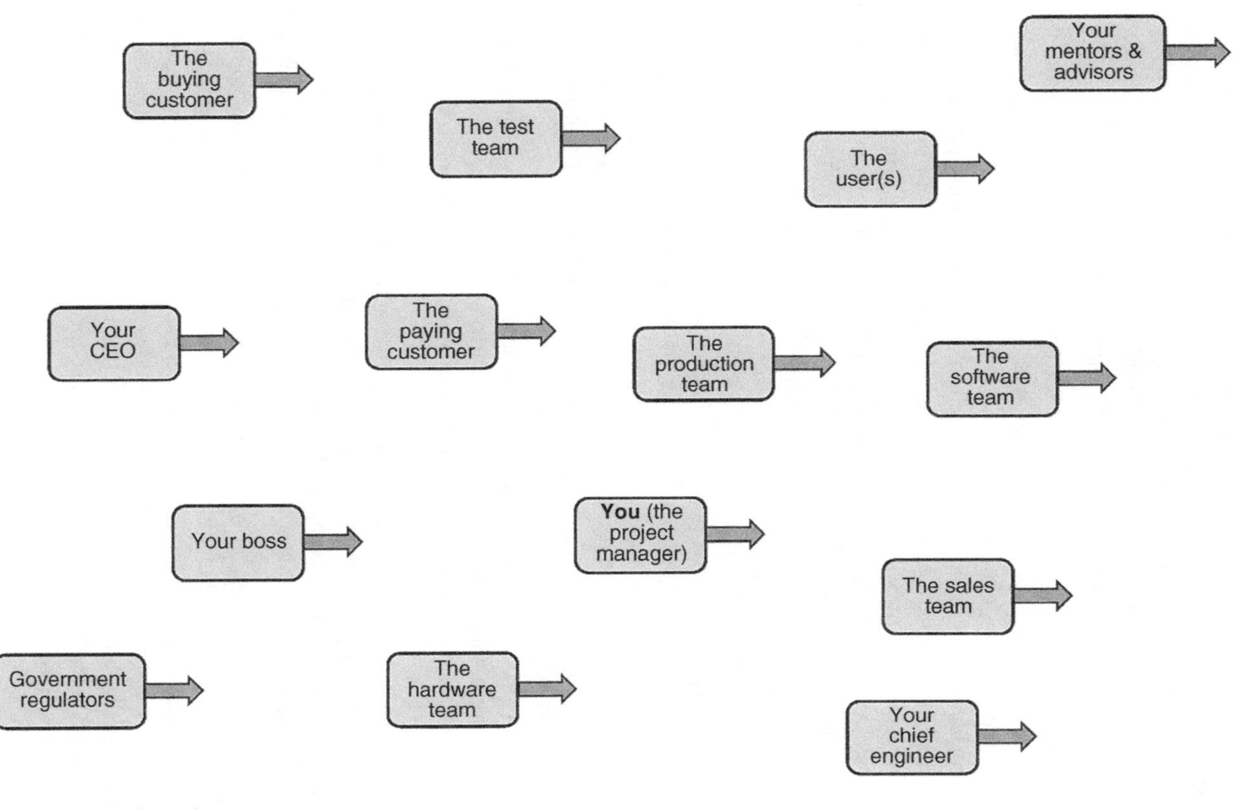

Figure 1.8 An aligned team.

activity to accomplish an end goal, or the person or people that perform such an activity. In this book, we will talk about both the process of management and the person who is the manager.

Our project is an engineering project, and we said that we call it that because engineering and technological considerations are central to the success of the project. But only a portion of the people working on your project are engineers performing engineering tasks; there are many other people with other skills as well. You manage *all* of them, not just the engineers. Let's look at your responsibilities.

- You work with your customer(s) to establish the *goals* of the project (e.g. what we are trying to build) and the *constraints* within which you and the team must accomplish them (time, money, laws, company rules, and so forth).
- You lead the *planning* of the project, by which we mean determining in advance what you will do, in what sequence, who will do each portion, how many people with what skills will be required, how we will obtain access to those people, where the work will be performed, what methods and tools will be employed, what records will be kept, and many other matters. These must mostly be determined in *advance* of the commencement of the project, and must be committed to writing.
- You lead the effort to agree upon methods for *measuring progress* of the project, in every applicable dimension: technical, schedule, staffing, cost, risk, and many others.
- You select the *key people* for the project, and agree with each of them regarding their role (responsibility and authority), committing these agreements to writing.
- You *motivate* those people to work effectively, to work well as a team, to believe in the importance of the project and its objectives. As we will discuss later on, it turns out that motivated people are far more productive than unmotivated people, which is important in achieving success in your project.
- You *monitor* the progress of your project as it proceeds, by which we mean you periodically compare results to date against your plans, and if there are material differences between those results to date and the plans, you take actions to resolve those differences.
- When the project reaches its conclusion, you take actions to *close-out* the project, which includes finding new work assignments for all of your personnel, returning facilities to the control of the company for other uses, archiving documented materials, properly dispositioning materials, closing the accounting books, and so forth.

In summary, you as the manager of an engineering project will direct the application of knowledge, skills, tools, and techniques to a set of activities that are designed to meet the needs and goals of your project. To achieve success (and to do so safely, legally, and ethically) is your *responsibility*. In order to execute those responsibilities, your company or organization will grant you certain *authorities*, that is, you can make certain types of decisions, commit certain types of resources, and so forth. But you do *not* have to invent all of this by yourself, or from scratch either.

1.4 Engineering Processes Can Help You

Other people have been managers of engineering projects before you, so you can learn from their experiences (both the positive and negative parts of those experiences!). Most organizations that undertake to perform engineering projects have captured the lessons learned from previous projects, in the form of written guidance about *how to do each of the different types of activities involved in an engineering project*, ranging from planning to execution, monitoring, dealing with personnel, and close-out.

We call this type of written guidance about how to conduct the steps of an engineering project *engineering processes*. They might take the form, for example, of written step-by-step instructions, or the form of checklists, or be in other formats. But they are *always* in writing and are always intended to tell you and your team how your organization expects your project to perform each type of project activity, to show you what artifacts that activity needs to produce, and what is expected on the content and format of those artifacts.

In addition to a library of such engineering processes that your company or organization might have, there are a variety of other organizations that create and publish guidance for how to conduct an engineering project. Examples include the Project Management Institute,[3] the Institute of Electrical and Electronics Engineers (usually known by its initials, IEEE),[4] and the US Department of Defense. In fact, your company's engineering process library most likely draws upon information from some of these sources.

So, there is a rich body of guidance available to you. Furthermore, there are actual examples of previous artifacts; most large companies build libraries of these too, for the use of new projects. In such a library, you can see an actual project plan, an actual risk register, an actual software test plan, and so forth. You can even use these as a model for your own project, adapting these existing artifacts to the specifics of your project.

Good companies also require that their engineering projects, after they are completed (whether successfully or not), contribute data to the library of past project artifacts and data and also a written "lessons-learned" report; these reports should also be made available to new projects. This is a chance for your peers to learn about what worked and did not work on your project.

In this book, we will frame many of the discussions in terms of "this is what the typical engineering process guidance says about how to perform this step," but we will also include a lot of discussion based on my own personal lessons learned from having been the manager of several large, complex, and successful engineering projects. What I have found is not that the books and the engineering processes are wrong, but that they are often woefully incomplete, and don't tell you what is actually vital to your success. In fact, that lack is exactly why I have decided to write this book!

As we noted above, although engineering is central to the success of your project, many of the people who do make important contributions to your project's success are not engineers. I have depicted a representative sampling of the range of people and skills with whom you will have to work in Figure 1.9.

As the project manager, you must manage and work with all of these people. Because of this, you have to learn a bit about each of those specialty skills – contracts, law, human relations, configuration control, quality management, and many others – so that you can interact with those people, reviews their plans and progress, and provide guidance to them, just like you do for the engineering staff. The way I like to think about it is that you need to know how to *talk to* each of these experts, but need not *be* an expert yourself.

---

3 https://www.pmi.org.
4 https://www.ieee.org/index.html.

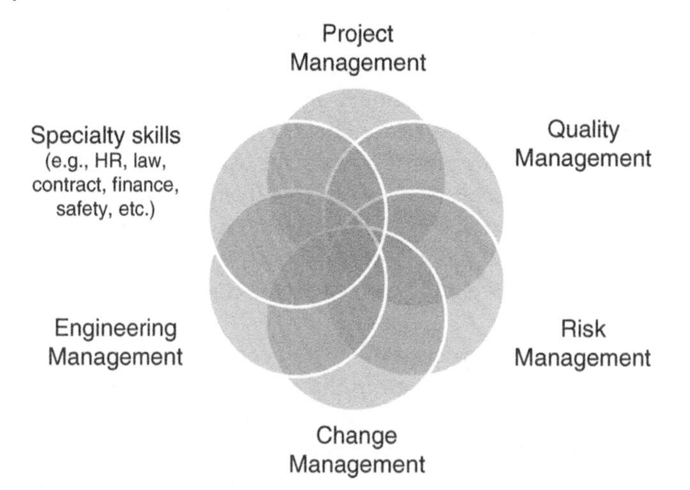

Figure 1.9 Engineering project managers must be able to interact with many different specialists.

We will therefore talk quite a bit about these non-engineering domains in this book. Your company probably has short courses that you can take in some of these fields, and certainly has manuals that describe the roles, processes, artifacts, and controls for each. My experience is that it is well within the capacity of an engineering project manager to learn all that he or she needs to know in order to work effectively with these specialists. Of course, you must devote time and effort to that learning. We will do some, but by no means all, of that learning over the course of this book.

1.5 The Engineering Project Manager Mind-Set

I have found that a certain mental orientation helps make for a better engineering project manager. You must:

- Be willing to spend time and effort to create realistic planning documents, in each of a number of domains (which we will name in a future chapter).
- Be committed to the success of the project, and be able to motivate your employees to share that commitment.
- Be able to persuade people with the requisite skills to come and work on your project.
- Be willing to spend time learning about your customer, what they value, and what they consider success (this is among my favorite items on this list; we return to this topic in Chapter 4).
- Be willing to spend time to build relationships of trust with many different people – your customers, your management, your other stakeholders, your employees, and your suppliers.
- Collaborate with your team to create a clear statement of scope and objectives for your project, and be willing to spend the time and effort necessary to obtain alignment of your team and stakeholders so that they agree with those descriptions.
- Motivate your team so that they believe in the importance of the project, and in the feasibility of the plan for implementing it.

- Be willing to accept that things will go wrong; spend time to create effective risk management and quality systems so that you can spot them and correct them.
- Be willing to solicit and act on advice; you cannot do this complex and demanding job without help.
- Be willing to insist on (and invest in) appropriate tools, technologies, methodologies, and techniques.
- Always be thinking ahead, looking for what might go wrong, and looking for opportunities.

This is a hard list, but I have found that most good engineers can learn to do it. All of these items are discussed somewhere in this book.

An engineering project, however, is *not* a science project; it is important to understand the difference between science and engineering. Look at Figure 1.10. *Science* seeks to study and understand the world; it asks questions about the underlying mechanisms and principles. *Engineering* does something different: it is concerned with using that knowledge (and other skills) in order to create a practical application, or a practical effect. Engineering is *not* knowledge for knowledge sake; such knowledge is important, but it is a job for scientists, not for engineers and engineering projects. We, in order to be effective engineering project managers, must keep our focus on achieving those *practical results.* But our projects must be feasible too; therefore, we may well have to consult with scientists in order to make sure that we are not promising more than we can feasibly deliver.

A presentation by an Englishman named Chris Wise (who, at the time, was a professor at University College, London) that I attended in 2013 got me thinking about what one might call "the tao of engineering project management"; *tao* is a Chinese word meaning approximately "way" or "path." I believe that the path to successful engineering project management encompasses much more than engineering. I have adapted what Professor Wise said to arrive at Figure 1.11.

As engineers we must base our work on facts, measurement, and rigor; it must actually achieve a practical result in the real world. But that is not enough; we must exercise judgment about what our product *should* be, create a compelling vision of what it *could* be, and use craftsmanship and artisanship to *build it well.*

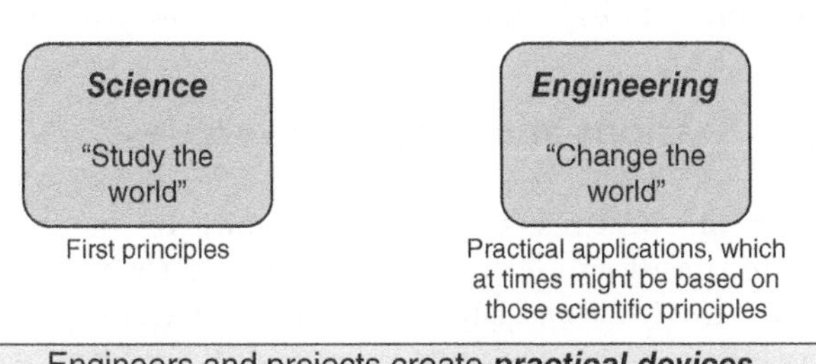

Figure 1.10 Science and engineering have different objectives.

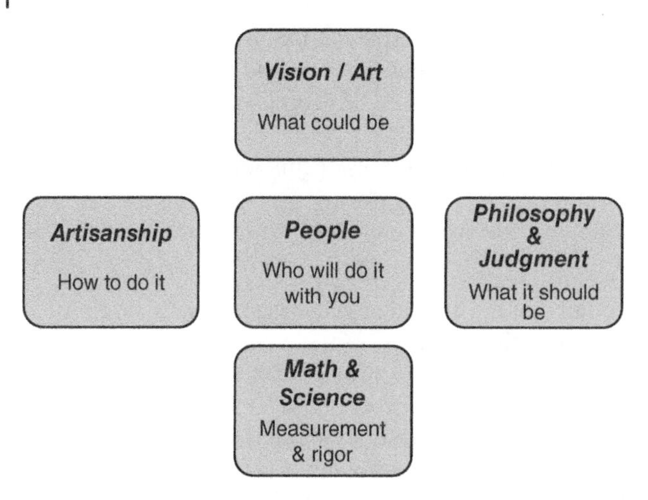

Figure 1.11 "The tao of engineering project management."

Lastly, we must meet the ever-rising expectations of our users and the general public. If you are a baseball player, you have the luxury of being considered a great success if you get a hit 30% of the time; that means you are allowed to "fail" 70% of the time! Not so in the world of engineered products; those are expected to work essentially 100% of the time. That may not seem fair, but it is a fact. We must also achieve the desired benefits of our product, but minimize the *unintended adverse consequences* (we will say a *lot* more about this in later chapters). Actually, our track record as engineers is pretty good (think of that doubling of human life expectancy that we talked about at the beginning of the chapter), but expectations are *very* high, and continuously getting higher. What was good enough yesterday will not be good enough tomorrow.

1.6 Next

Having set the stage, we are now ready to spend two chapters (Chapters 2 and 3) discussing how we actually do engineering on projects. We will then use that knowledge to optimize our project management procedures in light of those engineering processes.

1.7 About Facilitated Lab Sessions and Practical Exercises

Engineering is all about achieving practical results, and in that spirit of "hands-on" practical results, my class does *not* consist entirely of lectures. I use nearly half of the class sessions for what I call *facilitated lab sessions*. During these facilitated lab sessions, I take a short amount of time to explain a technique, and then provide the students with a problem which they work on (in teams) during the class session. They are allowed to consult with each other, to look at their class notes and books, to ask me questions, to show me their in-progress work, and get feedback on the spot. What I find is that they seldom actually finish the work during this class session (and therefore, they still have to do that work as homework), but by the end of each facilitated lab session they understand the method properly, and are able to solve the problem correctly.

Therefore, my course on engineering project management – and this book – are laid out as two parallel, week-by-week tracks: one of a progressive set of lectures and one of a progressive set of practical techniques and team exercises. As the students' knowledge and practical techniques are built up and mastered, this leads to a set of gradable course materials and an integrated set of learnings (via a combination of their reading, the lectures, the hands-on facilitated sessions, individual homework assignments, and team homework assignments) for each student.

At the end of the chapters that correspond to the weeks that have these facilitated lab sessions, there will be a section that addresses the topic, technique, and assignment for that week's session. This is the first of those sections.

1.8 This Week's Facilitated Lab Session

This week, we will explore the motivation for engineering projects and their contribution to the world. We will start by expanding on the contribution described at the beginning of Chapter 1: engineered systems that doubled the human lifespan as the most important human accomplishment of the last 2000 years. We will then branch out a bit and look at some other contributions of engineering projects to the world.

1.8.1 Exemplars

Consider the following six exemplary engineers.

1. *Thomas Edison*. The inventor of a practical electric light bulb (i.e. one that would last more than a few minutes, put out a reasonable amount of light, was reasonably rugged, not outrageously expensive, and could be manufactured in large volume), so that humans could have safe and effective light at night. Note that in some real sense Edison did not solve this problem via improved science, but much more by "tinkering"; he tried more than 1000 combinations of materials and preparations before he found one that he considered practical. He was obsessed with achieving a practical effect.
2. *George Westinghouse*. The inventor of alternating current, so that electricity could go a long distance. Edison's plan was to use the conceptually-simpler direct current, but direct current fades very quickly over distance as it travels through wires. We would have needed an electric power generating station on every block if we used direct current! Westinghouse had not only to solve the engineering problems so as to create a practical alternating current power distribution system, but also to overcome the fame and credibility that Edison had already established, in order to get people even to consider his approach based on alternating current.
3. *Nathaniel Wales*. The inventor of the first practical electric refrigerator, which has contributed significantly to humanity through its ability to preserve food against spoilage for long periods of time. Of course, if Westinghouse had not first invented a practical electric distribution system (based on alternating current), Wales could not have invented his refrigerator.
4. *Norman Borlaug*. Often called the creator of the "green revolution" in farming. Through the use of practical plant breeding, soil science, and engineering techniques to model growing patterns for crops in areas other than their native locations, he led the world to a massive increase in food production.

5. *Simon Ramo*. By 1949, both the United States and the Union of Soviet Socialist Republics (USSR[5]) had nuclear weapons. Europe had a 1000-year legacy of major wars almost every 10 years (there was a gap of only 21 years from the end of World War I to the beginning of World War II, and this gap was itself filled with many smaller wars). The leadership of neither the United States nor the USSR wanted a nuclear war, but what was needed was a way to use the potential power of nuclear weapons to *deter* the other country from attacking. Missiles that could fly quickly and reliably from one country to another carrying a nuclear warhead were recognized as a way to create such deterrence; since they could not be stopped, neither nation could attack the other without risking that the other nation would shoot its own nuclear missiles at it. The United States had a design for such a missile, but was failing in its attempts to build a practical missile out of that design. Since those in charge were certain that the missile would be of fairly poor accuracy – only guaranteed to fly within a few miles of its selected target – they concluded that giant, very powerful warheads (which would be very heavy) would be required. But carrying these heavy warheads required gigantic missiles, that turned out to be fragile and unreliable. An unreliable missile does not create deterrence! In desperation, in the mid-1950s the US Government turned to a pair of brilliant young engineers named Simon Ramo (see Figure 1.12) and Dean Wooldridge, and asked them to solve the problem. They decided that the design was wrong; we had to make a more accurate

Figure 1.12 The author with Dr. Ramo, 2011. On the occasion of the author receiving the Simon Ramo Medal for systems engineering. Source: Photo by Northrop Grumman. Used with permission.

5 The largest portion of the USSR became the Russian Federation in 1991, upon the dissolution of the Union of Soviet Socialist Republics. The Russian Federation inherited most of the former USSR's nuclear weapons and missile systems.

missile, which would enable the use of a less powerful and lighter-weight warhead; the lighter weight of the warhead would enable the use of a smaller missile, and this smaller missile could be made very reliable. The result was a credible deterrent, and that – so far! – the United States and the USSR have *never* gone to war against each other, even with conventional weapons. Most analysts credit this long period of peaceful coexistence to the deterrence achieved by Dr. Ramo's and Dr. Wooldridge's small, reliable intercontinental missiles.

6. *Judith Love Cohen.* As noted above, one of the key problems in creating a viable nuclear deterrent was improving the accuracy of the missile guidance system, so that a nuclear warhead that was far smaller (and therefore, far lighter in weight) could be used; therefore the missile needed to carry that warhead could be far smaller. My mother, Judith Love Cohen, was one of the members of the team of engineers who designed and built the first all-electronic missile guidance computer, which was an essential ingredient (along with a highly accurate gyroscope, and other components) in achieving the necessary accuracy and reliability (Figure 1.13). She started this work in 1952, and continued working on it (and other electronic guidance systems, including the one for the Apollo moon missions) into the 1960s. As noted above, they succeeded in creating a revolutionary new missile guidance computer, one that was small enough to fit in a missile and rugged enough to withstand the vibration of a missile in flight.

1.8.2 Points for Discussion

In each case cited above, the engineers created a *better design* than the preceding one. They looked at a wider range of options, including options that were not even considered

Figure 1.13 Judith Love Cohen in 1959. Source: Photo by Space Technology Laboratories. Used with permission.

previously. Identify and discuss the design selected, and how it differed from previous design attempts.

Their inventions had unintended consequences too. This is a vital observation, to which we will return in later chapters. Identify examples of such unintended consequences and discuss the implications. For example, increasing the human lifespan increased the number of people alive at any given time, requiring more production of food. Fortunately, Mr. Borlaug helped to solve that problem. But each accomplishment cited above, while solving one problem, had the potential to create other, unintended, problems.

The topic of the intercontinental ballistic missile and its nuclear weapons forms an example of an accomplishment that solved a then-current, urgent problem, but created serious unintended consequences afterwards. Nuclear weapons were created by the United States during World War II because there was evidence that Nazi Germany itself was working on such weapons; if the Nazis succeeded and the United States did not have such weapons, the Nazis would likely have won the war. But having been created for a reasonable purpose (e.g. ensuring the defeat of the Nazis) did not prevent nuclear weapons from creating serious adverse unintended consequences. The hardest part of creating such nuclear weapons turned out to be realizing that it could be done; the actual engineering, while difficult, could over time be accomplished by any country with a modern industrial base. This meant that the United States could not expect to retain a monopoly on these weapons; other countries – not all of which were benign[6] – could be expected to create them after the war as well. The United States elected to adopt a policy of *deterrence*; that is, present the world with a credible capability to strike back even *after* a nuclear attack by any foreign country upon the United States, with the expectation that the existence of such a potential retaliatory capability would discourage any other country from attacking the United States.[7] In fact, deterrence has worked; in combination with other US policies (such as the Marshall plan to rebuild a Europe devastated by war, and full of people who were literally on the verge of starvation; the basing of US troops in Europe after the conclusion of the war; and the formation of NATO[8]), Europe has enjoyed the longest period in its history without a major war between European nations, and the

---

6 During World War II, and for several years after the conclusion of the war, the USSR was governed by Ioseb Dzhugashvili (better known by his adopted revolutionary name of Joseph Stalin), recognized as the second largest mass murderer in all of human history (sources are cited at the end of this footnote). Those who study these matters agree that Stalin killed significantly more people even than Adolf Hitler, the head of Nazi Germany. By 1948, shortly after the conclusion of World War II, China was governed by Mao Zedong (formerly spelled in English transcription as Mao Tse-Tung); those who study these matters recognize Mao Zedong as *the* largest mass murderer in history. It was not unreasonable to see the acquisition of nuclear weapons by such people as a significant threat to the world. Sources: https://www.quora.com/What-are-the-worst-genocides-and-mass-murders-in-the-history-of-humanity, https://www.theepochtimes.com/the-worst-mass-murder-of-all-time_2134122.html, and https://about-history.com/list-of-dictatorships-by-death-toll-the-top-10-biggest-killers-in-history.

7 There were alternative policies considered at the time too. To oversimplify a bit, these consisted of (i) having the United States abandon nuclear weapons and hope that Stalin and the others would follow this moral example, or (ii) attacking the USSR before it had a significant nuclear weapon capability of its own. Both of these alternative policies were rejected (in my view, correctly) as presenting untenable risks to the United States and to the world (and in the case of a pre-emptive attack on the USSR, also entailing poor morality). Deterrence was selected, with all of its risks, as being the least-bad alternative available. As we will also discuss later in the book, there is not always a perfect approach to solving a problem. Deterrence, of course, works both ways: the USSR presumably believed that its own nuclear and missile capabilities would deter the United States from attacking them.

8 The latter two policies were intended to break the 1000-year cycle of continuous war between European nations. So far, they have worked.

United States and the USSR fought no war of *any* sort – nuclear or conventional – directly with each other. In my view, deterrence worked because it was implemented not just as a weapons program, but concurrent with efforts to understand the culture and decision processes of potential adversaries that might acquire nuclear weapons (starting with the USSR), so as to understand what actions would create a deterrent effect with those nations. A new risk to the world is that as additional nations elect to acquire nuclear weapons, we may not have the concomitant understanding of the cultural and decision processes of these new nuclear-armed nations.

When we get to Chapter 9, and I introduce my approach for dealing with low-probability but high-impact adverse events, you will likely conclude that deterrence is not a sufficient policy. We can discuss potential alternative and/or supplemental policies when we get to Chapter 9. Ronald Reagan introduced the only significant supplemental policy since the original adoption of deterrence when he advocated adding the ability to defend against intercontinental ballistic missiles as an additional component of deterrence. As moral choice, adding such an ability to defend is without flaw. Missile defense is technically quite difficult, however, and it may be considered an open question of whether or not we have achieved a level of technical missile defense capability that can yet be considered a *credible* addition to deterrence.

Deterrence nearly failed in 1962, in the event known as the Cuban missile crisis. Most assessments that I have read agree that leadership – and the characteristics of the specific individuals involved – played a big role in this near failure of deterrence. I like to think of this incident as showing the power of deterrence: it held, even in the face of specific individuals who displayed weak leadership.[9]

Leadership does in fact matter a great deal, as we will discuss many times in this book (most specifically, in Chapter 13).

---

9 In my view, the Cuban missile crisis was avoidable, and could have been settled by diplomacy, without recourse to threats of military action by the United States; trying to settle the crisis using only diplomacy was the course of action recommended by both the US Secretary of State and the US Secretary of Defense at the time: *The March of Folly*, Barbara Tuchman, 1984. It was the US President (John F. Kennedy) who insisted on threatening the use of military force, against the advice of his cabinet. The previous year (1961) had featured an incident known as the Bay of Pigs. Kennedy had openly stated that he was personally embarrassed by the poor way in which he handled the Bay of Pigs situation, and that he was not going to let himself be embarrassed again (for example, "After the Bay of Pigs ... Kennedy was determined to stand fast," *Thirteen Days*, Robert F. Kennedy, 1969). Notice that Kennedy did not say that he would stand fast when justified by the facts of a situation, but instead he said that would stand fast so that Nikita Khrushchev (Chairman of the Communist Party of the USSR, and therefore the de facto head of the Soviet Government) would not think that Kennedy could be pushed around. Robert F. Kennedy was President Kennedy's brother, and also the Attorney General of the United States.President Kennedy also used frequently to brag that he intended always to show himself as "tough" – this was his own term. For example: "Kennedy asserted that 'I have to show him that we can be as tough as he is'" ("Bearing the Burden," Thomas Patterson, https://www.vqronline.org/essay/bearing-burden-critical-look-jfk%E2%80%99s-foreign-policy); "Kennedy seemed eager to prove his toughness once in office" (ibid.); "Journalist William V. Shannon, after reviewing the first few months of the new administration, concluded that it had established a cult of toughness" (ibid.); "... the policy of toughness became dogma" (ibid.); "Toughness was the tone" (*The March of Folly*, Barbara Tuchman, 1984); "One does not even have to rehash his relationship with Joseph McCarthy to show how JFK willingly played the 'tough on communism' issue in all his campaigns" ("John Kennedy and the Cold War", http://mcadams.posc.mu.edu/progjfk5.htm); "... while running for President in 1960, JFK appealed to the 'tough on the Soviets' issue" (ibid.). There are literally dozens of additional quotes from Kennedy about his desire to appear "tough."The desire to appear tough and the desire to avoid embarrassment are not (in my view) mature, appropriate bases for effective leadership; we will discuss better attributes of leadership in Chapter 13. Even despite this weak leadership (on both sides; Nikita Khrushchev had his own weaknesses of leadership too), deterrence held.

2

Performing Engineering on Projects (Part I)

How do we do engineering on projects? Engineering projects are different from other projects, so learning to be an effective manager of an engineering project starts by understanding how we do engineering on projects. We accomplish this engineering through the engineering life-cycle. In this chapter, I summarize key aspects of how we do the initial stages of the engineering life-cycle, which are called "requirements analysis" and "design." We will complete our overview of how we do engineering on projects in Chapter 3.

2.1 The Systems Method

2.1.1 Motivation and Description

In Chapter 1, we introduced the idea of a project life-cycle – a series of steps or stages through which a project progresses toward completion.

Each stage can and should be governed by a set of guidelines – which we termed *processes* – whose level of rigor and detail should be adapted to the needs of each particular project.

A discipline called *systems engineering*, which is in many ways closely related to engineering project management, has developed what we might call the *system method*. This is intended to increase the likelihood that a system development effort will be successful, and achieves this increased likelihood of success by placing the focus on the behavior of the system as a whole, rather than exclusively on the parts and components.

Since the objectives are so similar, we can transfer much of this thought process from systems engineering to managing engineering projects. In fact, I tend to think of engineering project management as systems engineering supplemented by a specific set of management and people skills.

As I stated in the introduction to this book, I have come to believe – through long experience – that engineering projects are quite different from other projects (such as construction projects and artistic projects – and I have managed those too). The difference centers in something that later in this book we will call *risk*; engineering projects, since they are inventing something new and technological, have more risks, more profound risks, and risks that tend to be centered in engineering matters.

Engineering Project Management, First Edition. Neil G. Siegel.

Companion website: www.wiley.com/go/siegel/engineering_project_management

All of this has led me to the conclusion that one cannot be an effective manager of an engineering project without employing methods for project management that account for this specific nature of engineering projects. What I have learned is how to tailor and adapt the methods of ordinary project management to the specific needs and challenges of engineering projects, in many cases by drawing upon the methods and insights of systems engineering.

In this chapter (and continuing into the next), therefore, we will summarize how one performs engineering – in particular, systems engineering – on our engineering projects. With that background, in later chapters, we will be able to draw upon that knowledge to create those tailorings, adaptations, and new features that allow us successfully to address the challenges specific to managing *engineering* projects.

I will start by describing what I consider the *systems method.*

There is a natural tendency for engineers to focus on the *parts* and *components* of a system, such as computers, radios, motors, mechanical structures, and so forth. After all, they are visible, tangible artifacts; they are of a size and complexity that one can readily grasp; and some important portion of the actual system development effort involves the specification, selection, acquisition, and integration of these parts. Focusing on them is conceptually easy and comfortable.

But … it is usually the case that the reason we are building a complex system to accomplish a mission is that we desire or need something *more than what is provided directly by these parts*; that is, we aspire to create some *emergent behavior*, some "1 + 1 = 3" effect, wherein the new system will do something *more* than what is accomplished by the individual parts.

Let's consider an example or two. I am old enough to remember the first mobile telephones; they pretty much just had a numerical dial pad and two buttons marked "call" and "hang up." They did not store phone numbers, but they were still a breakthrough; you could make a call while you were out and about, without having to depend on finding a phone booth.

But you needed to know the number that you were going to call. At one time, we all had little booklets into which we wrote the telephone numbers of friends and business associates. To use that mobile telephone, I would have to first look up the name of the person that I wanted to call in my little booklet, look next to their name where I had written their phone number, and then punch their phone number into my mobile phone. In the "mobile communications system" of those days (not so long ago!), storing phone numbers – and relating those stored phone numbers to names – was one function; making the actual phone call was another, separate, function.

At some point, this was improved through the introduction of a little electronic device into which I could enter the names and phone numbers of my friends and business associates. This was a big improvement (my handwriting is terrible).

But there were still two separate devices, each of which implemented a separate function.

Then, someone came along and realized that one could put both functions onto a single device. Even with no electronic integration between the two functions, this was an improvement, as I had to carry only a single device, keep only a single battery charged, and so forth.

But then something very different was introduced: Since those two functions (1 = storing phone numbers and relating those stored phone numbers to names; 2 = making the actual phone call) were both on the same device, it was now possible to allow these two functions to *interact electronically*. I could now just find my friend or business associate by name, and indicate that I wanted to call them. The actual phone number was automatically transferred from the first function to the second. This was a radical improvement in simplicity, ease of use, and reduction of errors!

It is also an example of what we mean by the term "emergent behavior." The people that did this in essence created a capability that one might call "dial by finding a name on the list." The capability to do that is not inherent in either of the "parts" (e.g. the list of names and numbers, and the phone itself), but instead "emerges" from the carefully controlled union of those parts. I like to think of such emergent behavior as a "1 + 1 = 3" effect that we are striving to create; or to use an old-fashioned phrase, the "whole is greater than the sum of its parts."[1]

Exercise: Think of two more examples of emergent behavior in today's products and systems.

In the space below, write a short summary of each.

Emergent behavior is *critical* in systems engineering and engineering projects. Almost *every* engineering project today is undertaken to achieve such emergent behavior. Once you realize this, then you realize that your focus as either the designer or the project manager needs to shift from being exclusively or primarily on the *parts*, to instead being on the *system as a whole*. This is the start of the systems method.

Of course, when we put parts together so as to create the *desired* emergent behavior, we are highly likely unintentionally to create other, *undesired, unplanned* emergent behavior. Such unplanned emergent behavior can range in impact from a nuisance to a serious safety hazard. As we will see later in this chapter, good designers therefore not only design their systems to provide the emergent behavior they want, they *also design their systems so as to prevent other types of emergent behavior.*

There is another motivation for the systems method: parts – even large parts that may look like they are complete systems – may not be usable without the complete system. Imagine having cars, but without roads, without filling stations, without traffic signals, without insurance, without spare parts and repair facilities and repair technicians, and even without traffic police. It would not work. It is the role of the systems method to

1 This phrase is usually attributed to Aristotle.

help us figure out what are *all* of the parts that we must have; sometimes, the need for a part is obvious; sometimes, however, it is not. I have seen many systems that "forgot" what turned out to be a critical part; perhaps some data set that the computer program needed before it could operate, perhaps some special piece of test equipment, and so forth. The systems method provides a basis for actually determining early on in the project development cycle all of the parts that are needed. It does this through a technique called *decomposition*, wherein we conceptually break the system into segments, and then break those segments into smaller subsegments, and so forth. We in fact use *decomposition* throughout the systems method, decomposing different aspects of the system: the requirements, the design, the test program, and many other aspects of the problem are all analyzed through decomposition. We will say a lot more about this later.

There is yet a final motivation for the systems method: not only do we need all of the parts, but those parts must *match* and *be in balance*. Cars must be of a size that fits appropriately in the marked road lanes. The weight of the cars must be limited to that which the road can bear. The surface of the road must be reasonably smooth, but there is no reason to make it smoother than is necessary; that would certainly increase cost, and probably decrease reliability. Bumpers on all cars must be at around the same height. When we transitioned from leaded to unleaded fuel we had to ensure different sized nozzles at gas stations, sized so that people could not accidently put leaded fuel into a car that was designed for unleaded fuel.[2] We had to figure out for how long we needed to sell both types of fuel, as not everyone could afford to buy a new car that used unleaded fuel right away (and the car companies could not have built that many cars in a single year, anyway). And so forth. Parts must match, and *be in balance*. We want the car to drive smoothly, and with only a reasonable level of noise. How much of that noise control is to be achieved by the road surface? By the tires? By the suspension? By the shock absorbers? These must not only be in balance in order to achieve the desired effects; we must avoid putting effort into improving a single part if the effect of that improvement at the system level is either not worth its cost, or if other factors prevent there from being any improvement at all!

Here's an example to consider: society will have another very complicated system transition as we start introducing driverless cars. For many years, perhaps for decades, we will have both driverless and human-driven cars sharing the roads. How will insurance and liability work? Will we, at some point, require driverless cars to be connected to a network, so that we can route such driverless cars so as to even out traffic congestion? Will we at some point *ban* human-driven cars? After all, human-driven cars kill tens of thousands of people every year in the United States alone; once the driverless technology is really mature, such driverless cars are likely to be far safer than human-driven cars. Will we then abandon lane markings, and let the driverless cars crowd together, so that we can fit many more cars on the same roads? Doing some of these things would seem to be essential in order to achieve the real benefits of driverless cars – many fewer accidents and fatalities, and the ability to fit far more traffic on the same roads with less congestion. Yet there will be significant social resistance to all of these changes. These are all examples of the questions we consider when using the systems method.

2 Putting leaded gasoline in a car designed for unleaded gasoline would damage the engine.

The *systems method* guides us through all of these issues (see Figure 2.1).

Figure 2.1 points out that one of the key characteristics of the systems method is the use of *hierarchies* and *decomposition*. You will see that we use *hierarchies* and *decomposition* over and over again in project management (and in this book). We describe the product or system we aspire to create by decomposing it into a hierarchy of parts. For example, in Figure 2.2, we depict and describe the world air transportation system by decomposing it into a set of constituent parts – airports, airplanes, and air traffic control – which form level 2 of our hierarchy. We can then decompose each of these level-2 components further into smaller parts; in the figure, I have decomposed airplanes into fuselage, engines, avionics, and passenger equipment. The decomposition can then continue to further levels. Such decomposition into a hierarchy helps us understand complex entities.

We will discover that we can use such *hierarchies* and *decomposition* to represent and analyze many other items, not just the systems and products themselves. In particular, we can use them to represent and analyze our methods and intermediate work products (e.g. requirements, design, approach to testing, and so forth).

Motivations for using the systems method:

- Managing emergent behavior: obtaining that emergent behavior that we want, while preventing other, unplanned emergent behavior
- Identifying and characterizing all of the parts that we will need
- Helps us to achieve balance among all of the parts
- Helps us to plan transitions, especially from the existing system to our new system and its improved capabilities
- Helps us to identify all of the social factors that we must overcome in order to be allowed to implement our new system, and helps us create the data that will be used to justify the transition
- Helps us optimize at the systems level, rather than just optimizing each component

Key characteristics of the systems method:

- Work through hierarchies, using a technique that we call *decomposition*
- Transparent, credible connection of technical decisions to operational performance predictions and parameters
- A life-cycle model, usually separating development, operational, sustainment, and disposal, and within development, separating requirements, design, implementation, integration, validation – usually iterated at multiple levels of the hierarchy
- Iterative progress through a set of life-cycle stages
- Use of systems-level performance prediction, in order to achieve the desired optimization at the systems level
- Use of written processes to guide the work
    - To capture other people's mistakes and lessons learned
    - To operate at scale, and to do so consistently
- Use of reviews and decision gates
- Standardized engineering views, to help us communicate
- Identification of the desired *emergent* characteristics that result from the system as a whole
    - Obtaining these emergent characteristics is why we build *systems*!

Figure 2.1 Motivations for employing the systems method; key characteristics of the systems method.

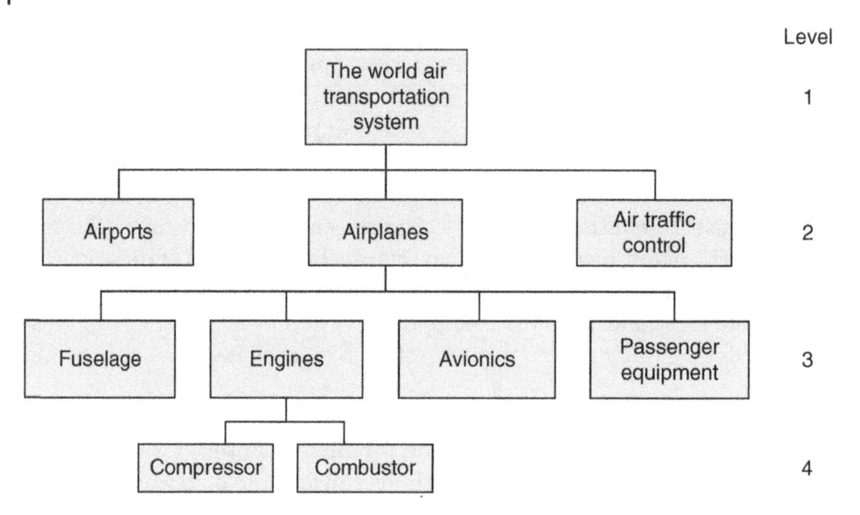

Figure 2.2 An example of decomposing a system into a hierarchy.

In Chapter 1, we introduced the idea of a project life-cycle, a series of steps or stages through which the project moves as it progresses. In Figure 2.3, I expand the list that we used in Chapter 1 to form my own version of the stages of the project life-cycle; other people use slightly different versions of this list.

Next, I will describe each stage:

- *The need and the idea*. In this stage, we try to understand the stakeholders, their needs, and their constraints. What is their mission? How do they do it? What is their product? How do they do their mission today? Why is that how they do it? What constrains the possibilities? What do they value? How do they measure that value? We also look at technology, techniques, and capabilities that are available to use in building the system. What can they really accomplish? What are their limits and

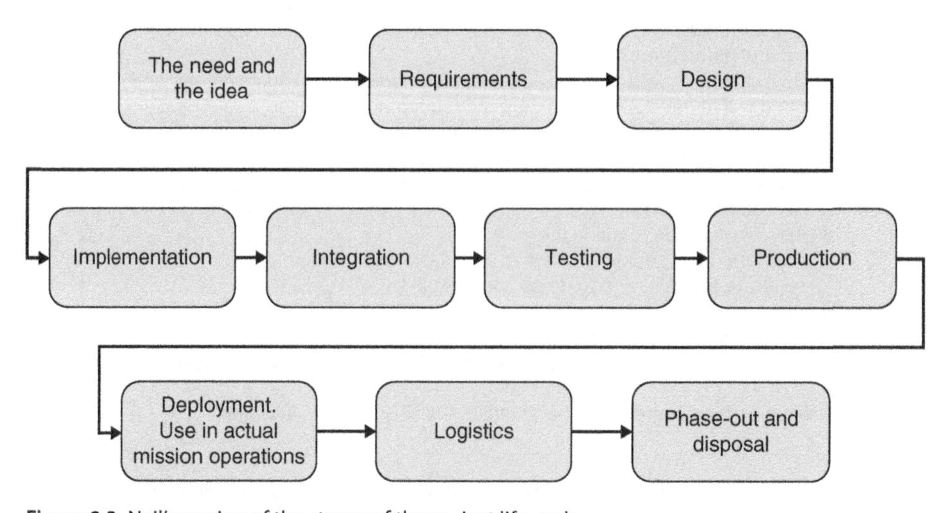

Figure 2.3 Neil's version of the stages of the project life-cycle.

side-effects? How mature are they (e.g. reliability, consistency, safety, predictability, manufacturability, etc.)? Can they be used productively in ways other than their original intent? In this stage, we are trying to understand broad needs and potential enablers, but not yet trying to define specific requirements.

- *Requirements.* The formal statement of *what* the new system is supposed to do, and *how well* it is supposed to do it. The *what* is a qualitative statement of a capability. For example, "The car shall be able to travel in both the forward and reverse directions, and the direction of travel shall be selectable by the driver." The *how well* is a quantitative statement. For example, "The car shall average 30 miles per gallon of fuel under the following conditions: X, Y, and Z," or "The noise level inside of the car shall not exceed 55 dB at speeds below 70 miles per hour, while traveling on road surfaces that meet the following conditions: A, B, and C." Requirements (of both types) are written down and placed into documents called *specifications.* Meeting the requirements, as documented in the specifications, is usually *mandatory*; that is, you may not get paid (or may get paid less) if the system you deliver at the end does not meet every single requirement in the specifications. Requirements – like so much else in systems engineering – are created in a hierarchy: we create requirements for the system as a whole, and then for the major functional elements of the system, and perhaps even a level below that; we do this using the decomposition process.
- *Design.* The requirements say what the system is supposed to do, and how well it is supposed to do it. The design, on the other hand, says *how* all of this is supposed to be accomplished; for example, will our car use a gasoline engine, a diesel engine, or a battery with an electric motor. The design will likely be quite technical at times, specifying, for example, particular algorithms, particular materials, particular structural methods, and so forth. The design, like the requirements, is also created in a hierarchy: for the system as a whole, for the major components of the system, and so forth. Note, however, that the hierarchies for the requirements and the design are seldom, if ever, the same: one is a decomposition of *what* and *how well,* the other a decomposition of *how.* The two hierarchies are related, but they are not the same. The top level of the design hierarchy describes how the system as a whole will be implemented, and in particular, will describe how we are to achieve the emergent behavior desired for our system. Lower levels of the design hierarchy describe successively smaller pieces of our system, and how those pieces interact. The lowest level of the design hierarchy describes how each of the smallest pieces are to work *internally.*
- *Implementation.* Through the decomposition process embodied in the hierarchy for the design, we arrive at the bottom of our top-down definition process (e.g. the left-hand side of the "U"; see Figure 2.4, below): the naming and describing of each of the little pieces into which we have decomposed our system through the design process. We now have requirements specifications and designs for all of these little pieces and, therefore, we are ready to go and build them. Any given such piece might consist entirely of hardware; or entirely of software; or entirely of data; or some combination thereof. We may not need a hierarchy for this stage; we may commission a set of independent teams each to build one or more of those little pieces.
- *Integration.* When people started building systems, especially systems with lots of software in them, their original concept was to finish all of the implementation, then put all of the pieces together, and then proceed to test the system. It was quickly discovered that this seldom worked in a predictable and consistent fashion. The

complete system, with its hundreds of separate parts and lots of software (nowadays, perhaps millions of lines of software code), turns out to be too complex for this sort of put-it-together-all-at-once approach to work. So, gradually, the need for a phase between implementation and testing was recognized, which we now term *integration*. The purpose of the integration phase is to put the parts together, at first in small subsets of the whole, and gradually working one's way through the integration of ever-larger subassemblies toward having the entire system. At this stage, we are not yet *testing* the system; instead, we are just trying to make it operate in an approximately correct fashion. Distinguishing integration from testing has been a gigantic boon to systems engineering and engineering project management; many projects today, however, still neglect the integration stage, and usually suffer greatly from that neglect. We will talk much more about the integration stage in the next chapter.

- *Testing*. Having conducted implementation (which builds all of the pieces of our system) and integration (which assembles all of the pieces of our system, and sorts out enough problems that the system operates in a reasonably correct fashion), we can then turn to the problem of testing our system. For most systems, we will do two different types of testing. First, we have our specifications that contain our requirements: the mandatory statements of *what* and *how well* for our system. We have to conduct some type of rigorous process to make sure that our system meets each and every one of these requirements. In this book, I call this first form of testing *verification*. Such verification, however, is not sufficient; I have seen plenty of systems which meet their requirements but were disliked by their intended users. Therefore, we must *also* assess a set of more subjective matters, such as "Can this system be used by the intended users, or is it too difficult for people with their education, experience, and training?" Or perhaps our systems is used by people who are in stressful situations (such as power plant operators, police, ambulance dispatchers, doctors, or soldiers); is the system designed in such a way that it can realistically be used by people under such stresses with only a reasonable number of errors? In this book, I call this second form of testing *validation*.
- *Production*. Our system development (e.g. all of the life-cycle stages through testing) yields us *one* copy of our system. Sometimes – such as for a satellite project – that is the only copy of our system that we will build. But more often, we then make *additional* copies of our system. This making of additional copies of our system is called *production*. Production may range in scale from making 10 copies, to 1000 copies, to 1 000 000 copies, or today, even 100 000 000 copies. The techniques used for production will need to vary significantly, depending on the scale of the production required. We must also perform some testing to ensure that our production copies are correct; this testing is usually far less in extent than the testing we perform on the first article, however.
- *Deployment*. Our system needs to be placed into service. That is often a complicated endeavor on its own. For example, a satellite needs to be launched into space. Or a new billing system probably needs to be operated in parallel with the existing billing system for a while, before we disconnect the existing billing system and switch over to the new one. Or our millions of new mobile phones need to be sent to retailers and sold. This process of placing our new system into service is called *deployment*.
- *Use in actual operation*. Once our system is in actual use (e.g. we have completed a successful deployment), it can finally be used by the intended users, and bring them the benefits for which it was designed. But those users need support: someone has to

create training materials for the new system, and perhaps even conduct actual training classes. Things break, and someone has to diagnose and fix them. To effect those repairs, we will need replacement parts; we need someone to make those replacement parts. It is likely that we will continue to find errors in the system – even after the test program has completed – and we will need to fix those errors. Most systems are operated for a long time, and our users expect us to design and implement improvements to the system over the course of the time that the system is operated. There are many other, related aspects of supporting our new system in effective operations.

- *Phase-out and disposal.* All good things come to an end, and someday – perhaps long after *we* have retired – our system will reach the end of its usable lifetime and will need to be taken out of service. This can be a very complicated and expensive activity on its own, and in such cases, methods to implement the retirement and disposal of the system should be *designed into the system from the very beginning.* A satellite might need a special rocket motor to deorbit the satellite, so that it burns up in the Earth's atmosphere. A nuclear power plant needs to be designed so that all of the radioactive materials can be taken out when they are expended, and then properly stored or reprocessed. Even mobile phones (and other consumer devices that contain batteries) need special disposal procedures, so that we do not create inappropriate dangers through pollution caused by old batteries. In other cases, we may have to figure out how to safely retire and dispose of a system where no such preparations were made. The world has many dams, for example; all have finite lifetimes, but methods, materials, and funds to dispose of them are seldom worked out until the dam is ready to be retired.

This is our basic system life-cycle; every project may have its own small variations and changes in nomenclature, but the general intent is usually very similar to that which I have described.

In each stage of the project life-cycle, we perform a mixture of activities:

- Technical
- Project management
- Agreements
- Planning and replanning
- Monitoring
- ... and so forth.

Each of these activities can and should be governed by a set of written guidelines – which we in this book will call *processes* – whose level of rigor and detail should be adapted to the needs of each particular project. An engineering process is simply a written description of the steps, guidelines, constraints, inputs, and outputs that we use to perform engineering activities. We will return to the subject of such *project processes* later on in this book.

2.1.2 Life-Cycle Shapes

An interesting insight is that engineering project life-cycles can have *shapes.* I will describe three of the most common such shapes: "U," waterfall, and spiral.

I start with what I call the "U" diagram (Figure 2.4). The concept is clear: we start our project in the upper left, defining our requirements, and then creating our design. As

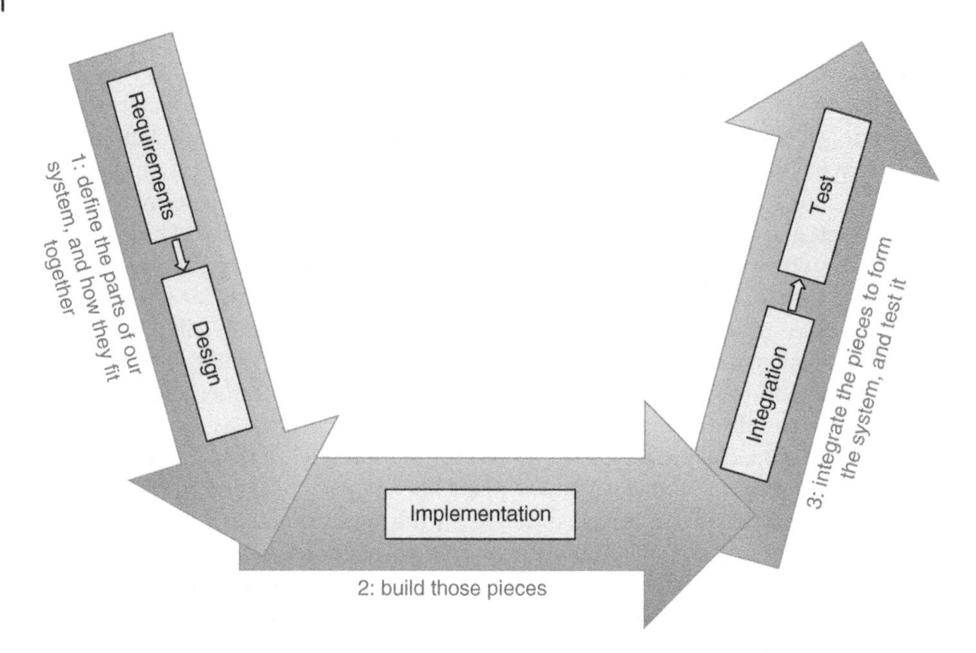

Figure 2.4 Neil's "U" diagram.

noted above, these stages proceed in a *hierarchy*; we do requirements at the system level, and then for each smaller segment of our system. We do design at the system level, and then for each smaller segment of our system. That is, we perform the requirements and design stages by starting at the system level and working our way "down" the hierarchy (where "down" in this context signifies from the system level to ever smaller pieces of the system).

To make it clear what we mean by such a hierarchy, consider the example that we presented in Figure 2.2:

- The *air transportation system* can be considered to consist of the following major parts: *airplanes, airports,* and *air traffic control.* These items – airplanes, airports, and air traffic control – form the second level of this system's hierarchy.
- An airplane consists of parts too, such as the *fuselage* and the *engine*(s). Those parts form a third level of our hierarchy.
- An airplane engine also consists of parts, such as a *compressor* and a *combustor.* The compressor and combustor form a fourth level of our hierarchy.

We say that the requirements and design stages are performed *top-down,* referring to the direction of motion through the hierarchies and through the "U" diagram.

Having completed our requirements and our design, we are ready to implement all the pieces of our system defined by our design. This portion of the diagram is drawn horizontally, because we may not use a hierarchy for this stage; we could just commission independent teams to implement each of the pieces we have defined.

We are then ready to move upward through the right-hand side of the diagram. Having implemented all of the pieces, we start putting those pieces together, at first in small subsets of the system, and gradually progressing to larger subsets, until we finally have the entire system interconnected and operating to some initial degree of correctness; as noted above, I call this process *integration*. We then perform *testing*: the verification that our system is *effective* (e.g. satisfies all of its formal requirements) and the validation that our system is *suitable* (e.g. meets the needs and desires of its intended users). On this side of the diagram, we show the arrow pointing upward, because we progress from small pieces and subassemblies, through larger subassemblies, to the entire system; that is, in this portion of the diagram, we say that we are proceeding *bottom-up*.

Many other people use a similar diagram, which they generally call the "V" diagram; it basically has the left and right sides of my "U." But the "V" version of this diagram either omits the implementation steps, or it places that actual building of the pieces on the right-hand side. I object to either of those approaches. Obviously, the implementation is important, and ought not to be omitted. Perhaps more subtle is the idea that the purpose of the left- and right-hand sides of the "U" diagram is to show that these activities proceed in a *hierarchy*. The actual *implementation* of the pieces that we have defined through the decomposition on the left-hand side, however, need not proceed in any sort of hierarchy; if we have decomposed our system into 500 little pieces, we might well build the 500 little pieces pretty much independent of each other. We resume working in a hierarchy (bottom-up, on the right side of the "U") when we start putting small numbers of those pieces together through the integration process. Hence my preference for the "U" shape over the "V."

The "U" diagram does not depict the latter stages of the project life-cycle (e.g. production, deployment, actual use of the system, retirement, and disposal); neither do the related "V" versions of this diagram. Instead, this "shape" concentrates on the stages that occur during the actual development of our system. Indeed, some of those later phases (e.g. use of the system in actual operations) are not actually "projects," as I defined that term in Chapter 1; instead, they are what I defined as "continuous business operations."

The next shape for a project life-cycle that I will discuss is called the *waterfall* (see Figure 2.5). The so-called waterfall method was introduced by Dr. Winston Royce in 1970.[3] Dr. Royce's purpose was to bring some order to what he perceived as the chaos that seemed to him to be a recurring feature of large software development activities. His recommendation was for a series of particular steps to be undertaken in a particular order, while endeavoring to complete one step before beginning the next. He believed that the need to perform all of these steps was not universally recognized – he actually said that some customers believed that doing some analysis, and then doing the coding, was all that would be required to deliver a software product – and therefore, part of his goal was to advocate the use of the complete set of steps. Although his terminology is in some ways specific to software systems, you can see that, in concept, Dr. Royce's list of steps is not very different from the list of stages that I presented above.

3 The figure is from Dr. Royce's paper "Managing the development of large software systems," Wescon, 1970. I knew Dr. Royce slightly. I later worked closely with his son, Walker – a very fine and original engineer and software architect in his own right – for many years in the aerospace industry. Walker kindly provided me with a copy of his father's original paper, and also gave me permission to use those materials in this book.

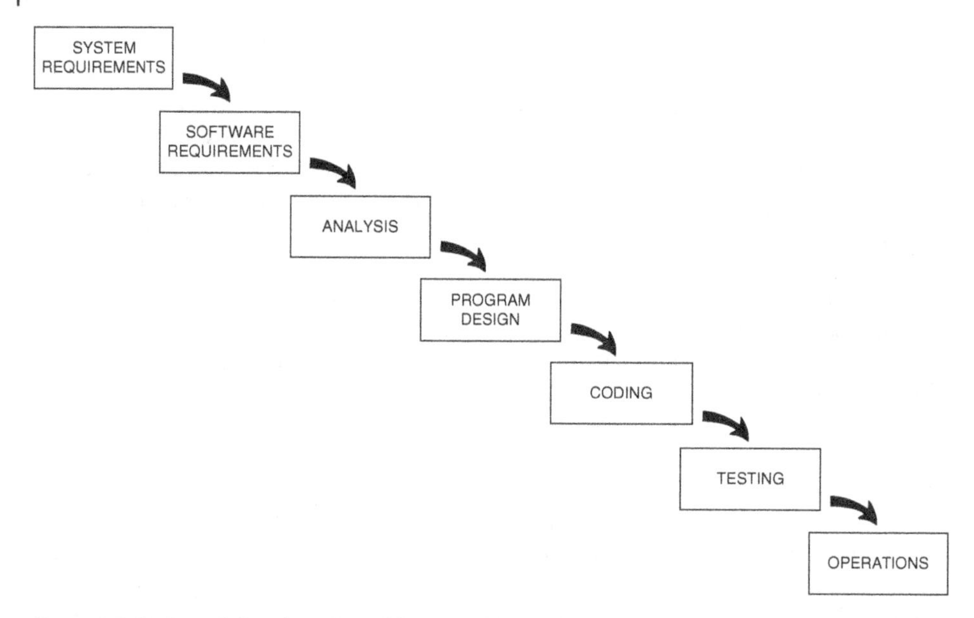

Figure 2.5 Dr. Royce's first depiction of the waterfall method. *Source:* Used with the permission of Dr. Royce's son, Walker.

It is difficult to overstate how important and influential Dr. Royce's work has been. Large companies, like TRW and IBM, created corporate software development policies that more or less adopted Dr. Royce's approach *in toto*. So too did the US Department of Defense, and through the Department of Defense's contractual terms that mandated that companies who were building software (and later, systems) for the US Government follow those standards, in fairly short order the entire world was following the waterfall method.

There is an additional insight that Dr. Royce provided: a recognition that things do go wrong, and one might at times have to go backwards (Figure 2.6).

The depiction of Figure 2.6 is often misunderstood. If one just looks at the drawing but does not read Dr. Royce's paper, one might get the impression that one is allowed – even encouraged – to back up as many steps as wished. This is not so! Dr. Royce is very clear that the desired approach is to be thorough enough at each step that one never has to back up more than a single step. His original caption for this figure actual reads "Hopefully, the iterative interaction between the successive phases is confined to successive steps."

The waterfall method contributed huge value through its "normalization" of the necessity for steps other than analysis and coding, and through its promulgation of the idea that there should in fact be a planned sequence of steps.

In short, the waterfall method was aimed at introducing some organization and structure into what was perceived to be an overly chaotic approach to systems engineering and engineering project management. But the waterfall method also implied a rigorous sequencing of (i) doing *all* of the requirements, then (ii) doing *all* of the design, then (iii) doing *all* of the implementation, and so forth; in fact, many of the corporate and

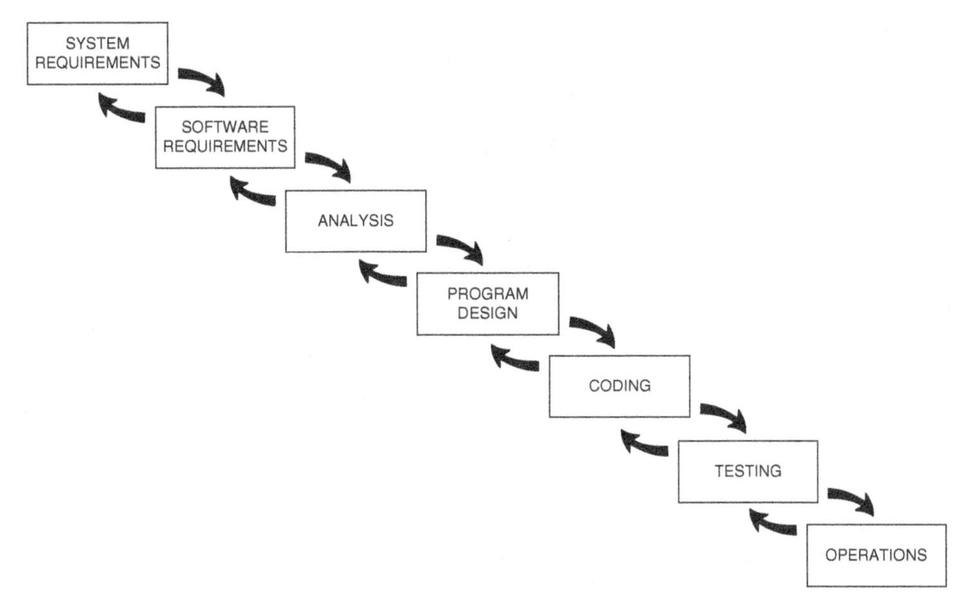

Figure 2.6 Dr. Royce's second depiction of the waterfall method. *Source:* Used with the permission of Dr. Royce's son, Walker.

government development policies that were created in the wake of Dr. Royce's paper said *explicitly*: finish one step before you proceed to the next, and go through the life-cycle exactly *once*.

Experience, however, soon showed that this was too constraining to be practical for many projects, especially those of larger scale and complexity. At times, people would achieve success through an incremental approach that involved a sort of successive approximation, by building a well-defined partial version of the system, operate that version for a while to gain additional insight, then build a second well-defined partial version of the system, operate that version for a while to gain additional insight, and so forth.

What to do? Live with the perceived inflexibility of the waterfall method, or return to the pre-waterfall chaos? Neither of those choices seemed very good.

Fortunately, someone came along and proposed a new method – and a new project life-cycle shape – that solved this dilemma, allowing the continued rigor and organization of the waterfall method, while creating a structured framework for successful development through incremental, successive approximations of the eventual system. The person who first put this forward as a candidate formal method was Dr. Barry Boehm,[4] in 1986;[5] he termed it the *spiral model*. The spiral model (Figure 2.7) forms a third shape for a project life-cycle.

4 I am fortunate indeed in being able to claim Dr. Boehm as my Ph.D. advisor! Dr. Boehm has kindly allowed me to use his materials in this book.

5 Barry Boehm, *A Spiral Method of Software Development and Enhancement*, ACM SIGSOFT Software Engineering Notes, Vol. 11, 4 August 1986, p. 14 ff.

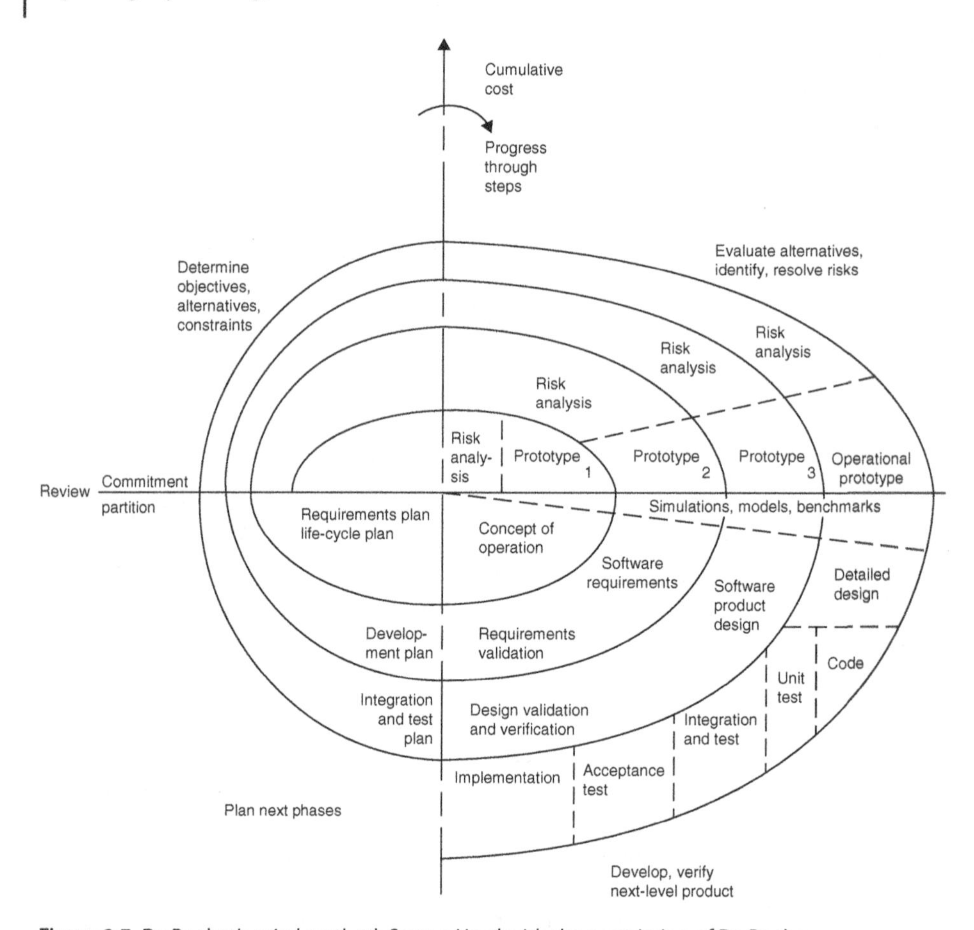

Figure 2.7 Dr. Boehm's spiral method. *Source:* Used with the permission of Dr. Boehm.

The point of the spiral method is that we develop a carefully thought out *partial* version of our system, and then we *deliver that partial version to the users* (Dr. Boehm calls this a *prototype* in Figure 2.7, but this partial version, ideally, is actually used to accomplish real work), who then use it for actual mission operations. The development team observes these operations, and thereby gains new insights into what those users actually need and want. The developers then incorporate those new insights into their plans for the next increment – which Dr. Boehm calls a *spiral* – of the system. This way, features or omissions that might cause the system to be unacceptable to the users get found and fixed along the way. It turns out (we quantify this in the next chapter) that fixing things *earlier* in the project development life-cycle costs *far less* than fixing them later. Fixing them earlier, of course, also increases the user's satisfaction with the system, and their confidence in the development team.

Again, it is difficult to overstate how important and influential Dr. Boehm's work has been. All of the same organizations that in the 1970s and 1980s created policies and directives mandating the use of the waterfall method modified those policies and directives in the 1990s to allow and encourage the use of the spiral method. As a result, some variation of the spiral method is used nearly universally on engineering projects today, especially engineering projects with lots of software (which these days is most of them). Legendary computer scientist Dr. Fredrick P. Brooks says of Dr. Boehm's spiral model, "I strongly believe the way forward is to embrace and develop the Spiral Model."[6]

Other shapes are possible, and have been propounded in the literature. But for the purposes of this book, we will limit ourselves to these three project life-cycle shapes.

2.1.3 Progress Through the Stages

Whatever shape your engineering project employs, you will use some type of orderly method to determine when you are ready to move from one life-cycle stage to the next (e.g. from requirements to design, and so forth). This method is centered around a *review*, which is the data-gathering and data-analysis exercise that forms the basis for a formal decision process (e.g. are we ready to move to the next life-cycle stage or not?). If we determine that we are *not* yet ready, we then determine what are the things that remain for us to accomplish, so that we are ready to move to the next life-cycle stage. Generally, you want a large segment of the stakeholders for your project (the development team, your company's management, the buying customer, the using customer, the paying customer, and so forth) to participate in these reviews and decisions. We include all of these people, both in order to get a full range of opinions to inform your decisions, but also to build a social consensus about the correctness of that decision.

This process is often referred to as *decision gates*. We call these reviews *gates* because we may be allowed to pass through them at this time ... or we may not; that is, the gate may be either open or closed. The reviews are intended to determine the adequacy of the system to meet the known requirements (specifications and process guidelines) and constraints. Reviews become progressively more detailed and definitive as one advances through the program life-cycle.

In toto, reviews provide a periodic assessment of your project's maturity, technical risk, and programmatic/execution risk. Equally importantly, they help one improve/build consensus around the go-forward plans; this is why you involve so many of your stakeholders in the review process. Reviews provide you – the manager of this engineering project – with the data you need to make the decision about whether to proceed to the next phase, or return to the previous one in order to resolve some issues.

As noted above, an important characteristic of the systems method is that we strive to optimize at the *systems* level, not at the *component* level. Even if we make each component of our system the very best possible, they may interact in a fashion that provides less than the best possible performance; since we are striving to make the *system* as a whole the best possible, we must look at the *interactions* of the components, in addition to the performance of each individual component.

This may have other benefits. For example, I have frequently discovered that I could make do with a less capable version of some component in a system and still get the system-level performance and capacity that I needed. It would be a waste of money to have paid for a better version of that particular component; limitations that arise out of

6 Fredrick P. Brooks, Jr., *The Design of Design*, Addison Wesley, New York, 2010, p. 58.

the interactions of the components might prevent the improved version of that component from having a positive effect on the system as a whole.

This quest for system-level optimization can lead to useful insights too. I once had a radio vendor on a military command-and-control system come to me and say that they had figured out (for a price!) how to make the signaling rate of their radio become twice as fast as it was at present. In the coordinate system of value for a radio designer, being twice as fast is of high value indeed. The radio signaling rate is an example of what I will later call a *technical performance measure*; this is an objective measurement within what I like to call the *engineer's coordinate system of value*. But before making a decision to pay that money for faster radio, I had my system-level modelers insert a model of that faster radio into our system-level performance model. We had created a set of metrics that measured the performance of the system not only in technical terms, but also in terms of the *operational benefit to the intended users*; this is what I will later call an *operational performance measure*, and it forms an objective measurement within what I like to call the *customer's coordinate system of value* (Figure 2.8). For example, we had created a metric that predicted the level of casualties on each of the opposing sides in a battle scenario; if our system design was going to add value to the customer (in this case, the US Army), the ratio of the number of enemy casualties to the number of US Army casualties should be higher than for the same battle scenario when our system was not used. We called this the *loss-exchange ratio*; a higher loss-exchange ratio indicated a better design, because that was a metric that the intended users of the system valued.

When we plugged the radio with the doubled signaling rate (a highly favorable technical performance measurement, in the engineer's coordinate system of value) into our system model, the result was *no improvement at all* in the system's overall performance, as measured by the loss-exchange ratio! Therefore, in the customer's coordinate system of value, the *faster radio had no value.*

<table>
<tr><th>The customer's coordinate system of value</th><th>The engineer's coordinate system of value</th></tr>
<tr><td>Usually non-technical in nature</td><td>Usually technical in nature</td></tr>
<tr><td>Characterizes something about the mission</td><td>Characterizes something about the technology that implements the system</td></tr>
<tr><td>Captured in what I call operational performance measures</td><td>Captured in what I call technical performance measures</td></tr>
<tr><td>Measures the goodness of the design for the users</td><td>Measures whether or not the design is feasible, and helps us estimate the schedule and cost needed to build the system</td></tr>
<tr><td colspan="2">The two types of metrics should be linked; that is, we need to credibly and transparently show how changes in design (which change the technical performance measures) cause changes to the operational performance measures

Our degrees of design freedom are located in the engineer's coordinate system of value, but we measure the goodness of the resulting design in the customer's coordinate system of value</td></tr>
</table>

Figure 2.8 The two coordinate systems of value, and the two types of objective measurements.

But we did not stop there; I asked the radio vendor and the modeling team to work together to figure out *why* the performance at the system level did not improve. They discovered that there was a subtle bottleneck, and this insight led to an idea for a *different* improvement that the radio vendor could make: leave the signaling rate the same, but decrease the time it took for the radio to acquire the channel and synchronize the encryption process. This was a far less expensive change to make than doubling the signaling rate, but led to a major improvement in system-level performance.

My lesson: You should always ask *why*! And you cannot trust your intuition; the interactions inside of a complex system make it very difficult to intuit the system-level impact of a change to a component.

We – not the Army – created the loss-exchange ratio metric. But clearly, this metric would have more value to our design team if the customers and eventual users agreed that this metric actually reflected their coordinate system of value, that is, a better score on this metric would indicate a system that the users would actually find to be better. So, you need to take the effort to *socialize* your ideas for operational performance metrics with your customers, users, and other stakeholders. I like to achieve what I call the *transfer of emotional ownership* to the customers. This signifies that in some real sense, the metric has become *theirs*, rather than *mine*. For example, after socializing the loss-exchange ratio metric with the US Army for several months, I learned that the Army modeling organization had started using that metric as the primary output of *their own* system performance models; our metric was now driving internal Army decision-making about the future of our system. That is a transfer of emotional ownership! When you achieve that, you are building credibility with your customers, users, and other stakeholders.

Another aspect of the systems method is that we employ written guidance regarding our methods. We call such guidance *processes*. We have processes for every aspect of our engineering project: engineering, but also finance, hiring, managing our people, configuration control, contracting, procurement, testing, quality, safety, and many other items (many of which we will cover over the course of this book). Each process will specify in writing what is to be done, when, by whom, what the products will be, what artifacts will be created, how the work is measured and quality ensured, who approves the work and the artifacts, and many other aspects.

Figure 2.9 summarizes the range of processes that we use in an engineering project.

Why go to the trouble to create and employ such written process? Because engineering a complex system is *hard*. Dr. Eberhart Rechtin[7] said that:

- Success comes from wisdom
- Wisdom comes from experience, and
- Experience comes from mistakes.

When possible, it is best to learn from the mistakes of *others*, rather than learning only from mistakes that one makes oneself. And *that* is the role of engineering processes – they allow us to learn from the mistakes of others. They are the "lessons learned" from past activities.

---

7 Eberhart Rechtin, *The Art of Systems Architecting*, CRC Press, Boca Raton, FL, 2002. I knew Dr. Rechtin too.

Planning	Written guidance for how each aspect of our project will be performed
Contracting	Written guidance for how all aspects for making and documenting binding, legal commitments with our customers is performed
Acquisition and Purchasing	Written guidance for how we acquire materials and services from outside organizations, whether by subcontract (usually used for items that involve invention or adaptation) or by purchase order (usually used for items that are being purchased without any adaptation)
Finance	Written guidance for how we will allocate budgets and funding to people and tasks, keep financial records, and all other financial aspects of the project, including compliance with customer requirements, laws, regulations, and company standards
Project Start-up	Written guidance for how we will get the project started, including acquisition of necessary facilities, equipment, people, cash, intellectual property, and so forth
Technical Management	Written guidance for how we will oversee the engineering, development, integration, testing, production, and other technical aspects of the project
System Requirements	Written guidance for how we will create and validate the design for our system
System Design	Written guidance for how we will create and validate the design for our system
System Implementation	Written guidance for how we will build the components of our system, including hardware, software, and data
System Integration and Test	Written guidance for how we will assemble the components of our system into a functional whole, and how we will then check that the system complies with the requirements and incorporates other desired features
Making Decisions	Written guidance for how we will make decisions, and keep records about them
Delivery	Written guidance for how we will deliver, install, and bring into actual operation the completed system
Property Management	Written guidance for how we will keep track, protect, and utilize material objects on the project, with particular emphasis on items that belong to our company, that belong to our customer, and to those items that must be delivered to the customer at specified time points in the project
Technical Evaluations / Quality	Written guidance for how we will achieve the levels of quality specified in our contract, our requirements, and in our organization policies
Safety	Written guidance for how we will keep our employees, our users, and our community safe while we perform the work on our project
Configuration Control	Written guidance for how we use control changes to items (documents, parts, subsystems, data, and so forth), and ensure that we know what is the state of each item, and that we are always using the correct version of each item
Human Resources	Written guidance for how we will acquire, motivate, and retain the people we need to do the work entailed in the project, make appropriate plans about the time-phasing of needed personnel, and provide for the orderly transition of people to another project when their contribution to this project is complete
Project Termination	Written guidance for how we will ensure that all items specified are delivered to the customer at the conclusion of the project; that all contractual obligations are fulfilled and documented appropriately; and that people, materials, facilities, and other resources are properly handled and accounted for as the project comes to a close

Figure 2.9 Examples of the range of processes that we use on an engineering project.

There is a *caveat*: processes are *necessary* – they help us be repeatable, and operate at scale. But processes by themselves are *not sufficient* to ensure a good design! We need *both* good processes and a good design. Good designs come from *good designers*, not from good processes.

Some companies have gone through a phase of assuming that good processes are in fact sufficient to ensure good design; the result was a series of expensive project failures.

We discuss how to achieve a good design later in this chapter.

Lastly, the systems method involves a lot of *planning*. We formulate and write plans about how we are going to perform each of the various aspects of the project, ranging from how we will validate the technical requirements, to how we will acquire the people with all of the specialized skills we need (and at the right time, and in sufficient quantity), to how we will keep our people (and the general public) safe as we perform this work.

2.2 Requirements

One of the first stages in the project life-cycle is the process of creating and validating the requirements for our system.

Let's start by considering *why* we think about requirements. It is a fact that people often just do things, without having given a lot of consideration to the detailed nature of the problem, or without having spent a lot of time considering what is the best approach to use. Is that how successful people attack a problem?

I have been told that someone once asked the famous physicist Albert Einstein how he would allocate his time, if he had only an hour in which to solve a problem. His answer?

45 minutes to understand the problem
5 minutes to formulate a solution
10 minutes to implement the solution

(He left off verifying the solution.)

That is, Dr. Einstein would allocate 75% of his time just to the task of *understanding the problem*, before he started doing any actual work to formulate or create a solution.

On real engineering projects, we generally cannot allocate 75% of the time to this single task, but the point I take away from this (potentially apocryphal) story is that this particular highly successful person[8] believed that the path to success entailed allocating a significant portion of time to the question of *understanding the problem to be solved.*

In systems engineering and in engineering project management, therefore, we try to understand the problem, and then we write down what we have decided.

Requirements is the term we use for the *formal, written statement of the problem* that we are trying to solve by building an engineered system. Requirements are a statement of *what* the system is supposed to do, and *how well* it has to do it. But requirements are not a statement of *how* the system does it; *how* it does it is the *design* (which we will discuss next). See Figure 2.10.

- Requirements are a statement of *what* the system is supposed to do, and *how well* it has to do it
- But requirements are not a statement of *how* it does it; *how* it does it is the *design*

Figure 2.10 Definition of the terms *requirements* and *design*.

8 Dr. Einstein's fame is well deserved. He had four separate breakthrough insights into physics, each of which would have been a good lifetime accomplishment for anyone else. He won the Nobel Prize in Physics in 1921 for his explanation of the photoelectric effect; the other three breakthroughs he made included his explanation of Brownian motion, his special theory of relativity (which won him worldwide fame but *not* the Nobel Prize), and his general theory of relativity.

Let us illustrate this definition with an example. What is a car supposed to do? We might say that a project to create a car "shall provide a separate physical entity that is capable of moving under its own power over a paved road from one location to another, under the control of a human being." That is a statement of *what.* We might also say that "The car shall be able to reach a speed of at least 50 miles per hour, sustained for 2 hours without needing to stop for refueling or any other purpose." This is a statement of *how well.* The question of whether our car uses an internal combustion engine and gasoline, or an electric motor and a battery – or a hamster in a cage – is a question of design, that is, *how* we accomplish the *what* and the *how well.* We choose the design *after* we have specified the requirements; the *what* and *how well* requirements stated above do not tell us what type of engine to use. We might have a *how well* requirement about limitations on the pollution generated by the operation of our car, but ideally, that requirement does not tell us what type of engine to use either.[9] Nor do either of our requirements determine whether the car should have three or four wheels, how large those wheels and tires should be, and other considerations of *how*; those are *design* decisions.

Here's another example. One of the US Army's short-range air defense weapons – called the Avenger – has a missile operator in a turret on the back of a small truck (see Figure 2.11). There are eight missile tubes located on the top of this turret. A radar located somewhere else sees objects flying in the sky, and a computer makes a preliminary assessment about which ones are friendly aircraft and which ones might be hostile aircraft. Information about both types of aircraft are sent by a data radio to the depicted unit, which receives that information and displays it on a computer screen with a map. The weapons operator may select an aircraft that he thinks he may want to shoot down. But ... before he is allowed to shoot, he is required by US Army policy visually to look at the aircraft through a magnifying optic, and make a determination, based on the training that he has received, that in fact this is a hostile aircraft (and *not* a friendly aircraft). Only *after* making such a visual identification may he press the button to shoot a missile. All of this, by the way, takes place while the depicted unit is moving, driving either on a road or cross country.

But how is the operator to perform this visual identification? When he selects the aircraft that he thinks he may want to shoot down, the turret on the back of the truck turns and elevates so that the magnifying optic is pointing at the correct aircraft, and the turret continues to adjust its position automatically, so as to keep that aircraft in the field of view of the optic. This process is called "slew-to-cue," and there are written requirements defining exactly what the slew-to-cue process must accomplish, both *what* (e.g. "Upon designation of a candidate target aircraft by the operation, the computer shall compute and issue the appropriate commands to the turret's positioning motors so as to adjust the position of the turret in both azimuth and elevation, and to continuously update these positions as both the aircraft and the weapon continue to move") and *how well* (e.g. "Upon slew-to-cue, 90% of the time, the designated aircraft will be in the narrow field of view of the launcher's optics"). Note that neither of the

---

9 An electric car is not pollution-free! Not only must the car be manufactured, but the electricity must be generated (and the generation equipment manufactured); all of those steps cause pollution. In most parts of the world today, given the methods used to generate electricity, careful studies have concluded that electric cars actually cause slightly more pollution than gasoline-powered cars. This will change only as we change the way we generate electricity. Electric cars do, of course, change *where* and *when* the pollution occurs.

Figure 2.11 Avenger air defense weapon.

statements of *what* and *how well* says anything about *how* we will accomplish this; the choice of computer, programming language, algorithms, servo-motors, rate sensors, and so forth is left to the design activity.

But suppose we were at the beginning of our air defense project. You do not yet have all of the information contained in the previous two paragraphs. What might we have when we started the project? Probably, we have only some statements of objectives and constraints from the customer, such as

> Never shoot down a friendly aircraft
>
> and
>
> You may not shoot at any aircraft, even if you think that it is an enemy aircraft, until after a trained operator has conducted a visual identification of the target through an appropriate magnifying optic, and determined by that visual examination that it is actually an enemy aircraft

So, what do you do next? My approach is to start by gaining an understanding of the customers, especially the eventual users. What is their mission? What are the constraints placed upon them? How do they accomplish their mission today? What do they like about the way they do it today, and what do they think needs improvement? Out of questions like these, you can start to distill a statement of what I call the customer's *coordinate system of value* (please review Figure 2.8): What do they value? How do they measure it?

Some of this might already be written down (in policy manuals, training manuals, and so forth), but much of it will not be. You have to go and talk to people, listen to people, and equally important, watch them while they work.

In my experience, it is vitally important that this knowledge be acquired by the engineers on the project; it is not sufficient for the project engineers to depend on other people ("domain experts") for this knowledge. I find that we need to have *both* the domain knowledge and the engineering skills in a single brain. Only then are we able to create useful new insights.

Let's go back to our air defense example. After you read the manuals, talk to people, go on exercises, and watch air defense personnel train in their mission, and maybe even go to a theater of war and watch them conduct real operations, you can start creating a more detailed list of the key steps, functions, and attributes needed to accomplish the short-range air defense mission. It might look like Figure 2.12. Are we done? Are those requirements?

- Find objects in the sky
- For each such object in the sky, make a preliminary assessment: is this a friendly, neutral, or enemy aircraft?
- For each aircraft that appears to be an enemy aircraft, determine if it is engageable under the current rules of engagement
- For each aircraft that is engageable under the current rules of engagement, decide if "now is the time"that you want to shoot at it
- If now is the time that you want to shoot, select a weapon, and give it the pointing coordinates
- Using those pointing coordinates, perform a "slew to cue" operation, which puts the candidate target aircraft into the field of view of the optical sight
- Perform the visual identification of the candidate target
- If the trained operator, as a result of performing the visual identification, decides that in fact this is an enemy aircraft, conduct the engagement (press the button to fire a missile)
- Figure out if we destroyed the target, or not. If not, decide if we want to take another shot
- Do all of the above while on the move!
- Do all of the above at least a certain percentage of the time
- Do all of the above with the available personnel (e.g. don't make the system require skills that the available personnel do not possess)
- *Never* engage a friendly aircraft!!
- Be able to plan and execute movements of the units that comprise the air defense force
- Be able to monitor the location, status, and readiness of each element of the air defense force: equipment, personnel, supplies
- Be able to plan and execute resupply operations

Figure 2.12 Key steps, functions, and attributes of accomplishing the short-range air defense mission.

Unfortunately, no. Let us consider just the very first statement on that list: "Find objects in the sky." Before we can declare that we have a viable requirement about finding objects in the sky, we have to consider items such as these:

- How small an object do we need to see?
- Are there limits on the speed of the object that we need to be able to see?
- Are there limits on the materials of which the object is made? Can we be satisfied with seeing airplanes made of metal; do we not have to see airplanes made of plastic?
- Do we have to provide 360° azimuthal coverage, or can we be satisfied with looking only in a particular direction?
- Do we need to look at all elevation angles, from the horizon to the zenith, or can we be satisfied with only looking at some lesser set of elevation angles? It might be expensive to build a sensor that can see all the way to the zenith, for example.
- How far out do we need to look (e.g. to what slant range?).
- Is it sufficient if we report only the slant range and azimuth of the objects we find in the sky? Or do we also need to report the elevation angle?
- In what coordinate system do we need to make our reports? In latitude and longitude? In some circular coordinate system centered on our sensor? In some circular coordinate system centered on a weapon?
- Is it sufficient to report the object just once? It is an airplane, and therefore will continue to move; how often do we need to send subsequent reports?
- How accurate do the measurements contained in our reports need to be? How is accuracy specified?

For each functional statement on your list (e.g. "Find objects in the sky"), we need first to figure out all of the detailed questions that the requirements must address, and then we must go and figure out all of the answers to those questions. We then phrase what we learned in a particular fashion: we use the verb "shall" to indicate that a sentence is a mandatory requirement; we use the verbs "is" or "are" to indicate that a sentence is a supplemental description, rather than a mandatory requirement; we place each statement with a "shall" in its own paragraph and with its own paragraph number. When we have done all of that, then we have the actual requirements.

Notice that some of the resulting requirements will contain numbers; these form the *how well* portion of the requirements.

Requirements are written down in documents called *specifications*. Therefore, a specification is a document used in acquisition/development, which clearly and accurately describes the essential technical requirements of an entity (system, subsystem, component, etc.).

On most projects, we will have a *hierarchy* of specifications; that is, we will have a specification that defines the requirements for the system as a whole, but we will also have subordinate specifications that describe in more detail the requirements for individual subsystems and components. On a typical large engineering project, this hierarchy of specifications is likely to be three or four layers deep, moving from (at the top of the hierarchy) defining the requirements for the system as a whole, to defining the requirements for smaller and smaller pieces (Figure 2.13). This hierarchy is usually called the *specification tree*. We break the requirements into these separate specifications because each subsystem and component is likely to be designed and built by a separate team, and it is convenient, and reduces errors, to have separate requirements

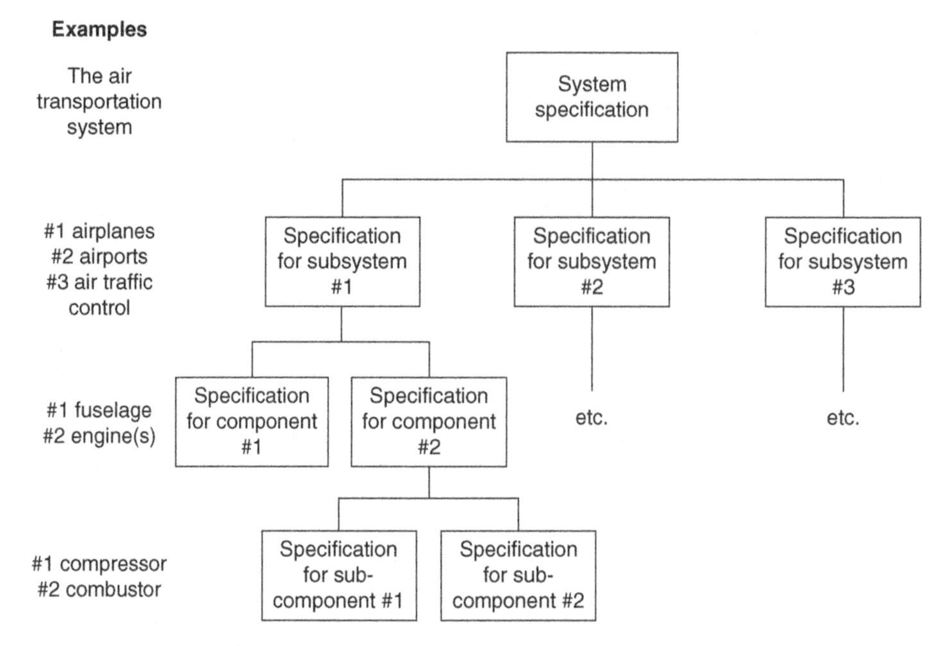

Figure 2.13 The specification tree.

specifications for each such team. By separating the subsystem and component requirements from the system-level requirements, we also gain the benefit of being able to focus separately on the system-level requirements, where so much of our desired emergent behavior will reside.

Remember what we said about requirements: they define the *what* and the *how well* for our system, but not the *how*.

Specifications contain a little more than just the statement of the requirements; since it would be useless to write a requirement that is for some reason impossible to test, it has become customary to include in the specification an indication of the basic strategy by which each requirement will be tested. We return to this subject in the next chapter.

Specifications usually form a *contractually binding* commitment; that is, if the system you eventually deliver fails to implement some requirement, your company may be paid less for their services, or there may be some other form of penalty. We will therefore pay a lot of attention to making sure that our design is in fact implementing every single requirement in our specification tree.

How do we go about creating the requirements? Figure 2.14 defines the steps that I recommend. There are a few new terms in this figure, each of which are explained in the paragraphs that follow.

We have already talked about some of these steps, but not all of them:

- *Identify who are the users and the other stakeholders.* We have already talked about this; we must know all of the users and the other stakeholders for our system. The buying customer is only one of those stakeholders, and it is likely that they are not *users* of the system at all. As we already discussed, there are likely many other stakeholders besides the users and the buyers.

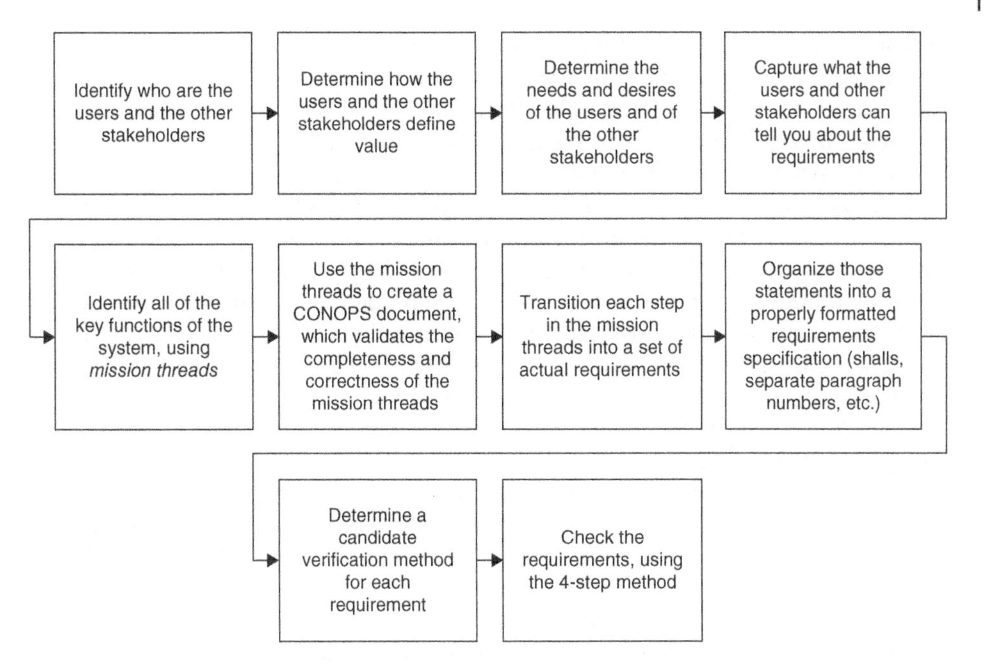

Figure 2.14 Creating the requirements.

- *Determine how the users and the other stakeholders define value.* We already talked about this one too; we must determine how the users and other stakeholders define value, in the context of our system. As noted above, we do this by reading, talking to people, listening, and watching the users perform the mission with their current tools and methods.
- *Determine the needs and desires of the users and of the other stakeholders.* Having determined how our users and stakeholders determine value, we are ready to determine what they need and want. This involves more reading, talking, listening, and watching. Many texts on requirements tell you only to consider documented needs, as you may not get paid for doing additional things (e.g. *desires*) that are not in your contract. I strongly disagree; the customers must end up being happy. This always involves figuring out what they want, in addition to what they need. You may well then try to get those items added to the contract (so that you can get paid for them), but you might elect to do some of them anyway, in the interest of establishing good relations and high credibility with the customer. When things go wrong (and things *will* go wrong at some point), you need the customer "on your side."
- *Capture what the users and other stakeholders can tell you about the requirements.* Usually, the buying customer will have some written description of what they want to buy, and of course you start with that. But even if they do not, you must elicit that information from them.
- *Identify all of the key functions of the system, using mission threads.* Consider Figure 2.10; here, we tried to capture a list of all the major functions of our notional short-range air defense system. How did we create that list? The method I advocate is to create *mission threads,* the major operational sequences of our user's mission.

What are all of the external stimuli that cause our users to start some sort of a task? What is the sequence of steps (this sequence is what I call a *mission thread*) that they move through in order to accomplish each task? What are the results of each such sequence of steps? You can read Figure 2.10 and create a list of what were the likely mission threads that I had in mind when I created the figure.

- *Use the mission threads to create a concept of operations (*CONOPS*) document,* which not only describes the mission threads, the inputs, and the outputs, but also summarizes how often each mission thread is likely to be exercised during actual operations, identifies any timing constraints (e.g., a particular thread might have to be accomplished in fewer than 10 seconds), identifies other constraints (e.g. the requirement to perform the visual identification step) imposed by outside authorities, and so forth. You also write down what you have learned about why each of these items is done in this particular fashion. The CONOPS document allows you to validate the completeness and correctness of the mission threads.
- *Transition each step in the mission threads into a set of actual requirements.* Recall that we said items like those in Figure 2.10 do *not* constitute actual requirements. We discussed how we would turn just one statement from the figure ("find objects in the sky") into actual requirements. Now, we must complete a similar process for every step along every one of our mission threads.
- *Organize those statements into a properly formatted requirements specification.* Over time, we have developed standardized formats, lexical conventions, and so forth for requirement specifications. These are often documented in government or company policy manuals, and include things such as the use of particular verbs for distinguishing actual mandatory requirements from explanatory materials (e.g. *shall* versus *will*), the use of a separate paragraph number for each individual requirement, and other conventions.
- *Determine a candidate verification method for each requirement.* Recall that earlier, I stated that in our requirements specification we also include a preliminary idea for how we will *verify* each requirement. It is possible to write requirements in a fashion that is difficult or impossible to verify, and we strive to avoid that (if for no other reason than we will not get paid until we have verified every single requirement!). In order to avoid such a situation, for each requirement we identify a candidate method (usually the candidate methods are *inspection, analysis, demonstration, simulation,* and *assessment by operation*) for verification, and include this verification method right in the specification.
- *Check the requirements.* We cannot consider something as important as a requirements specification complete until we have performed some type of check that it is complete and correct. I use a four-step process that I learned from Jim Hines and Scott Jackson in 2009[10].
  - *Editorial check* (format, grammar, spelling, punctuation, consistent use of terminology).
  - *Soundness check* (no missing items or "to be determined," quantitative values where required, positive statements [e.g. *shall,* rather than *shall not*], no statements of design [e.g. no statements of how, only statements of *what* and *how well*], no statements constraining external systems, validation of all assumptions).

10 Scott Jackson has kindly given me permission to cite his method (and the SMART acronym) in my teaching materials, for which I am very grateful.

- *Substance check* (complete, consistent, every requirement is necessary, appropriate level of constraints).
- *Risk check* (technical feasibility, consistent with budget and schedule, identify things that could go wrong, estimate likelihood and consequence of each, assess realism of available approaches to mitigate each risk should it actually occur).

Remember that we said requirements usually become a *binding part of the contract.* If you let the requirements say more than you intend, you may well have to foot the bill for building that extra capability!

Scott and Jim also have a cute acronym intended to remind us about the desired attributes of requirements, which I have modified slightly so that it now reads:

A requirement must be *SMART*

(***S***pecific, ***M***easureable, ***A***chievable, ***R***ealistic, ***T***estable)

These five attributes are essential, but there are other essential attributes of requirements too. For example, we have stated several times that the requirements are about *what* and *how well*, but not about *how*, because *how* is the design. *The requirements should not unnecessarily constrain the design.*

We also noted above that the requirements must be verifiable; and there are other important attributes of requirements too. But I have decided not to tamper with Jim and Scott's SMART acronym!

2.3 Design

2.3.1 The Design and its Process

The next phase in the development process of an engineering project is the design; we aspire to create a design that satisfies the requirements, but also one that is feasible and affordable to build.

We get a lot of "help" with the requirements; after all, our customers and our users understand well *what* they want the new system to do, and such *what* constitutes a major portion of the requirements. It is my experience that most systems eventually develop pretty good requirements, although it may take them longer to do so than they originally planned, and cost more money to do so than they planned.

But the design is a completely different matter; many systems simply have bad designs. Why might this be? For one thing, the customer and the users are generally *not* qualified to provide expert help with the design, in contrast to the way that they *are* qualified to provide expert help with the requirements.

How do I know that many systems actually have bad designs? More than once I have seen cases of two completed systems that do approximately the same thing, where one runs 100 times faster than the other. Similarly, I have seen cases of two completed systems that do approximately the same thing, where one is 1000 time more reliable than the other. I have then had the opportunity to examine these systems so as to find the root cause for the slower and less reliable performance, and therefore I can state with confidence that this gigantic gap in performance derived from specific (undesirable) features of their designs.

This finding has many interesting implications. First, having a 100× or 1000× range of outcomes for a critical parameter from an engineering project is shocking; mature disciplines do not have such large ranges of outcomes. Consider mid-sized family sedans offered for sale that meet US emissions control requirements; the variation from best to worst in, for example, gas mileage is no more than 25%, not 100× (10 000%) or 1000× (100 000%). Something is going radically wrong inside the designs of the systems that exhibit such poor performance on such an important metric.

Second, a lot of engineering projects turn out to be problematic, in the sense that they end up far over budget, far behind their delivery schedule, and a shocking number (some studies say more than half of all engineering projects) are canceled before they complete, because of customer and user dissatisfaction with progress. The people who study these problem projects nearly all assign the blame to poor requirements. But I spent many years of my career as a designated fix-it person for engineering projects that were in trouble, and I will tell you that they all had pretty good requirements. *What they all lacked was a sensible, feasible design.*

So, in light of the above, I have come to view the design as *the* critical portion of the engineering project development cycle. It is the stage that will likely make or break your project.

What is a design? In the previous section, we defined the requirements as the statements that tell us *what* the system is supposed to do, and *how well* it is supposed to do it; in contrast, the design tells us *how* the requirements are going to be accomplished. Consider a house: the *requirements* might tell us that the house needs to have four bedrooms and three bathrooms. The *design* tells us how we will satisfy those requirements: whether we will use wood or metal for the frame, whether we will use a raised structure or a concrete slab for the foundation, whether we will use casement or sliding windows, whether we will use wooden shakes or concrete tiles for the roofing materials, and so forth. We can build a house that meets the requirements – four bedrooms and three bathrooms – using either wood or metal as the framing material; both probably allow us to satisfy the requirements. But there may be other reasons for choosing one design approach or the other, reasons that have little to do with the requirements (e.g. "four bedrooms"). For example, if our house is going to be in a location with a really severe termite problem, we might not want to select wood for the framing material. But if wood is satisfactory, then using wood is probably a lot less expensive than using metal for the frame. A wood-framed house can probably be built in less time than a metal-framed one too. These are examples of *alternative designs.*

The process that we use to create a design centers around a method that we call the *trade study.* The trade study process helps us create a set of candidate alternative designs, helps us create a way to measure the "goodness" of each alternative, and finally allows us to select a preferred alternative, while also creating the data and the artifacts that will allow us to explain to our peers and stakeholders why we believe that it is the best possible alternative.

In the design process, there will seldom be a clear winner, in the sense that a particular alternative design is best in every category. That is why we call the process a *trade*; we make judgments (backed up by data and analyses) about which combination of positive and negative features achieves the best overall solution for our system. We strive not for perfection, but for a reasonable *balance.*

In this book, we are considering engineering projects, and therefore technology and technological concepts are *central to the success of those projects.* We therefore use *engineering methods* to guide project management decisions, and that in turn implies that we use *data* to help make decisions. The data that we use to measure the "goodness" of our candidate designs takes the form of two types of metrics: one that I call *operational performance measures* (OPMs), and one that I call *technical performance measures* (TPMs).

Why two types of metrics? Our degrees of freedom in creating alternative designs lie mostly in alternative technical concepts and approaches, and these are best measured using the technical performance measures. But we must also make sure that improved technical performance actually results in improved operational performance, as measured from the coordinate system of value relevant to the users, customers, and other stakeholders. We therefore also need to use the operational performance measures.

At first blush, one might believe that improved technical performance always leads to improved operational performance. That is simply not true. Remember the example a few pages back about the radio vendor who offered me a radio that sent and received data twice as fast as his current radio model? When we plugged twice the data rate into our system-level model, the operational performance measures did not improve at all! This, then, is a real-life example where dramatic improvement in a key technical performance measure did not result in *any* improvement in an operational performance measure.

In my experience, this is in fact a frequent occurrence; as a result, we must separately measure *both* technical performance measures and operational performance measures. We cannot abandon technical performance measures, as they are at the heart of our technical analyses that allow us to determine whether or not our design will in fact work. But we must somehow relate the effect of the technical performance measures to the operational performance measures. In the radio example cited above, we did that through a system-level model. In other instances, we have done this through actual benchmark measurements. But however you choose to do it, it must be done.

Furthermore we must convince our stakeholders – most of whom are *not* engineers – that our predictions about operational performance measures are credible. So, we must be able to explain the connection between the technical performance measures and the operational performance measures in a credible and transparent fashion, even to our non-technical stakeholders. This might be done by explaining the logic in the system-level model, and then showing that we have calibrated that model by using it to predict the performance for a set of situations for which we can go out and make actual real-world measurements. If the stakeholders understand and agree with the logic inside our system-level model, and see that in a set of real-world circumstances the system-level model makes accurate predictions, then they may accept that when we use the model to make predictions about circumstances for which we cannot yet make measurements (e.g. how the new system will perform), the predictions may be believed.

I use the method depicted in Figure 2.15 in order to create a design. This figure introduces a few new terms, which will be described in the following text.

Let's discuss each of these.

By a *pressure point* that the design must actually address, I mean that we must use our operational knowledge of the users and the mission to determine what are the real design drivers for the system. Do not depend on your customers to do this for you! They are not engineers and designers, and while they understand their mission well, they often have an imperfect understanding of how technology interacts with their mission.

1. Understand the *pressure points* that the design must actually address
2. Perform the actual *trade study*: create alternatives (each of which address every requirement), create measurement methods, assess the alternatives (and adjust the alternatives, as is usually necessary), select the preferred candidate, and gather the rationale showing why that is in fact the right candidate design to select
3. For the selected design, assess in some detail various important features of that design: the *performance* and *capacity* (via modeling and benchmarking), *stability* (that is, we want the design to be relatively insensitive to small errors in inputs and assumptions), *design margin* (we want the design to be relatively insensitive to small errors in implementation, e.g., some part weighs a small amount more than planned, etc.), and *avoidance of known design pitfalls*
4. Verify the *completeness* of your selected design; that is, make sure that it actually addresses every requirement. This is accomplished through a *traceability* analysis, where we map every requirement to a section of the design that implements that requirement
5. Finalize your selection of a candidate design, perhaps with alternative designs for a few selected features. Prepare a thorough description of the design itself (which should be detailed enough to serve as a guide to those who will build each of the pieces called for by the design), as well as of the assessment findings, including the rationale for the selection. Archive all materials, so that the process can be reconstructed and rerun. It is not uncommon later on in a project to have to reopen the trade study, add a new design candidate, and rerun the assessment

Figure 2.15 Steps to create a design.

Think of the short-range air defense system that I described earlier. If the Army were to notice that only about one-third of the missiles they fire at airplanes in the sky actually hit the target airplane, they might well conclude that they need a better missile. While this may sound reasonable, that conclusion may be completely incorrect! When I was actually designing such a system many years ago, we discovered that the Army gunners were actually taking very few shots; they were shooting at a target only about 10% of the time that they could have. Most of the time, the very short nature of the shot opportunity (a high-speed jet flying very close to the ground passes you by in just a few seconds) meant that they were *not even shooting 90% of the time.* Instead of building a better missile, we decided that the *pressure point* in the design was to help the gunner get ready, and to cue him when a shot opportunity was coming up soon, so that he would not miss so many shot opportunities. A few years later, after we had finished an automated system designed to help gunners achieve more shot opportunities, not only were they taking nearly 10 times as many shots per day, but *most of those shots were hitting the target airplane.* It turned out that by using automation in the system to help the gunner find and take his shot opportunities, we were not only creating *more* shot opportunities, but also creating *better* shot opportunities, ones where the target airplane was closer, or otherwise situated so as to make it easier for the missile actually to hit the target airplane. They didn't need a better missile at all!

Think about that: not only did we create a revolutionary improvement in the performance of the system (almost a 10× improvement!), but we did it *without making any changes at all to the item that the customers and users initially might have thought was the problem.* We had to discover what was the *real* pressure point in the design. In this case, that pressure point was improving the number of shot opportunities that complied with the rules of engagement (and also improving the quality of those shot opportunities), rather than improving the probability of kill once a missile was launched. This was despite the fact that the major observable of poor performance was that most of the time, the missile missed the intended target – which made it seem like the problem was with the missile itself.

In my experience, this sort of *focusing on the wrong aspect of the problem* takes place quite frequently. Therefore, my design methodology always starts with the assessment and analysis needed to determine where the actual *pressure point* is in the design of our new system. There may well be more than one, of course.

Once we know the pressure points, we can turn to the *trade study*. I use the following steps to perform a trade study (Figure 2.16):

- Use the knowledge and insight we acquired about the customer in order to create *operational performance measures*, and then discuss those with the customer. Of course, we actually started this process while we were creating the requirements.
- Use the operational sequences to firm up a list of all the independent stimuli that can activate a mission thread in your system, and also define what is produced as the output of each mission thread. We started this when we prepared the concept of operations document as part of the requirements, but now we need to do it in more detail.
- Use these items – the lists of stimuli and the partitioning of steps that could be in parallel and steps that must be in sequence – to define all of the *independently schedulable entities* within your system. These are the system activities that can be started in response any sort of asynchronous stimuli, and can therefore operate at the same time as other system activities.
- Use the knowledge and insight we acquired about the mission, together with the requirements, to create operational sequences that describe how they perform this mission, the *mission threads*. This is a mechanism that helps you to ensure that every requirement is addressed by the design. We started this while we were creating the requirements too. Now we have to do it at a finer level of detail, showing which steps on the mission threads can be performed in parallel, and which must be performed sequentially.

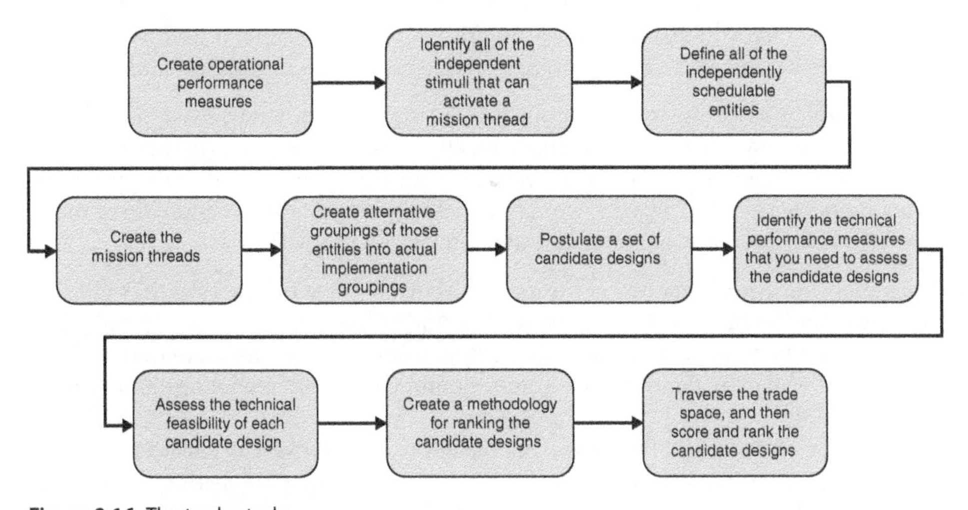

Figure 2.16 The trade study.

- Use the list of mission threads and the list of independently schedulable entities to create alternative groupings of those entities into actual implementation groupings (e.g. the items that will actually comprise the separately buildable elements of the system).
- For every alternative implementation grouping, postulate a set of candidate implementation methods (e.g. software, digital hardware, analog hardware, electromechanical, and so forth). Perhaps there will be more than one for some of these categories (e.g. several software-based approaches, each using a different algorithmic method). These form the set of candidate designs; we call this set of candidate designs the *trade space*. The candidates that you select for the trade space ought to be informed, of course, by your previous selection of a set of *pressure points* that the design needs to consider.
- Perform an assessment of the technical feasibility of each candidate design and eliminate from further consideration those that are deemed not feasible, or too risky.
- Using the technical performance measures and the operational performance measures, create a methodology for assessing the "goodness" of each candidate design, by which I mean how well this design balances the two key goals of the design process: (i) satisfying all of the requirements, while also (ii) being technically feasible, and at the same time also satisfying any other goals that may apply to this specific project (such as unusually high levels of reliability, if our system is safety-critical, and so forth). What will you measure? How will you measure it? How do you ensure validity of the measurements and the predictions that you make from those measurements? What are the desired values for each measure? What are the minimum acceptable values for each measure? How do you combine and process the measurements into an overall assessment of "goodness" for a candidate design? What are the relative weightings you will give to each measurement as a part of that overall assessment? I provide a list and description of common assessment methods in Figure 2.17.
- Use that measurement and assessment methodology in order to perform the actual assessment of the goodness of each candidate design. This will probably be done multiple times, eliminating a few of the worst-performing candidates each time, adding additional measures for the remaining candidates, perhaps using the insight from the measurements in order to create new candidates. Keep careful records documenting the rationale behind all decisions, even for those candidates that are discarded.

Since we will be assessing several candidates during the trade study, the depth of each assessment is necessarily limited. Once we reach the point, however, where we believe that we have our single selected design (perhaps with alternatives for a small number of selected points within that design), we can afford to assess that final candidate in more depth. This involves assessing the final candidate design in several ways:

- *The performance and capacity of the design.* Many systems fail because they fall significantly short of promised performance rate and/or capacity, so we try to avoid that risk by a careful assessment against those dimensions of our candidate design. This assessment will be done using some combination of modeling and actual benchmarking.
- *The stability of the design.* We want the design to be relatively insensitive to small errors in inputs and assumptions, so we assess our design by subjecting our system model to such variations in inputs, initial conditions, operating conditions, and so

Quality functional deployment	Diagrammatic method to relate user goals to requirements, and requirements to design candidates
Pareto analysis	Diagrammatic method for finding the dominant influences in a situation
Functional analysis	A method of decomposition that enables the definition of various system descriptors, such as states and modes, system functions, sequential dependencies, external interfaces, assigning quantifiable performance requirements to functional groupings, analyzing timing, analyzing off-nominal behavior and types of errors, analyzing how to detect faults and how to recover from error conditions, and so forth
Optimizing across multiple parameters	A quantifiable method for creating mathematically optimum combinations from pairs of parameters
Modes and states	A method to analyze and define operational segments within the systems operational sequence and operational life-cycle (modes), and the conditions of existence of a system (states); these start us thinking about the dynamic behavior of our system
Timeline analysis	A method to define a specific time dimension to functional relationships, where sequence and/or concurrency are important; again, this helps us think about the dynamic behavior of our system
N^2 chart	A graphical depiction of a set of elements and their interfaces
IDEF-0 analysis	A graphical depiction of a function, showing its *inputs*, *outputs*, *enablers*, and *constraints*
Swim-lane methodology (activity diagram)	A graphical depiction that assigns functions and tasks to people and/or organizations, (e.g. who among the users performs each of the depicted elements of a mission thread)
Schematic block diagrams	A graphical depiction of the functional decomposition down to the level of modular units, that is, small pieces of a candidate design that (a) perform a single, independent function, (b) have a single entry point and a single exit point, (c) display low external coupling, and (d) can be separately tested
Functional / physical matrix	A graphical map between the functional architecture (requirements) and the candidate physical architectures (design), intended to aid in the identification of gaps and conflicts
Decision tree	A graphic technique for finding an optimized choice between specific alternative decisions (including design decisions) that mathematically accounts properly for conditional probabilities
Pugh method	A tabular method for assessing alternatives against a baseline, against multiple assessment criteria that may be weighted differently
Constraint theory	A formal mathematical / graphical method for identifying and resolving conflicts (in the form of *over-constraints*) in the definition or characterization of a system
Modeling	A computer program that is designed to simulate selected aspects of the behavior of the eventual system. We use such a model to make predictions about how well the candidate design will work, and therefore, hope to make better design decisions. Using such a model can both help you improve each design candidate, and also help you select the final design

Figure 2.17 A list of candidate analysis methods for assessing the goodness of your design.

forth, and check to see that small variations in these factors lead only to small variations in the performance of our system, rather than to catastrophic degradations in our system. See Figure 2.18.

- *Design margin.* We want the design to be relatively insensitive to small errors in implementation too. For example, if some part ends up weighing a small amount

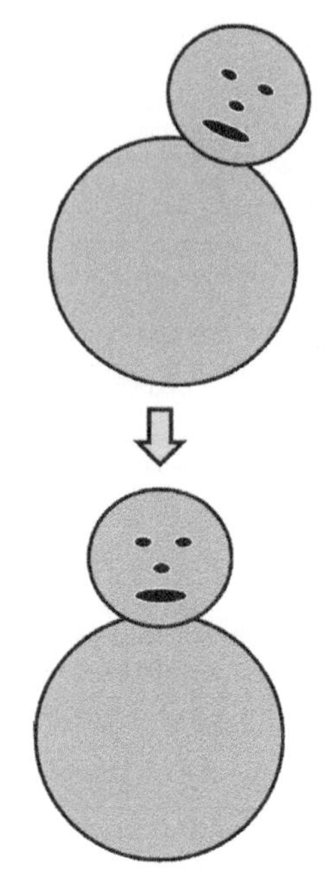

Figure 2.18 Stability in the design.

more than planned, we want the impact on our system to be small (e.g. a tiny decrease in fuel economy, etc.) rather than catastrophic (e.g. our satellite fails to reach the intended orbit). It is a *major* portion of our role as designers to provide design margin; this will enable the system to work, even if a few things don't work as well as planned. A classic example is managing the weight of a spacecraft; there is a hard limit for how much weight the launch vehicle can take to the designated orbit. It you allocate all of that weight, and something shows up late a few pounds over, you are in big trouble. So, instead you should keep some design margin (in this case, unallocated weight) in your "back pocket." As you get closer to delivery, you can allocate from your design margin to solve problems. This is analogous to the way the program manager allocates from her/his management reserve of funding (which we will discuss in a later chapter). But, at the same time, you cannot keep an unreasonable amount of design margin; it costs too much, and detracts from operational performance.

- *Avoidance of known design pitfalls.* I have seen a lot of system designs. Those that fail often share a small number of characteristics. So, I advocate checking that your candidate design avoids those known design pitfalls. See Figure 2.19.

Of course, we may learn something from this assessment that causes us to adjust the design, or even to have to return to the trade study, create additional candidates, and reassess.

- Systems are ***dynamic***; that is, they change over time:
  - Processing happens in defined sequences
  - Steps happen within defined time frames
  - Various stimuli (both asynchronous and synchronous) cause processing to occur
  - . . . and so forth
- Such dynamics have implications:
  - Physical parts move, get hot or cold, vibrate, etc.
  - Software parts get started and stopped
  - Data gets passed from one place to another
- You *design* how these interactions are to proceed, hence they constitute your *planned dynamic behavior*
- However, your system can have dynamic behavior for which you did *not* plan; things go awry during the dynamic execution:
  - Steps get out of sequence
  - The wrong activity gets started in response to a stimuli
  - Timing requirements are not met
  - Control signals get lost, or associated with the wrong data
  - . . . and so forth
- This is very common!
  - In fact, there is an entire vocabulary for such problems (e.g. "deadlock," "race conditions," etc.)
- The major design pitfall: most designers realize that their design must implement the dynamic behavior *they want*, but they do not realize that their design must also *prevent* the dynamic behavior *they do not want* (see Figure 2.19b)
  - Designing to prevent the dynamic behavior that you do not want is hard; you must first *envision* all of the ways in which your system's dynamics could go wrong, and then *create a corrective* to each
  - Understanding all of the implications of dynamics is hard; but most designers don't even try
  - I have invented a design methodology and design pattern to accomplish this; see https://cpb-us-e1.wpmucdn.com/sites.usc.edu/dist/a/54/files/2018/01/98909-vak73m.pdf for a description
- But it can be done!
  - Example: contrast the use of ***separate control and data signals*** (the near-universal approach) with the concept of having the data ***be*** the control signal. If the data are the control signal, the data and the control signal simply cannot get out of synchronicity with each other

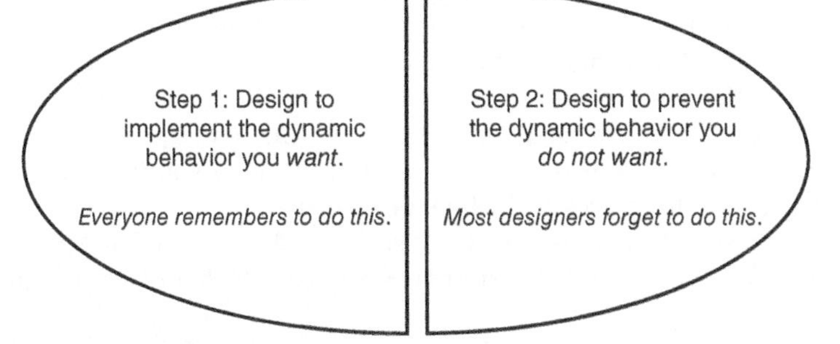

Figure 2.19 (a) Avoidance of known technical pitfalls in a system design. (b). Designing for the two types of dynamic behavior.

Once you have created a candidate design that appears to satisfy the requirements and meets your measures of goodness, then we need to do a more thorough analysis to show that this candidate actually meets all of the requirements. You inserted the functional requirements – the statements about *what* – into the mission threads, and used the mission threads to help create design candidates, so there is a reasonable chance that they are all addressed by the design. But we need to verify that. We do this by a traceability analysis, where we map every requirement to a section of the design that implements that requirement.

Furthermore, there are many other requirements: all of the *how well* requirements, which deal with reliability, capacity, timing, availability, and many other quantifiable matters. These are not usually addressed by a single step on a mission thread, but instead by a more subtle amalgam of elements within the design. Proving that they are addressed by the design is part of the traceability analysis too, but often the verification of the incorporation of the *how well* requirements must be verified through use of our system model and its predictions about the operational and technical performance measures.

2.3.2 The Design Hierarchy is Not the Same as the Requirements Hierarchy

A common mistake is to assume that the groupings created to organize the requirements – the functional groupings – are appropriate for organizing the design. My experience is that people who try to do this end up with bad designs. For the requirements, it is best to group things together that are functionally related, but we must group things differently for the design: when we design, it is best to group things together that support the implications of the steps that must be executed in sequence, and the freedom and performance improvement opportunities implied by steps that could be executed in parallel.

2.3.3 Modeling

Modeling appeared on the list in Figure 2.17, and in several other places I have made reference to the idea of a *system-level model* that we use for making predictions about the behavior of our eventual system. Let me say a bit more now about modeling.

Modeling of some sort is used in most large system development efforts. These days, such modeling usually takes the form of a computer program that is designed to simulate selected aspects of the behavior of the eventual system. We use such a model to make predictions about how well the candidate design will work, and therefore hope to make better design decisions. Using such a model can both help you improve each design candidate, and also help you select the final design.

A typical computer-based simulation model will simulate the components of a system, their actions and interactions, and the external interfaces (which provide stimuli to the system).

It is easy to code up a model; the first big issue is the *credibility* of the model. Why should anyone believe that the predictions from the model are reasonably accurate? Just because the computer says so is not good enough; computer models can at times make predictions which are wildly inaccurate. We use techniques such as the following to prove the validity and credibility of a model's predictions:

- *Analytic validation.* We check that the algorithms are coded correctly within the model.
- *Calibration against benchmarks.* We use the model to make predictions about things for which we already have actual measurements of a real system; we then believe that we can rely on the predictions of the model for additional situations.
- *Assessment of the accuracy and risks of extrapolation beyond the benchmark data.* Obviously, using the model to make predictions for situations where we do not have actual measurements is one of the primary purposes of employing a model. But such extrapolation necessarily entails some *uncertainty*; if you were doing an experiment where you were measuring the viscosity of water as you cooled it, and you made measurements at 60, 50, and 40 °F, you would conclude that cooling the water did not change the viscosity very much. You might therefore be tempted to extrapolate, and make a prediction about what would happen if you cooled the water to 30 °F.[11] Since such extrapolation is a major purpose of our models, we must make explicit efforts to look for the non-linearities, phase changes (such as the fact that water will freeze if cooled to 32 °F!), turbulence, queuing, and other major disruptions that would limit the range of validity of such extrapolations.

There is another big issue with models: How accurate is the model? We use the term *fidelity*: How faithful to the real world are the predictions of the model? Nor do we always need 10 decimal places of accuracy: How accurate is good enough for my purposes?

We start with the question of determining the necessary fidelity with an *error budget analysis* of the system. What is an error budget analysis?

Recall the air defense system that we described earlier in this chapter. We said that the gunner could not shoot a missile at an airplane until after a trained operator had conducted a visual identification of the target through an appropriate magnifying optic. The airplane, of course, is constantly moving. The vehicle with the missiles and the trained operator is moving, too. In order to get the magnifying optic on that vehicle to point at the correct airplane, there are a series of steps that must be completed, all within some degree of accuracy; if we do not achieve at least that accuracy, the optic may be pointing in the wrong direction. To achieve the necessary degree of overall accuracy, we must specify the degree of accuracy for *each step* along the processing chain, and then combine all of those individual errors correctly to arrive at an estimate for the overall accuracy. That is an *error budget analysis*; it tells you how accurate each component of your system needs to be in order for the system to meet its performance requirements (*how well*). Obviously, then, our model needs to be slightly more accurate than the actual system will be (there are ways of calculating exactly how much more accurate the model needs to be), in order to make useful predictions about our system.

One can then verify the achievement of that accuracy (at least for some scenarios and ranges of data) through benchmarking (that is, making measurements under controlled circumstances, such as in a laboratory, or under constrained field conditions). Furthermore, as you build the actual system, you can continuously re-benchmark the

11 Water, of course, freezes at 32 °F. Therefore, the viscosity of water changes materially between 40 and 30 °F.

model against the emerging actual system over a larger and larger range of scenarios and data.

Models are often *nested* or *chained*. An example of such nesting might be:

- A physics model of radio-frequency propagation, which feeds ...
- A model of an antenna, which feeds ...
- A model of the antenna mast height, which feeds ...
- A model of the received signal quality, which feeds ...
- A model of successful packet completion rate, which feeds ...
- A model of message completion delay (average and variance), which feeds ...
- A model of end-to-end completion time and accuracy for a specific capability, which feeds ...
- A measure of some system operational performance measure!

There are, however, many different ways to interconnect these models. Sometimes, they are all put together into a model-of-models, with fully automated interactions between each model. Other times, they are run separately, but the outputs from one are automatically fed into the next model in the chain (these are usually called *federated models*). Other times, the models are completely disjoint, and the outputs from one are manually transferred into the next model in the chain.

It is my experience that it is important that each model be maintained and operated by its actual creator; that creator is the expert who knows the limits of credibility for their own model better than anyone else, and having them maintain and operate that model, in turn, contributes to achieving better and more credible predictions. This motivates me usually to prefer the use of separate models with manual transfer of data! That sounds old-fashioned, but better accuracy and credibility in my view is more valuable than automated interconnection.

We use the models to analyze our system and its candidate designs; that is, to determine how well each of our candidate designs performs. Since we are concerned with whether it meets the needs of the *users*, the model must finally reach the level of being able to make predictions about the *operational performance measures*, not just about the technical performance measures. This is a common failing of system models; many are designed only to make *technical* predictions.

2.3.4 Design Patterns

In the building construction business, there is something called the "building code," which actually defines how certain portions of the design for a building must be accomplished. This lowers risk, in that these model design segments can be created by experts, and reused by average designers. This is an example of what I call a *design pattern*: a vignette for a small portion of a design that has been codified by some authority as working well, under the appropriate conditions.

The last bit about *appropriate conditions* is important; a bit of building design that is great for a single-story private residence might be a disaster for a 30-story office building (and vice versa).

We too can make use of design patterns in designing our engineered systems. There are plenty of such design patterns for us to consider: things like EtherNet, TCP/IP (the network transport standard for the Internet), various standards for electronic circuit cards so that they can plug into a standardized backplane, and both the client-server and HADOOP data storage architectures; these are all examples of such *design patterns*.

In the context of the engineered systems that we are considering herein, things can be much more subtle than when the local city building and safety department tells you that you must have studs every 18" in the subfloor for the new room in your house (which is, of course, another example of a design pattern). The problem lies not in being unable to understand client-server or HADOOP; rather, the problem lies in understanding when each is appropriate or inappropriate for our particular system. The range of design variation in buildings is far smaller than the range of design variation in engineered systems. This is in part due to the intrinsic complexity of today's engineered systems; something with 100 000 000 lines of software code is far more complex than an office building. But remember the examples cited above about staggering levels of variation in quality between apparently similar engineered systems; I cited ranges of 100× in speed and 1000× in reliability. A major cause of this variation is poor matching of design patterns to the specific nature of the system being designed (e.g. using client-server when that is just a poor choice for the nature of the system).

Good system design organizations keep libraries of past system developments (successful or otherwise), and you can peruse those in order to gain insight about which engineering design patterns are likely to be effective for your system.

I wish I had a better answer for selecting design patterns; we need to use design patterns in our systems (our customers will expect it), but if we choose patterns that are inappropriate for our particular system, we are likely to end up being the next example of a system failure. I advocate careful benchmarking and modeling of your design patterns in the context of your system before finalizing selection. There is, in my experience, too much reliance on the idea that "this is a standard approach, so it ought to work for our system, too" without understanding whether or not such use is appropriate to the specific nature of *your* system.

2.3.5 Do the Hard Parts First

There is a temptation to start design (and implementation too) with the easy, low-risk parts of the system. For example, there may be portions of a system that are reused from previous systems with only minor adaptations. Another favorite for early design activity is the user interface software.

These are among the easier aspects of the design, and some people like to start with them so that the team can make a lot of quick progress. This might be thought to help the team learn to work together, and to establish credibility with the customer.

While one can in fact make quick progress by concentrating at the beginning of the design (and implementation) phase on the easy parts, there are, in my experience, far better reasons to start the design and development phases with the *hardest and/or highest-risk portions of the system*. If nothing else, this approach provides you with more time to work on these harder/higher-risk portions, and it is often the case that having more time is a pretty effective element of a risk-mitigation strategy. Beware of those who would have you concentrate first on the "low-hanging fruit"!

2.3.6 Designs and Your Team

A former radio personality[12] used to make his audience laugh by saying "... and all of the children are above average." We laughed because we knew it could not be true – very few children can *actually* be materially above average. We also laugh because people recognize that they do tend to believe that too often about their own children.

There is actually a very important insight in this idea. You might be able to select the staff for a three-person project so that everyone is above average. You cannot, however, realistically do that for a 30-person project, nor for a 300-person project. Most people on such larger projects will be ... about average.

At the same time, it is often the case that a design requires that a small segment of very difficult and error-prone work is *spread across the tasks that many different people have to do*. This was true, for example, in early implementations of cyber-security techniques; every programmer had to implement a relatively small amount of very complex and highly error-prone software code. Not surprisingly, the result was a lot of design and coding mistakes, causing very poor cyber-security characteristics in the resulting system. If most of the designers and programmers on your project are average, you cannot expect them all to perform properly on very difficult tasks.

There is, however, a better alternative. It is possible to use the design process to *isolate* certain types of difficult and error-prone tasks into a *small segment of the implementation*.[13] You can then assign this isolated bit of work to a small set of experts. The overall difficulty of the entire project remains the same, but you have *partitioned* the design into (a small set of) difficult parts and (a large set of) less difficult parts. This matches the distribution of the difficulty of the work to the distribution of skills on your team: you have a small amount of difficult work, which is isolated into a small segment of the design and implementation, and you can therefore assign this difficult work to your small number of expert practitioners. You have a large amount of work of average difficulty, which you can assign to the remainder of the team. I have found that this approach significantly improves the outcome on real projects. This is a new and, I think, important design objective: matching the distribution of the difficulty of the work to the distribution of skills on your team.

A second example of how to use this technique involves the management of the system's dynamic behavior; I have found that this is a highly complex and error-prone task, and one that has a significant impact on the overall success of the project. Remember what I stated above: most designers remember to design to implement the dynamic behavior they desire, but do not do nearly enough to prevent other (unplanned and undesirable) dynamic behavior. This is all very difficult work, especially the latter portion.

On many projects, responsibility for designing and implementing such controls is (perhaps inadvertently) spread out to every software package; because this is difficult work that ends up being performed by almost every member of the software team (most of whom are, after all, average in skills), this leads to poor technical performance, low system reliability, and lots of schedule and cost over-runs. This was in fact the *most*

---

12 Garrison Keillor.

13 I invented this technique. For additional details, see section 4 of https://cpb-us-e1.wpmucdn.com/sites.usc.edu/dist/a/54/files/2018/01/98914-zif8ym.pdf.

common design flaw that I saw in all of those systems for which I was tasked to act as the fix-it person.

To address this problem, I adapted technical concepts invented by some of my colleagues[14] into a management and design methodology for isolating this complexity into a small segment of the design and software implementation, which allowed me to assign this difficult, error-prone, but highly significant work to a small team of experts.

Good project managers have always tried to match people to assignments, but this quest was often confounded by the phenomena noted above: really difficult and risky tasks can creep into every single part of the system. Using the design process to avoid this problem by the sort of partitioning described above gives the project manager a new and powerful tool to improve the probability of a good outcome for their project. The design stage of your project is where you can do this.

2.3.7 Summary: Design

As I stated at the beginning of this chapter, in my experience, design is the make-or-break activity for most engineered systems. As you can tell from what we have discussed in this chapter, there are a lot of established practices that can get you going, but there is a lot that depends on judgment (which comes from experience) and art.

Design is also one of the most fun aspects of systems engineering. Learning to do it well will distinguish you from your colleagues.

In Figure 2.20 I summarize some of the key points we have discussed about the design activity in a format called IDEF-0.

2.4 Interaction of the Requirements and Design Processes with Project Management Processes

As I stated in the introduction, one of the important characteristics of my approach to engineering project management is my emphasis on the *uniqueness* of engineering projects (as distinguished from other types of projects) and the resulting need to have the actual engineering activities influence the way we perform the project management activities. Figure 2.21 illustrates this.

This figure is in a format called an *N^2 chart.* The N^2 chart format was invented by Robert Lano.[15] The diagonal entries (in the shaded boxes) are a set of interacting activities; the purpose of the chart is to examine and depict the interactions among these activities. In this N^2 chart, there are five interacting activities, two engineering life-cycle activities (*requirements* and *design*), and three project management activities (*risk/opportunity management, creation and management of operational performance*

14 Let me name Walker Royce, David Bixler, and Peter Blankenship as the main people whose technical ideas I turned into this design and management method. Thanks to them, and the others who worked with them! We employed this resulting method on many projects in a variety of domains, all of which were resounding successes.

15 Robert J. Lano, *The N^2 chart.* TRW Software Series SS-77-04, TRW, Inc., 1977. I met Dr. Lano too! Thanks to Northrop Grumman (successor company to TRW) for providing me with permission to us the N^2 chart format and Dr. Lano's work in my teaching materials.

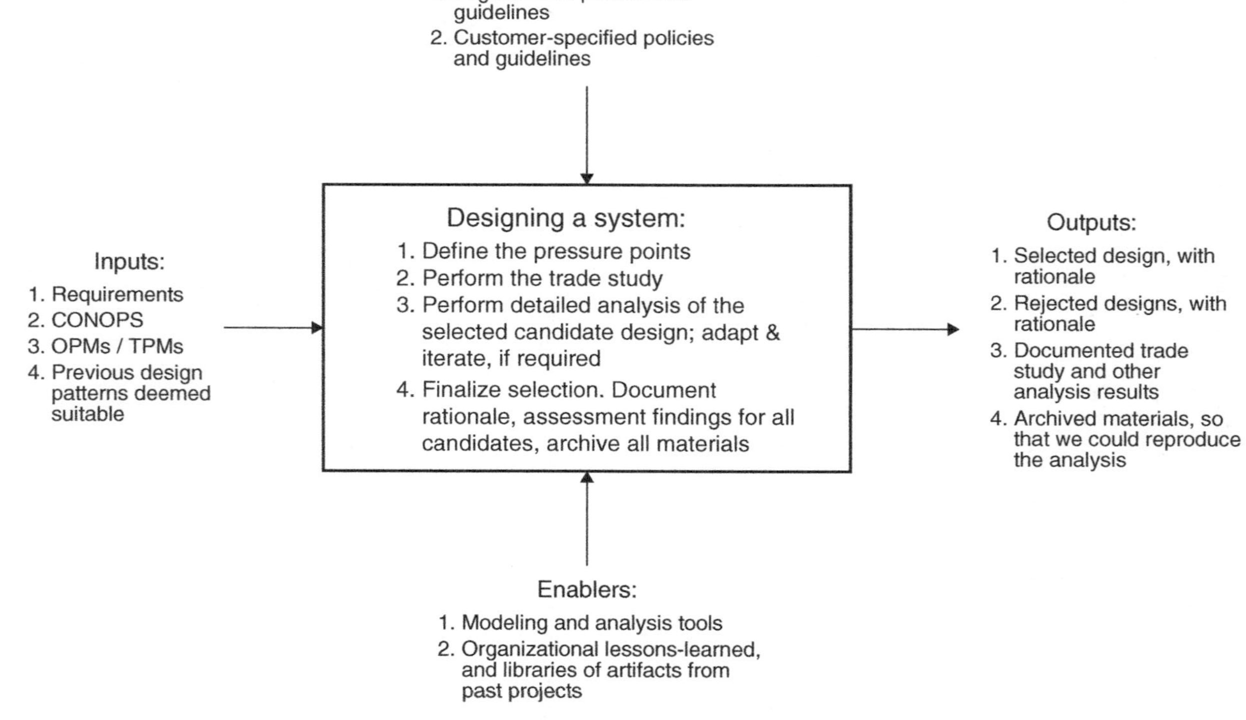

Figure 2.20 IDEF-0 representation of the design process.

Requirements	Updated &/or candidate requirments	Updated &/or candidate requirments	Updated &/or candidate requirments	Updated &/or candidate requirments
	Design	Anticipated levels of design margin; features that provide stability in the design. Design alternatives.	Predictions of the performance of the design, and of the alternatives.	Design alternatives
Proposed changes to requirements	Decisions to execute backup design alternatives; decisions to change which alternatives is considered the baseline	**Risk / Opportunity Management**		
Proposed changes to requirements	Required / desired levels of technical performance	Predictions of operational and technical performance against the thersholds established by the OPMs and TPMs (early indications of problems). Predictions of customer satisfaction with the sytem.	**Creation and management of operational performance measures (OPMs) and technical performance measures (TPMs)**	
	Design-to-schedule and design-to- cost guidelines	Predictions of schedule and cost (lagging indications of problems)		**Schedule & Cost Prediction**

Key to entries on the diagonal:

Engineering Activities

Project Management Activities

Diagonal elements of the matrix

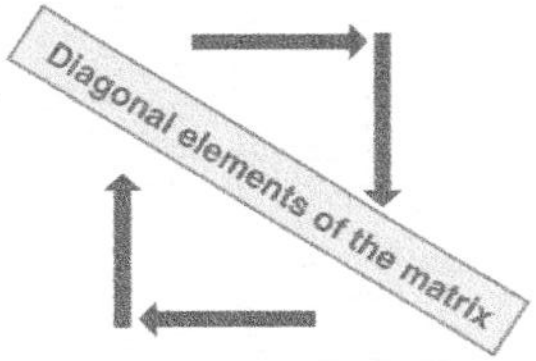

The diagonal entries are the interacting activities;
the off-diagonal entries are the data that flow between the interacting activities

Figure 2.21 An example of how we tie engineering activities to project management activities.

measures and technical performance measures, and *schedule & cost prediction*). As shown by the off-diagonal elements (which show the data flowing between the interacting activities on the diagonal), there are extensive interactions between the two engineering activities and the three project management activities. In fact, as we will discuss in later chapters, we actually change the nature of these three (and other) project management activities due to the fact that this is an engineering project. For example, the analysis that we perform in those project management activities is adapted to the specific nature of this particular design. There are other ways too in which the project management processes are changed because this is an engineering project. We will return to this subject in more detail in later chapters.

2.5 Your Role in All of This

As the manager of an engineering project, you are *not* the chief designer; you have someone working for you who has that responsibility, and it is important that you allow them authority to match *their* responsibility. But, because the design is so important to the success of the project, driving the team to a good design is one of your most important tasks. Your role includes emphasizing the importance of creating a good design, ensuring that the metrics (operation and technical) created are suitable (and accepted/liked by the customers), visibly participate in the process as a way of continuously indicating its importance, motivate the right behaviors by the participants, and push for a thorough trade study guided by suitable metrics and performed with suitable rigor. Of course, the design also needs a good set of requirements, and you must do many of the same things in order to ensure the creation of a good set of requirements. But the design is where my experience suggests that projects succeed or fail. Allocate your time and emphasis accordingly.

I have come to see engineering activities as tending to fall into one or the other of two categories: routine tasks and those that are *not* routine tasks.

The *routine tasks* are those that require real work in order to be completed, but no original technical breakthroughs are needed. For example, perhaps our system has a computer screen that the users view. Various information needs to be displayed to those users on that screen. You decide to do this through a series of forms. The forms need to be designed and coded. No dramatic technical breakthroughs need to be made, but there is work that needs to be done. It might even be possible to plan and estimate this work. For example, if there are 100 such forms to be created, and similar forms have been created for previous projects, you might have data that says it takes about four hours of work for a single person to design, code, and unit test each form. Since there are 100 of them, you can estimate that creating the entire set will take around 400 hours, and since you have two people to do the work, the entire task will take about five weeks of calendar time. This is genuine and value-added work – it truly contributes to the completion of the customer's mission – but it does not require the creation of new concepts, new approaches, new insights, or new techniques.

The second type of engineering activity is those that are *not* routine tasks; these are characterized by the fact that something has to be invented or created. Or the solution requires an insight or breakthrough of organization or structure, rather than routine application of known techniques. Here's a trivial but real example. It used to be the case that a person could type too fast on their computer's keyboard, and an occasional character would "drop"; that is, would not appear on the screen or in the document that you were typing. Even in the days of slower computers, this was not a problem caused by slow computers, nor was it solved by the advent of faster computers. Instead, it was solved by the realization that typing is an *asynchronous* activity; that is, each key-press comes at a fairly random time. The sequential set of keystrokes made by a competent typist may seem like a fairly even stream of key-presses to a human, but to computers that (even in the bad old days) operated in terms of milliseconds and microseconds, the keystroke-to-keystroke spacing has a significant *random element*. That is, there is a level of uncertainty about when the keystroke will arrive. What solved the problem was not *faster computers*, but instead the realization that whereas the computer operates internally on a completely *synchronous* basis – it essentially does exactly one operation per time period (in a modern computer, that time period is far less than a microsecond) – the typing of keystrokes is inherently an *asynchronous* operation; that is, the keystrokes are not evenly spaced in time, but instead the interval between keystrokes has a significant random element. What solved the problem was the realization that there was a *conceptual difference to be bridged* between the human's asynchronous action of typing and the computer's internal operational methodology of exactly even, synchronous processing. The solution would be described by a computer architect by saying that we realized that we needed the external facing interface to operate on an asynchronous basis, and then to queue up the received keystrokes into a storage area where they could be processed by the computer's synchronous internal structure. In technical jargon, we would say that the external interface was changed from a polling architecture to an interrupt-driven architecture.

This example is in some sense a "small" discovery but was real enough at the time. The point is that a conceptual breakthrough was required to create a true solution.

My experience is that the creation of this type of solution – an insight, the creation of a solution by reconsidering the structure and nature of the problem, an actual new discovery (or, more likely, the creation of a way to adapt someone else's discovery to solve a new problem) – is *not schedulable* in the same way that the creation of those 100 computer-screen forms is schedulable. You cannot *predict* when the discovery will be made, nor can you estimate how much time and labor it will take to make it.

These problems, in my experience, are solved by a *method other than scheduling and estimation*. One solves these problems by getting a set of people to spend time thinking about the problems and doing tinkering/experimentation; in some real sense, the more people are thinking about it, and the better they understand the problem statement and its constraints, the more likely it is that one of them will create a solution.

As a project manager what you do, therefore, is achieve that *alignment* that we discussed in Chapter 1 (and that we will discuss in more detail in Chapter 13); this *alignment* creates an environment where (i) many people are aware of the non-routine problems that must be solved in order to build your system, (ii) many people understand that you believe that spending time thinking about these problems is a legitimate portion of their job, and (iii) these same people understand that the project has provided tools, data, and other helpful infrastructure that can help them refine their thinking, and conduct their tinkering/experimentation.

Alignment can include – *should* include – the *identification of the problems* that need to be solved. The risk register (Chapter 9) does some of this for your project. When I was vice-president and chief technology officer of a large organization, I used to publish annually what I called an *unsolved problems list,*[16] that is, a list and description of what I considered to be the principal unsolved problems of the customers and the markets that we wished to address. This motivated a lot of thinking by a lot of engineers about those problems. Not surprisingly, this led to an occasional breakthrough. It itself also contributed to *alignment,* by providing a topic for hallway and lunch-time conversations! And it provided me a clear and understandable way to connect our company's system of *rewards* to clearly articulated organizational goals. All good stuff; as the manager of an engineering project, you should do the same sort of thing.

The result is *not* that these problems get solved on a schedule; the result is that lots of smart and/or motivated people (the *alignment of the team* helps to make them motivated, for example, by allowing them psychologically to participate in, and contribute to, the socially valuable mission of your customers) spend many of their waking hours thinking about your problems. And at some point, solutions are created. But not on a create-a-form-every-four-hours sort of basis.

The project manager has many ways to contribute to this process:

- The project manager creates the *alignment* that allows the people on your team to be *motivated,* usually through believing that they have the opportunity to participate in the socially valuable mission of your customers, and providing them the shared vision of an approach, a sequence, and so forth.
- The project manager makes sure that the problems are *identified.* Do not worry too much that you may not have identified the problems correctly; having lots of motivated people thinking about your list will result in the list being constantly improved. All you have to do is create an initial version, and then *listen.*
- The project manager creates the supportive *environment,* the psychological mindset that allows people to spend time at work thinking about the unsolved problems, and the tools, data, and other infrastructure that will help those people actually move from thoughts to experiments.
- The project manager also creates the *culture* that makes it safe for people to step forward with their ideas.
- The project manager creates a *reward system.* This should include financial rewards for especially good and/or important work. It must also go beyond money, and provide institutionalize thanks, recognition, pats-on-the-back, and so forth.

My experience is that engineers *like* to create and to tinker. To some reasonable degree, you should encourage them to do so. I am told that Google does this by allowing some portion of its staff to devote some portion of their time to work on anything they want. To me, this approach seems needlessly pedantic; if your people are aligned and made aware of the key problems that need solving, and if you have also created the culture and environment that makes problem-solving safe and rewarded, my

16 The idea of an *unsolved problems list* is one that I borrowed from the German mathematician David Hilbert, who published such a list (of what he considered the principal unsolved problems in mathematics) in 1900. As I stated earlier, a lot of discovery is actually the adaptation of someone else's discovery to a new problem domain.

experience is that many people on your team will spend not just some allocated time slot thinking about your problem, but many of their waking hours! And having a large body of motivated people thinking about your key unsolved problems seems to me to be the best way to get those problems solved, even if it will not be done according to a schedule.

You should be thinking about these things yourself too. Play out scenarios in your head. Hold practice conversations with customers in your head. Brainstorm with yourself about potential additions to the unsolved problems list, and potential methods toward a solution. Chat with your chief engineer and other senior technical staff (you can invite them to lunch once a month), and pass your thoughts on to them. It is no longer your personal responsibility to design the system, but you can create ideas, and pass them on to those whose job it is to create the design.

I have spent a lot of time in this chapter warning you about the problems to your system and to your project that *unplanned dynamic behavior* might cause. Please also realize, however, that *some* unplanned dynamic behavior might be *good*. If you find an example of such *positive unplanned dynamic behavior*, cultivate it. We will provide you with some specific methods to accomplish this cultivation in Chapter 9 when we talk about *opportunities*, the mirror image of *risks*.

2.6 Next

In this chapter, we started our discussion of how we actually do engineering on projects. We conclude that discussion in Chapter 3. Starting with Chapter 4, we will then use that knowledge to optimize our project management procedures in light of those engineering processes.

2.7 This Week's Facilitated Lab Session

This week is all lectures; there is no facilitated lab session.

3

Performing Engineering on Projects (Part II)

In this chapter, I continue our summary of the key aspects of how we do engineering on projects, covering the remaining stages of the engineering life-cycle, from "implementation" all the way through to "phase-out and disposal."

3.1 The Remaining Stages of the Project Life-Cycle

In Chapter 2, we introduced the system method and then discussed the initial stages of the project life-cycle, concentrating on the requirements and the design.

In this chapter, we discuss the remaining stages of the project life-cycle, from implementation to phase-out and disposal (see Figure 3.1).

3.1.1 Implementation

Through the design process (described in the previous chapter), we decomposed our system into a set of pieces, each of which might involve physical structures, electronics, software, data, and other elements. Each piece might consist of only one sort of constituent (e.g. be entirely software, or be entirely a physical structure, or be entirely data, etc.), or might be a multiple of these in combination.

In the implementation phase, we build each of these individual pieces. We use the design created (and written down in the form of formal design documents) during the design phase as guidance for each piece of implementation. If we have done our decomposition well, there is relatively little interaction among the pieces as we implement them, and therefore the implementation of the pieces can proceed in parallel with each other.

Although testing of our system comes a couple of stages later, there are some basic features and characteristics of each individual piece that can and should be checked as we complete each piece; this is often called "unit testing." I will, however, defer our discussion of this aspect until we get to the test stage, below.

3.1.2 Integration

When I was a child, I liked to build things from kits: model cars, clocks, and so forth. I would assemble all the parts into the entire entity, and then try it out and see if it worked.

Engineering Project Management, First Edition. Neil G. Siegel.

Companion website: www.wiley.com/go/siegel/engineering_project_management

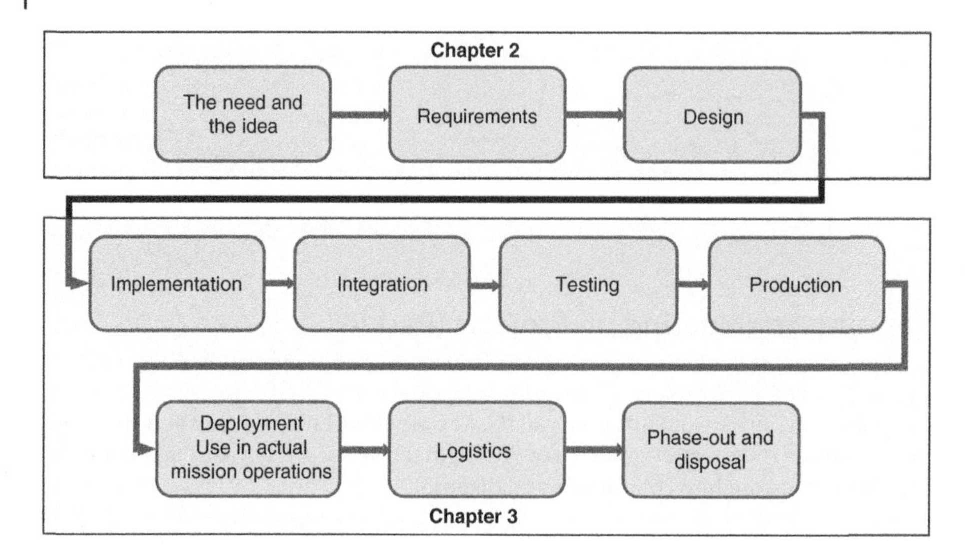

Figure 3.1 Showing in which chapter we discuss which stages of the project life-cycle.

For something simple like a model car, this "put it all together and only then try it out" approach might work. But for the big, complicated societal systems that we aspire to create, that approach does not work at all; there are just too many parts (especially software), and too much complexity. If you put the entire system together and only then started to try things out, when something did not work, it would be very hard to find the root of the problem; there are just too many parts, and the root cause would be hidden in all of the size and complexity of the entire system.

When people started building systems, especially systems with lots of software in them, the original concept was in fact to finish all of the implementation, then put *all* of the pieces together, and immediately try to conduct the formal testing of the system. It was quickly discovered that this seldom worked in a predictable and consistent fashion. The complete system, with its thousands of separate hardware parts and lots of software (nowadays, perhaps tens of millions of lines of software code), turns out to be too complex for this sort of put-it-together-all-at-once approach to work. So, gradually, the need for a phase between implementation and testing was recognized, which we term *integration*.

The purpose of the integration phase is to put the parts together, at first in *small subsets* of the whole, gradually working one's way through the integration of *ever larger subassemblies*, and only stepwise reaching the goal of having the entire system.

During the integration stage, we are *not yet* formally testing that the system meets its requirements; instead, we are only trying to make it operate in an approximately correct fashion. We take a small subset of the system, input a representative set of system stimuli, and see what happens. We compare the outputs received to the outputs expected, and when there are differences, we go and track down the source of the discrepancy (which we can do relatively easily, because we are working with only a small subset of the entire system, and therefore we are not overwhelmed by scale and complexity) and make the necessary corrections to the implementation. We then try out our stimuli again, repeating this process until we consistently get the outputs we expected for each type of stimuli.

While we are working on integrating our small subset of the system, other teams are similarly working on integrating *other* small subsets. When we each seem to be getting the correct outputs from our first small subsets of the system, we put a few such subsets together into a slightly larger portion of our system, and try more stimuli. We will likely find additional errors and inconsistencies, which we can trace to a root cause and correct (we still only have a few small subsets of the system, so we are still not yet overwhelmed by scale and complexity). When our combinations of subsets seem to be working, we continue the process, reaching larger and larger scale and complexity, but doing so only after having worked out a pretty large portion of the major problems at a much smaller scale.

In essence, we are now starting *upward* on the right-hand side of the "U" diagram (Figure 2.3; see the excerpt from Figure 2.3 below, labeled Figure 3.2).

Because the direction of flow in the diagram is now upward, from small pieces, through larger pieces, to the entire system, we say that integration proceeds *bottom-up*.

In this fashion, we can spot and correct lots of problems in the design and implementation far more easily than if we wait until the entire system is assembled; it is just much easier to see the problems, and trace those problems back to a root cause, when we are dealing only with small portions of the system.

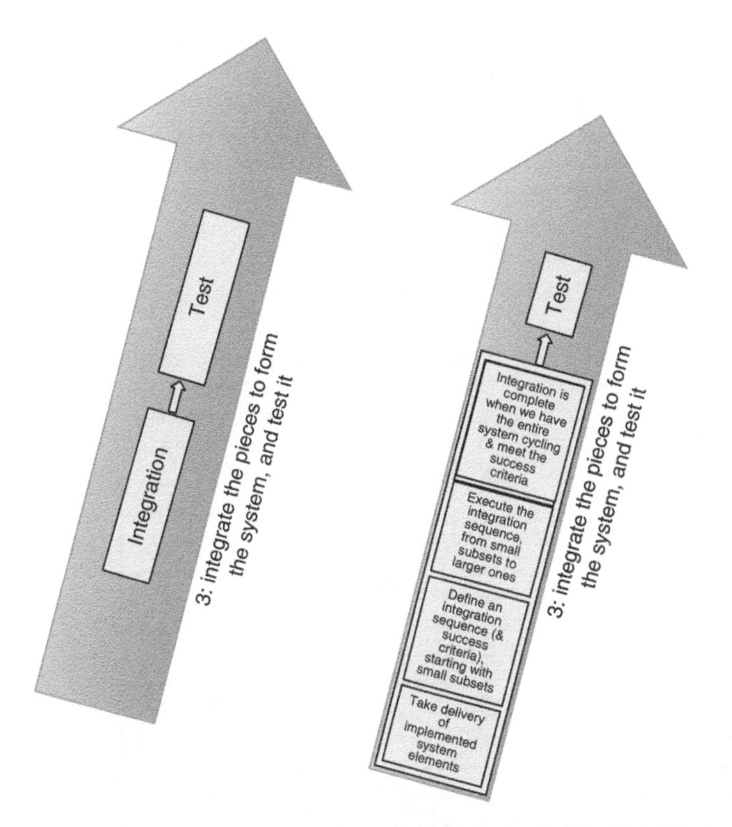

Figure 3.2 The right-hand side of the "U" diagram. On the left, the original version (from Figure 2.3). On the right, an expanded version showing details for the integration stage.

Distinguishing integration from testing has been a gigantic boon to systems engineering and engineering project management. Ironically, even though integration has proven to be important for project success, it is still often under-resourced, in terms of time, money, people, and tools. Far too many projects still neglect the integration stage, and usually suffer greatly from that neglect.

Finding errors earlier in the process is not only easier (due to the smaller scale, as noted above), it is also far less expensive. I have seen data which says that the cost to find and fix a problem increases by about a factor of five in each successive project stage (e.g. requirements, design, implementation, integration, test, deployment, operation). This means that it costs about 1000 times as much to find and fix a problem in the deployment stage as in the requirements stage! This is a big part of why good projects are willing to spend a lot of time and money on those early project phases, especially requirements and design. It also explains why good projects are willing to spend time and money on integration: it costs five times less to find a problem during integration than during the very next stage (testing).

Another interesting and subtle aspect of integration is that the sequence you select for integration matters a lot. Figure 3.3 depicts two alternative integration sequences (A and B) for integrating nine components, which are labeled components C1 through C9.

In each alternative integration sequence, we first integrate the three components inside each gray box, and only after we have completed those initial integration steps do we put all nine components (e.g. all three of the gray boxes) together for a larger integration step.

In the sequence on the left, we integrate C1, C2, and C3 first; another team is integrating C4, C5, and C6; and a third team is integrating C7, C8, and C9. When all three teams have completed their integration of their three components, we then put the three subassemblies together, and have all nine components together in the integration activity.

You will recall that in the design process we generally desire *loose coupling between elements*; similarly, in selecting the order of integration, we want to select an order of integration that mostly involves loose coupling between elements. That is, we want to

Candidate integration sequence A

1. Integrate C1, C2, and C3 to form A1
2. Integrate C4, C5, and C6 to form A2
3. Integrate C7, C8, and C9 to form A3
2. Then integrate A1 + A2 + A3

C1		X			X			
	C2	X	X				X	
X		C3			X			
	X		C4	X			X	
	X			C5		X		X
X		X	X	X	C6			
				X		C7		X
	X		X				C8	
X				X		X		C9

A1	A2	A3

Candidate integration sequence B

1. Integrate C1, C3, and C6 to form B1
2. Integrate C2, C4, and C8 to form B2
3. Integrate C5, C7, and C9 to form B3
2. Then integrate B1 + B2 + B3

C1	X	X						
	C3	X						
X	X	C6		X		X		
	X		C2	X	X	X		
			X	C4	X			
			X	X	C8			
			X			C5	X	X
						X	C7	X
X						X	X	C9

B1	B2	B3

Figure 3.3 An example of the importance of how we select a sequence for integration.

exercise the most demanding interactions *within* the boundaries of our initial integration structures.

In both drawings, the off-axis Xs show where there are component-to-component interactions.

In candidate integration sequence A (on the left in the figure), the off-axis interactions fall largely *outside* the initial 3 × 3 integration steps (labeled A1, A2, and A3). In candidate integration sequence B (on the right in the figure) – you can check that the interactions defined between the components C1 through C9 are exactly the same as in the left-hand candidate – many more of the interactions between the components fall *within* the initial 3 × 3 integration steps (labeled B1, B2, and B3). That way, we have the chance to find and fix more problems during the simpler initial integration step. In my view, the right-hand candidate is a superior choice for an integration sequence.

I call this representation a *coupling matrix*. It can be used as a tool to help you select an appropriate integration sequence.

Integration needs specialized tools: drivers, simulators, data sets, recording and analysis tools, and so forth. Make sure that you allocate time, people, and money at the beginning of the project to build these things, so that they are ready in time for your integration stage.

3.1.3 Testing – Verification and Validation

When we first put together a complicated engineered system of the sort we are considering in this book, it will certainly contain errors. This seems to be just a fact of life when we are dealing with something as complicated as engineered systems, with human beings as the designers and builders. We therefore need a process that allows us to find and fix a reasonable number of these errors, because we want to provide our users with a good product that is safe to use.[1] We also need to convince our customer that the system is complete, and that we are ready to deliver it and place it into operation (and incidentally, to get paid for our efforts). And finally, we need to be sure that the system is going to be liked by, and valuable to, our users.

The process we use for all three of these purposes – (i) driving down the number of errors that remain in our system (the term for these remaining errors is *latent defects*), (ii) convincing our customer that the system is complete, and we are therefore ready both for delivery and payment, and (iii) ensuring that the system is going to be liked by, and valuable to, our users – is called *testing*, and it forms the next stage of our engineering project life-cycle.

We will discuss testing in two ways; first, we will introduce the two different types of testing. Then, we will go on to discuss in some detail the methods for accomplishing these two different types of tests.

---

1 Notice that we said "fix a reasonable number of these errors," rather than "fix all of these errors." Human contrivances are likely always to contain some remaining errors; the Navaho people have an insightful philosophy of deliberately introducing a few errors into the pattern of their woven rugs, as a way of acknowledging that no human creation can be error-free. We will quantify the likely extent of these remaining errors in Chapter 14. Of course, we not only want to minimize the number of such errors, but also to make sure that none of them seriously compromise the capabilities and safety of our system.

Verification – Does the system meet its formal and documented requirements? Are we therefore ready to deliver it, and request final payment?
Validation – Does the system provide value and utility for its intended users?

Figure 3.4 The two types of testing: verification and validation.

The approach I advocate is to use *two different types of testing*; the first – called *verification* – is aimed at the first two of our three purposes for having a test program. The second type of testing – called *validation* – is aimed at the third of our three purposes for having a test program. See Figure 3.4.

People sometimes like to poke fun at government regulations as being all useless "red tape," so I am glad to be able to mention a bit of a US Government regulation that I consider profoundly insightful. In the US Government document called the Federal Acquisition Regulations,[2] it says that a system may not be fielded until it has been demonstrated to be ***both*** *effective* and *suitable.*

The fact that it uses both of the terms *effective* and *suitable* implies that the writer thought that they were *different*, and therefore the possibility exists that a system could be effective without being suitable, or suitable without being effective.

In the usage that I advocate, *effective* means that the *system meets the formal requirements* contained in its specifications. This is an *objective* assessment that all mandatory requirements have been successfully implemented, and is proven through an appropriate *verification* program.

In the usage that I advocate, *suitable* means that the system is appropriate for its intended purpose, and for its intended users. This is inherently an assessment with some *subjective* aspects. Do the users like it? Does it fit the operational context for which it is intended (e.g. the confusion and stress of combat, of a nuclear power plant control room, of responding to a fire or a crime, and so forth)? Can it be operated by the intended personnel (e.g. does it require a Ph.D. in computer science in order to make full use of the system, when the intended operators of the system have 10th-grade educations)? And so forth. Suitability is demonstrated through the *validation* activity; since suitability usually has a significant subjective aspect, a validation activity is quite different from a verification activity (which is objective, in the sense that verification checks off one requirement after another, until all requirements are checked off).

In principle, one could argue that since we took the trouble to understand the users, their value system, their needs, and their desires *before* we wrote the requirements, all of these suitability factors ought in fact to be incorporated into the formal requirements, and verifying those formal requirements will automatically ensure that the system is suitable, as well as effective. I have seen, unfortunately, many instances where a system meets all of its formal requirements, and the users still dislike the system – and at times, have even refused to use the system. This is not a good outcome for your project!

How does this occur? In my experience, it is simply difficult to capture all of the factors that relate to suitability in the formal requirements, and therefore conducting only verification does not always lead to satisfied users and stakeholders. Therefore, conducting *both* verification (which checks if the system is *effective*) and validation (which checks if the system is *suitable*) is good practice.

2 Available on-line at https://www.acquisition.gov/browsefar.

In summary:

- Verification determines if the system is *effective*. Verification is essentially an *objective* process.
- Validation determines if the system is *suitable*. Validation is usually a process with a significant *subjective* component.

Verification and validation are complementary processes; you generally need *both*.

Note that not everyone uses these terms – *verification* and *validation* – exactly this way, so be sure to check the definitions used on your project. If you are working in the aerospace industry, or building systems for a US Federal Government agency, what I described is likely exactly what they expect. Commercial contracts sometimes use the terms differently.

Let's define verification and validation in a little more detail. System *verification* is a set of actions used to check the *correctness* of any system, or portion of a system, where correctness in this context means it *satisfies the written requirements*. The verification process consists of the following steps:

- The items to be verified are *every mandatory requirement* in the entire project's specification tree (e.g. every sentence that contains the verb *shall*). Define a *verification action* for each mandatory requirement in the specification tree – that is, what will we inspect, measure, execute, or simulate? How will we do it? What portion(s) of our system will be involved? (We can often perform some parts of verification using only a portion of our entire system) What tools, recording mechanisms, and data will be involved? What types of personnel and expertise will we need? How many times will we need to do it? (If we have a requirement that is expressed statistically – and this is very common – we will need to collect a statistically valid set of samples.) And so forth.
- Define the *expected result* for each verification action, and define *pass–fail criteria* for each verification action. That is, when we perform the verification action, what result do we expect? What range of results will constitute a pass? What range of results will constitute a fail?
- Create a *procedure* (that is, a step-by-step checklist and methodology) for executing the verification actions, with all of the associated supporting artifacts: test tools, recording devices, input data, and so forth.
- Execute the procedures, obtaining and recording/logging the results.
- Determine which requirements pass, based on an *objective* assessment of whether they satisfied the pass–fail criteria assigned to that requirement.

Verification does *not* occur solely at the end of the project. Planning for verification takes place during earlier project phases, generally in parallel with the definition of the requirements and the realization of the design. The actual verification activity can take place at various times and levels of the hierarchy too. We will talk more about this aspect in a moment.

In contrast, system *validation* is a set of actions used to check the *appropriateness* of any system against its *purposes, functions, and users*. The validation process consists of the following steps:

- Create a *list of items to be validated*. These usually include each major system thread and each major system operation (as described in the system concept-of-operations document).

- For each item to be validated, define a *validation action* – that is, what will we inspect, measure, execute, or simulate? How will we do it? What portion(s) of our system will be involved? (For validation, generally, the *entire system* is involved.) What tools, recording mechanisms, and data will be involved? What types of personnel and expertise will we need? How many times will we need to do it? And so forth.
- Define the *user's goals and expectations* for each validation action, and describe what you believe will constitute conformance [e.g. how do we demonstrate that the user's (and other stakeholders') needs will be met].
- Create a *procedure* (that is, a step-by-step checklist and methodology) for executing the validation actions, with all of the associated supporting artifacts: test tools, recording devices, input data, and so forth. Note, however, that below I advocate that many important validation activities take place *without* the use of a procedure.
- Perform the indicated validation actions, obtaining and recording/logging the results.
- Make a *determination* (which will be at least in part *subjective*) of whether each result constitutes a pass or a fail. This will likely require having some of your users and your other stakeholders participate in creating the list of items to be validated, the validation actions, the list of user goals and expectations, and having them participate in the actual validation activities (or at least witness them). Ideally, the users participate in each decision about whether a validation step has been passed or not.

In contrast to verification, the actual validation activity *does* in fact generally take place largely at the end of the project. Planning for validation, however, like planning for verification, takes place during earlier project phases, generally in parallel with the definition of the requirements and the concept of operations document.

Some limited validation processes might take place earlier (e.g. assessment of the human interfaces of the system), but these will generally be *repeated* in the context of the entire system at the end of the development cycle

Figure 3.5 provides a side-by-side comparison of verification and validation.

Facet of comparison	Verification	Validation
Why are we doing this activity?	Detect defects; ensure compliance with processes and methods: assess quality (development-contractor oriented)	Acquire confidence: ensure that the system produces a desired effect for the users (end-user oriented)
Assessment criteria	***Truth***: yes or no (objective & unbiased)	***Value judgment*** (in part, subjective)
Scope	Detail and local	System-wide
Metaphor	***Glass box*** (how it runs inside)	***Black box*** (inputs cause the expected effect)
Method	Detailed, scripted use	Unscripted use; realistic conditions; off-nominal conditions
Baseline reference for correctness	System requirements	Stakeholder needs and desires
Order of performance	First	Second
Organization of activity	Verification actions are defined and performed by the development contractor	Validation actions are defined and performed in *partnership* by the development contractor and the users

Figure 3.5 Comparison of verification and validation.

3.1.4 Testing – Planning, Procedures, Test Levels, Other Hints About Testing

During the requirements phase, we will not only define the requirements; recall how we said that we must also identify *how* we are going to test each requirement (we called this the *test method*). We do this in order to prevent accidently writing requirements that cannot be tested.

Hence, we select a *test method* for each requirement. Typically, these are one of the following:

- Inspection
- Demonstration
- Analysis
- Simulation
- Assessment via operation.

These terms are defined in Figure 3.6.

We capture this information in what is termed a *test verification matrix*. This is a fancy term for the simple idea of a two-column chart, with the *left-hand column* being the paragraph number of each requirement (recall that we give each separate requirement – indicated by use of the verb *shall* – its own unique paragraph number in the specifications, so each paragraph number identifies a unique requirement) and the *right-hand column* being the name of one of the five test methods. See Figure 3.7 for an example.

The test verification matrix usually becomes an appendix to each requirements specification document.

Inspection:
- We look at something, but we do not operate it or manipulate it
- For example, we look to see if a physical label is attached to a unit, in order to show conformity with the California Proposition 65 labeling requirements

Demonstration:
- We operate a portion of our system, and observe a visible indication
- We do not measure anything; we only take note or record the visible indication
- For example, we might look for a particular message on a display screen, in response to pressing a certain key or button

Analysis:
- We do not operate the system
- But we subject a portion of it to a mathematical analysis
- For example, we perform an accuracy assessment of an algorithm, based on the mathematics and logic of that algorithm

Simulation:
- We build (and confirm the accuracy of!) a computer model of the system (or of an aspect of the system), and use the predictions from that model to show compliance with a requirement

Assessment via operation:
- We operate / manipulate the system or item under test, we record data and results, and then process those data to check performance against pre-established pass/fail criteria
- This test method is often called "test," but I dislike using the term "test" for both the overall process and for this particular method

Figure 3.6 The principal test methods.

Requirement paragraph number	Selected test method
3.2.5.6.11a	Demonstration
3.2.5.6.11b	Assessment via operation
3.2.5.6.11c	Analysis
3.2.5.6.12	Assessment via operation
3.2.5.6.13	Inspection
3.2.5.6.14	Simulation

Figure 3.7 Example excerpt from a test verification matrix.

Having selected a test method for each requirement, we are ready to start *planning* our test program, and ready to create an important artifact called the *test plan*. Here's how to do that.

A typical system will have 1000 or more requirements … or 10 000. Usually, multiple requirements can be verified through a set of related actions. Doing so provides efficiency, and thereby reduces the cost of the test program.

So, we need to identify those groupings of requirements that can be verified through a set of related actions and arrange the test program to suit. My approach is as follows:

- Prepare the test verification matrix (e.g. select a test method for each requirement).
- Define *test levels* for each requirement (which I discuss below).
- Make a preliminary grouping of requirements into *test cases* (where each test case is a set of requirements that can be tested together, through a set of related actions).
- Define scenarios, test data, stimulators/drivers, instrumentation, analysis tools for each test case, and then look for similarities across the test cases. Use this information to improve the groupings of requirements into test cases.
- Develop the sequencing of the test cases and their interdependencies (perhaps one cannot start until another has completed, etc.).
- All of the above items together constitute the *test plan*.

In the above, I used the term *test levels*. Here's why we introduce this concept.

Like many other aspects of systems engineering, we use a *hierarchy* in our test program. Whereas in creating the requirements and design we traversed those hierarchies in a *top-down* fashion, in testing, we will generally traverse that hierarchy in a *bottom-up* direction (as indicated in the "U" diagram of Chapter 2).

What does it mean to test in a *bottom-up* direction? It means that we first test small pieces, then small integrated assemblies, then larger integrated assemblies, and only then do we attempt to test the entire system. You will see that testing therefore is following an approach similar to that we defined above for integration.

We do this because it is easier and less expensive to test smaller pieces than big pieces (just like it was for integration) … and it turns out that some requirements can be completely tested and signed off as being formally verified *at a level below the system level.* This is possible because one can show that nothing that gets added/integrated later will perturb that verified performance. When it is possible to test something at a level below the system level, that's what you want to do; just as for integration, it is less expensive to test a requirement at a lower test level than at a higher one.

So, we must assign the test of each requirement not only to a *test method* (as described above) but also to a *test level*. In doing so, we identify:

- The *lowest level* in the hierarchy at which we can test each particular requirement.
- Which requirements can be tested once (at that low level) and which requirements might need to be tested at more than one level. We might do this because we believe that we can test a requirement at a given low test level, but there may be some uncertainty about whether interactions with other components will invalidate the results of those low-level tests. In such a situation, we will test the requirement at the lower level (where it is easier and cheaper to find and fix the problems) and also include that requirement in a test case at a higher level of testing, so as to prove that no result from that lower-level test was invalidated by the other components.
- The goal is always to test at the *lowest level possible*, because (as was the case for integration) it is easier to find and fix problems in small-scale settings, and it costs less to find and fix each error in small-scale settings.

You will recall that I mentioned during our discussion of the implementation stage that there are some basic features of each piece that we should check out right as we do the implementation. This is, as we can now see, simply an assignment of those particular features to the lowest of the test levels, that of a single piece of the implementation. In software, this lowest level is often called a *unit*, and this type of testing is therefore called *unit testing*.

We cannot select test levels until the design is complete, so we do not include the assignment of requirements to a test level in the requirements specification.

Once test planning is complete, once can turn one's attention to the preparation of *test procedures*. A test procedure is the detailed, step-by-step instructions for conducting each test. It must include all of the supporting tools and information needed to conduct the test, such as scenarios, data, configuration instructions, stimulators, and test drivers (if you are testing only a portion of your system, you might need some special tools to operate that portion by itself, to inject data into that portion, and to capture the outputs of that portion), and pass–fail criteria.

Test procedures can be big, complicated documents, and getting them right will usually require a significant number of practice runs (some people calls these *dry-runs*); during and after each practice run, you can make corrections and improvements to the test procedure (a process we call *red-lining* the test procedure).

One can often work out significant portions of your test procedures during the integration activity. But don't expect to get all of your test procedures to be sorted out during integration; the purposes of integration and testing are different!

The natural tendency is to hope that all goes well during your testing. But I have found that it is actually important to push your system past its nominal limits, and past the expected operating conditions; you *want* to make your system fail, so that you can find its boundaries of effective operation and (this next item is really important!) *understand what happens when it is operated outside of those boundaries*. Does it just slow down when overloaded, or does it actually lose valuable data? Does it slow down or shut off when overheated, or does it let the system actually sustain damage? You need to know these types of things, and at times you will decide to make changes to the design and implementation so that the system behaves better under such off-nominal and beyond-boundary conditions. Systems that pass their requirements but behave poorly when pushed to their limits are often rejected by users as "not suitable." Don't let this happen to you!

The methods above test *effectiveness* (e.g. verification); we still need to test *suitability* (e.g. validation). The following summarizes the methods that I use for this purpose. Notice that they are quite different from the methods described above for verification:

- Unscripted use of the system
- Realistic operating conditions
- Off-nominal operating conditions.

These are each described in the subsections below.

3.1.4.1 Unscripted Use of the System

You might be amazed at how often users find ways to use your system differently than you expected it to be employed. In fact, many systems are used for missions and capabilities other than those for which they were originally designed; think of a screwdriver being used as a chisel. In fact, there are billion-dollar systems that are used *only* for a mission other than that for which they were originally designed; the users discovered something more valuable to do with the system, and decided to use it *only* for that more valuable mission. The US Air Force Airborne Warning and Control System (AWACS) is a good example of this; intended to manage air *defense*, during testing the users decided that the AWACS was fantastic to manage *offensive air operations*, and the system was repurposed for that mission.[3]

Such valuable insights can only be obtained by allowing the users to have *unscripted* access to use the system. During verification, only the carefully designed specific sequences contained in the test procedures are executed; no deviations at all from that script are permitted. During validation, we at times do the opposite: I have found it essential to allow the users unscripted access to use the system. That is, we train the users in the operation of the system, and then let them experiment with it, while watching and recording what they are doing. From such observations, we learn where the system seems to confuse them or slow them down; they learn what the system can do, and might invent new or additional uses, as described above for the AWACS.

Here is another example of such a repurposing of a system, based on insights learned during validation. I was project manager for an army system that was intended for use during close-combat operations; however, during unscripted validation experiments, the users discovered that the system was in fact also valuable during maneuver, planning, and supply/support operations. Those roles were then incorporated into the contract as formal capabilities of the system, and in future versions we enhanced the capabilities of the system so as to be even more suitable for the users during those additional mission operations. Support for missions that originally were *outside* the scope of our contract became some of the main and most valued features of the system!

3.1.4.2 Realistic Operating Conditions

Many of the systems that we build are designed to support *people in stressful or emergency conditions*. Examples include police, fire, and ambulance dispatch and management;

3 A big "thank you" to my friend and former colleague US Air Force Lieutenant General (retired) Bruce Brown, who first told me about this repurposing of the AWACS.

military systems; controls for nuclear power plants, oil refineries, chemical plants, and other complex industrial sites; and many others. In other cases, it might be the *system* that is under stress, rather than the operators: periods of very high input loads, periods of computational stress, periods that require exact timing or accuracy requirements, and so forth. Another common situation is *teamwork*: multiple people (perhaps at different locations) have to collaborate in order to accomplish a task. Most systems have opportunities for *interruptions* to a task in progress, and opportunities to *multiplex* operator attention between more than one task. Lastly, it might be the *operating environment* that is under stress, rather than the people or the system: electronic jamming, cyber attack, problems with electric power, and so forth.

It is unlikely that the verification activity is going to explore these sorts of situations in adequate detail. So, we do this type of exploration during validation. This requires different scenarios, and sometime additional stimulators/simulators, so as to create the loads and stresses.

3.1.4.3 Off-Nominal Operating Conditions

Your system probably has a requirement to operate up to a specified maximum ambient temperature, and similarly down to a specified minimum ambient temperature. But what happens to your system if the temperature is 10 °F above the specified maximum? Do things catch on fire? Do things sustain permanent damage? Or is the effect more transitory, and the system will operate again according to its specifications when things cool down a bit?

There are many such *off-nominal conditions* to explore: input data presented at a faster rate than allowed for in the specifications, more physical shock and/or vibration, under- and over-voltage conditions on the power lines, recovery from power outages, incorrect actions by users, and so forth. You and your users need to understand how the system behaves under such circumstances.

Summary for testing Usually, because they are so different, verification and validation are *separate* events. You must pass *both* verification and validation! You can learn from them both too.

The following are what I have found to be typical errors in the design of a test program:

- Not investing enough in scenarios, drivers/stimulators, instrumentation, and analysis tools. These are little development projects in themselves, and require schedule, budget, and appropriately skilled people.
- Not allowing enough time and money for the integration stage. Without a good integration phase, the chances of success during the test stage are significantly reduced.
- Not allowing enough time and money for a lot of refining and rewriting of test procedures. As noted above, your test procedures will need to be tried out (we use the term *dry-run*); you will find things that are wrong, incomplete, or ambiguous; you will need to fix them, and then do another dry-run. You will almost certainly do more iterations through this process – for every single test procedure – than you likely anticipate. Get data from previous projects to help you estimate the size of this activity.
- Not pushing your system until it breaks (e.g. what above was called using *realistic operating conditions* and *off-nominal operating conditions*).
- Not investing enough in configuration control and a change control process. The test program is fast-paced, and a lot of things happen at once. It is very easy in such a setting accidently to run a test using the wrong version of a piece of software, or use the

wrong version of a data file, and so forth. You need good configuration control (e.g. the mechanism to ensure that a version of an article is the one you intend to use) and good change control (e.g. the mechanism you use to authorize and document the decision process about making changes to an article) in order to prevent making such errors.

We will examine – in a *quantitative* sense – how well testing works to remove defects later in Chapter 14. The short answer is "not as well as you would like." This is primarily because most systems today have lots of *software* in them, and large software products are simply so complicated that they always have *lots* of latent defects. There is a very large range for the number of such latent defects, however: some systems have 20 times more defects per line of software code than other, similar systems. The techniques in this book will help you end up at the good end of those ranges.

3.1.5 Production

For some systems, you are only going to build one copy of the resulting system. For other systems, you might be under contract to build 10 copies; or 1000; or 100 000 000. Alternatively, the number to be produced may *not be completely defined in advance*; you will produce them until people stop buying them.

We call the project life-cycle stage where we make these additional copies of our system *production*. The techniques for making these additional copies vary significantly depending on the scale of the planned production:

- If you are only going to build one item (e.g. a specialized satellite), you may be content to have your engineering staff do portions of the production and assembly.
- If you are going to build 10 (e.g. an entire constellation of a particular type of satellite), you may decide to hire some *specialized production staff* to build units 2 through 10, and stock a warehouse in advance with all of the necessary parts.
- If you are going to build 1000, you may invest in some *automation* to do some of the assembly: pick-and-place machines to build electronic circuit cards, for example. You will probably also have a production staff that is completely disjoint from your engineering staff.
- If you are going to build 1 000 000 (or 100 000 000), you will probably invest in a lot of automation to do the assembly: *robots* to do the work at each assembly station; machines to move items from one assembly station to another, and so forth. You will probably manage your parts differently too: instead of just amassing all of the parts you need in advance, you might establish a sophisticated *supply chain* to provide the parts you need on a *just-in-time* basis.

Why not use the best production techniques (robots, a complicated just-in-time supply chain, etc.) for every project? You will likely find that it is too expensive to set such mechanisms up for smaller production runs. We must therefore *design* our production processes based on an estimate of how many copies, over what period of time, we will be making of our system.

Another important decision about production is what your company will do themselves versus what you will buy from other companies. We call this a *make/buy decision*.

Companies used to aspire to do almost everything themselves, in their own facility; this approach provided a great deal of control. The Ford Motor Company was famous for a plant that it used to operate just west of Detroit, where they claimed that iron ore and raw rubber went in one end, and finished cars came out the other.

No one builds cars this way any longer; companies have found it too hard to be good at everything. This has led to a strategy of picking the things that you want your company to be good at, and buying other components from companies who you believe are good at that kind of work. Car companies, in fact, generally no longer even buy small parts; they buy large subassemblies (transmissions, front-end chassis assemblies that already combine suspension, brakes, and steering, and so forth).

What this all means is that the *design of our production process* is part of our responsibilities, just like the design for the rest of our system!

Consider the following:

- There will be a big difference in up-front versus recurring costs, depending on what you *make* versus what you *buy*. It requires more up-front investment to make a part than to buy it, but the recurring cost may be lower. A big reason to select *buy* is that you can pass a lot of the risk (which we will discuss later in Chapter 9) and inventory cost to your supplier. You must balance all of these factors, and select *make* or *buy* for every part in your system.
- For those items that you elect to *buy*, for each you must decide whether you want to buy individual parts (e.g. shock absorbers and brake calipers) or complete subassemblies (e.g. entire integrated front-end chassis assemblies).
- There will be a big difference in up-front and recurring cost, depending on how you do the production, as described above. You must select a production approach that balances these two types of costs.
- Different designs will turn out to be easier or more difficult to produce in quantity; if your system is planned to have significant production, you must include *producibility* as a goal of your design process.
- There will also be a big variation in the *defect rate* induced by different assembly methods; generally, more automation (which, of course, costs money) results in lower assembly defect rates. You can also influence the defect rate significantly by a good design for the assembly process.
- Errors arise not just from defective parts (whether hardware or software), but also from defective assembly (e.g. parts inserted upside-down, over-tightened, wrong part in a particular location, wrong assembly sequence, missing or left out, and so forth). You can design the parts and the assembly sequence so as to minimize the likelihood of such errors.
- You must also design the assembly process to match the skills of your intended assembly personnel.
- You must account for these sources of error in your system reliability, availability, and mean-time-to-repair predictions. These parameters are likely to be contractually binding, so satisfying them is not optional.

Since parts can be defective, and since the assembly process can introduce errors, we must do a small amount of testing on the items we produce. Sometimes, you conduct an abbreviated test (called *subsequent-article testing,* because these are items produced subsequent to the first copy of the system) on every copy. In other circumstances, you may elect to test only some of them (e.g. every tenth item, etc.).

3.1.6 Deployment: Use in Actual Mission Operations

Let's consider a scenario where our project is building a satellite. It is complete, and sitting in our factory or test facility. It is not providing much value to the user at that location; it needs to be shipped to a launch site, mated with a launch rocket, launched into the correct orbit, checked out in orbit, and then connected to the ground station that will send it control signals and receive the product from the satellite's sensors. These latter steps are an example of a *deployment* activity, our next stage in the life-cycle.

Let's consider another example, one where our project is building a new automation system for a large chain of hospitals, a system that will combine accounting, billing, insurance processing, patient case management, and electronic patient records. As you can see, our system is going to be both safety-critical and very important to the hospital (and their patients). Our new system may not need to be launched into space, but the deployment stage is still quite complicated: we must disassemble the system in our factory, pack it, and ship the appropriate parts without damage to each of the various locations where they will be installed. We must physically and electronically install it, without disabling the existing systems that will continue to perform these same functions until our system is ready to take over. We must train the staff who will use the new system. We must do some sort of trial operations, so as to verify that our system works properly in the actual context of the hospital. All kinds of things can go wrong: for example, one of our computers may be located close to a medical machine that periodically emits large bursts of electromagnetic energy, which causes our computer to crash! We cannot *go live* – that is, transition our system into actual mission operations – until we are sure that we have worked out all of these types of problems; the hospital cannot be "down" for even a minute, as lives are on the line. We therefore probably have to operate our new system in *parallel* with the old systems for some period of time (by which we mean that both the old and our new system receive all data inputs and process them, but only the old system is used for actual hospital operations) – perhaps for several weeks, and then organize a switch-over that does not leave the hospital without service for more than some agreed-to period of time (which might be very short indeed!). We must make sure that we have transferred all of the data from the old system to the new system, and that we have done so without violating any privacy provisions of the law. We must somehow confirm that all of the data (such as patient records) have been transferred correctly to our new system.

As you can see from these examples, the deployment stage can be quite complicated. The typical steps involved in deployment include:

- Packing, shipping, installation
- Check-out at site
- Initial training of operators
- Load actual mission databases
- Interconnection with real interfaces
- Configuration management:
  - Have we loaded the right software?
  - Have we loaded the right data?
  - Have we correctly configured all of the settings?
  - ... and so forth.

- Transition into operations ("go live"):
  - Dry-runs
  - Operate in parallel with legacy system
  - Then cut-over
  - Be prepared to roll-back!
  - Lots of rehearsal will be required
  - Instrumentation and analysis tools are required – the problems are not always obvious.

Once our system is in actual use (e.g. we have completed a successful deployment), it can finally be used by the intended users, and bring them the benefits for which it was designed. But those users need support: someone has to create training materials for the new system, and perhaps even conduct actual training classes. Things break, and someone has to diagnose and fix them. To effect those repairs, we will need replacement parts; we need someone to make those replacement parts. It is likely that we will continue to find errors in the system – even after the test program has completed – and we will need to fix those errors. Most systems are operated for a long time, and our users expect us to design and implement improvements to the system over the course of time that the system is operated. There are many other, related aspects of supporting our new system in effective operations; these are discussed in the next two life-cycle phases.

3.1.7 Non-project Life-Cycle Stages

The remaining life-cycle stages are usually *not a project* (following the definition of a *project* that we introduced in Chapter 1), but your project must create the ingredients of success for these stages too, even if someone else then receives responsibility for executing these stages.

3.1.7.1 Logistics

Typically, 80% of the cost of ownership of a system takes place ***after*** deployment. That's right; the entire cost of designing, building, and testing our system probably accounts for only 20% of the cost of operating the system for its full intended operating life.

Because of this, designing our system so that we minimize overall cost *over the entire time period for which the system will be used* (called the *life-cycle cost*) is usually very important to our customers.

Furthermore, it is not always obvious what is actually going to incur the biggest portion of those life-cycle costs. Consider a data center: a building full of computers, disks, and communications equipment. Perhaps this data center is going to be used for 20 years. Over those 20 years, what will likely cost the most? It might seem that the information technology (IT) equipment – those computers, disks, and communications gear – is likely to be the most expensive. But take a look at Figure 3.8, which graphically depicts a notional view of the relative magnitude of the life-cycle costs over that 20-year period.

In the figure, you can see the small box that represents the cost of the information technology equipment. The next larger box represents the cost of the building and the heating/ventilation/air-conditioning (HVAC) equipment; over the 20-year life-cycle, those items likely cost more than the computers and disks!

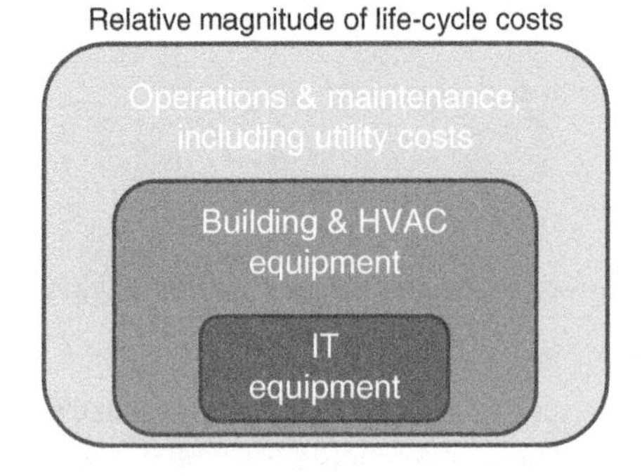

Figure 3.8 Notional relative magnitude of life-cycle costs for a data center.

But the box representing the building and the HVAC is not the biggest box in the figure; the biggest box is the one that represents maintenance and utility costs. Most of this box is just paying for the electricity to run the data center! For many data centers, electricity is the dominant cost over the 20-year life-cycle.[4]

What does this mean to us? It means that, as designers, we ought to think about items such as:

- The power efficiency of the IT equipment: are there computers and disk drives that use a lot less power than others?
- The power efficiency of the HVAC: is there some type of air-cooling method that uses a lot less power than other types?
- Are there investments we ought to make in power management controls for the computers and disk drives?
- The location where we place the data center may be more important than we think: a cold location could allow the use of ambient air as part of the cooling strategy (which might save a lot of electricity) versus a sunny site that might allow for the use of solar panels to generate the electricity needed to power the air conditioning.

Life-cycle costs are a major design driver … in fact, often *the* major design driver.

Therefore, even though the post-deployment activities are *not* usually part of our project, I will describe what takes place during those life-cycle stages so that you understand the *design objectives* that arise.

What happens after we deploy our system? Activities such as the following:

- Operation of the system for its intended purpose by the users
- Sustainment
- Spares
- Training
- Support to operations
- Data-gathering and record-keeping.

I will describe each.

4 Most analyses that I have seen assert that *data centers alone* now use between 2% and 3% of all the electricity in the United States; some assert even higher levels! See, for example, https://eta.lbl.gov/publications/united-states-data-center-energy.

Operation of the System for its Intended Purpose by the Users Trained personnel are assigned to operate the system. The intended missions are performed. Some process and people are assigned to monitor operations, manage the assignment of personnel, and look for problems with the system. Problems found generally fall into one of three categories: something might have broken; some consumable item might have run short; or some *latent defect* that was not uncovered and corrected during the test stage might have been encountered. The development contractor (us!) likely has a role in monitoring, assessing, and correcting such problems, especially those in the third category.

Sustainment The system must be kept in operation. The system likely has some preventative maintenance actions that have been identified by the development contractor; for example, the manufacturer of your car probably suggests that you change the oil at specified intervals of time or distance traveled. A team of people have been trained and assigned to perform these *preventative maintenance actions*, and provided with the necessary equipment and parts to do so. In addition, as noted above, problems will be encountered that are determined to be caused by something breaking, or some consumable item (e.g. toner in the computer printer) having run short. A team of people has been trained and assigned to perform these *corrective maintenance actions* when such problems arise. In both cases, someone has to make a decision about the balance between having lots of such maintenance personnel at every location 24 hours per day, with lots of spare parts and consumable items right on hand (which allows corrective actions to take place promptly, but costs money), or some less intense strategy of assigning personnel and parts to locations and work-shifts. Both types of maintenance (preventative and corrective) are tied to metrics of system availability, reliability, and allowed down-time, which are often contractually binding commitments by the development contractor; you must *design the maintenance activities so as to meet the contractually binding quality commitments*. You must also match the difficulty of the maintenance tasks to the skill levels of available personnel, and to the training provided to those personnel. You must develop concepts for managing repair turnaround times (no one likes to wait for repairs!), including what will be done on-site versus at a maintenance station or depot located elsewhere, and what will be done by the customer's personnel versus what will be done by the manufacturer and their suppliers. You may have to develop concepts for formal record-keeping about failures, the certification or qualifications of the person performing the repairs, and maintenance logs; in some systems, these records are necessary to obtain and keep special certifications like flight-worthiness and safety certifications. You must decide what tools (e.g. on-line diagnostics/prognostics, on-line manuals, etc.) will be cost-effective for this system; you can spend more on diagnostic and prognostic tools and likely decrease mean time to repair, but that costs money too.

For us as the development contractor, the focus of sustainment planning is not just figuring out how many spare parts to stock. Rather, it is the *determination of the balance between inherent reliability and maintenance* – how are we going to make this system be operationally suitable, given its intended use, the nature of the mission, and the nature of the operators. We can design the system to be inherently more reliable (and thereby need fewer repairs and fewer repair parts), but that costs money too.

We must also balance the above consideration against the life-cycle cost of the system; that is, within the trade space of the ways to balance between inherent reliability and maintenance that are operationally suitable, which of them have the most advantageous cost profile?

The sustainment planning trade space therefore often includes:

- Inherent reliability – designs that have higher inherent reliability often have lower maintenance costs but higher acquisition costs. Which balance is right for this system and this customer?
- Which personnel can/should do which type of maintenance/repair actions? Which are performed at operational locations, and which are performed at various sorts of depots and repair facilities? For example, most of the military systems that I have built have a three-level maintenance concept: (i) some repairs are performed in the field, (ii) some repairs are performed at a military repair facility, and (iii) some items are returned to the contractor's facility for repairs.
- Most systems have an operational life far longer than the anticipated lifetime of their parts. How do we accommodate that? How do we deal with parts that become obsolete and/or unavailable? More flexible designs are possible, but usually have higher development costs.
- The system will experience errors and failures. What level of system capability must we continue to provide? How much of that is accomplished through design features, and how much is accomplished through diagnosis/repair/sparing strategies?
- Increasingly, software, rather than hardware, is what drives system reliability rates; how are we going to account for that during our design and test activities?
- Mean-time-to-repair and mean-time-between-failure, which combine to drive system availability, are often tradable with each other, and are each driven by design and maintenance plan decisions. What values do we choose for target mean-time-to-repair, mean-time-between-failure, and other related parameters?

Spare Parts Repairs, of course, cannot be effected without spare parts. It is nice to have lots of spare parts right at hand, but having such a large inventory of spare parts costs money too. So, you must develop a strategy for spare parts: where are they kept (at the user's site, at a customer depot, at the factory, or some combination)? Furthermore, your system may be in service for years (even decades), but it is likely that at some point, the manufacturer of a part will stop making . Do you build up a big supply before that part is discontinued? Or do you redesign a portion of your system to make use of a newer part? You will also need an inventory control system to keep track of parts, where they are located, what is ordered, and when it will arrive, and to help you make predictions about when you need to order more parts.

Training We have mentioned training of users, repair technicians, and others many times already. But people come and go, they retire, they move to another organization, they move to a different assignment within the same organization, they get promoted, etc. So, we must be training people all of the time! We must therefore predict and plan when we need to train people, how many people at which locations, select and qualify people to serve as the trainers, prepare training materials, and so forth. In some systems, we must keep formal records about who has received what training, as important personnel and system certifications may in part depend on records about people having completed periodic refresher training.

Support to Operations All kinds of activity takes place behind the scenes of a large system. Upgrades, patches, and minor upgrades must be applied with minimal disruption to regular operations. Larger repairs and upgrades that require actual outages and down-time must be scheduled and coordinated with the users. Training materials and documentation must be updated and corrected. If versions of the system are sold to users in other countries, materials need to be translated, culture differences need to be assessed and these effects incorporated in the systems operational concept, and so forth.

There is also an important opportunity for *learning*: the operation and use of the system can be observed by the development team, in order to create ideas for what improvements can be made to the system, and when those improvements should be made.

Data-Gathering and Record-Keeping All of the above involves a great deal of data-gathering and record-keeping; we need good data in order to make good decisions. When should we order additional spare parts? How many parts should we keep at each operational site? How many maintenance personnel should we have on-site at each location on each work shift? The number of questions that we must continually ask and answer during the post-deployment stages is very large. We must always be ready to reassess; perhaps the load on the system has changed, or the relative cost of stocking parts in various locations has changed, or it is now cheaper just to throw away and replace a part rather than repair it. We make these decisions based on data. So, we must be gathering and analyzing data continuously, and keep well-organized records.

Also, as noted above, sometimes our system and/or our system's operators require special certifications, and we must gather data and keep records to support those needs too.

3.1.7.2 Phase-Out and Disposal

All good things come to an end, and eventually our system will be obsolete; or for some other reason, the users will be ready to retire it and remove it from service. We call this the *phase-out and disposal* stage of our life-cycle.

This stage can be difficult and complicated. For example, the system might include radioactive materials (even smoke detectors have radioactive materials in them, and must be properly disposed of). Or there might be other materials that require specialized disposal; for example, the lithium-based batteries in most consumer electronics and computers.

Today's systems almost certainly contain data that cannot just be placed in a dump where others might get access to it. Properly erasing data is quite complicated, and is at times subject to laws and regulations about how it must be done; even the method of erasing a disk drive may be subject to law and regulation – just reformatting a disk drive is usually not enough.

This means that the phase-out and disposal stage should be considered in the design as well. For example, it makes disposal easier, safer, and less expensive if those special materials can be easily removed from the system.

Lots of systems and devices require special disposal. Our systems can be designed to make this disposal easier. And don't forget about handling private data during disposal!

3.1.7.3 Summary for the Post-Deployment Stages

Training and post-deployment support are big drivers of user satisfaction (or dissatisfaction!). Resource these activities appropriately, and start working on your plans for how to accomplish them right at the beginning of the development program. All of the above activities need to be planned, budgeted, staffed, and monitored.

Remember, most of the costs of a system are incurred *after deployment*. We minimize and control those costs in large part through our *design*. So, even though the post-deployment life-cycle stages may not be a formal part of our project, we designers are the ones who determine whether the customer will be happy with the system after they start using it. We need to understand these post-deployment life-cycle stages because our design is an important factor in their success.

3.2 Next

We have devoted two chapters to discussing how we actually do engineering on projects. We will next use that knowledge to optimize our project management procedures in light of those engineering processes.

3.3 This Week's Facilitated Lab Session

This week is all lectures; there is no facilitated lab session.

4

Understanding Your Users and Your Other Stakeholders

We introduce the two coordinate systems of value, and we also discuss how we engineer the user experience. Engineering projects often create products and/or services that never existed before. Under these circumstances, it is easy to lose sight of what aspects of the new item are essential, and which are less so. We solve this dilemma by rigorous and continuous focus on our eventual users and customers. What are they trying to accomplish? How do they do it now? What are the shortfalls? What are their needs and desires? At the same time, our degrees of engineering freedom are usually entirely within the technical domain: choices about materials, parts, algorithms, mechanical structures, and so forth. In this chapter, you will learn how to understand your users, how to relate that understanding of your user to the engineering choices that are your degrees of design freedom. We then extend this focus on our users to all "stakeholders" of our project. We end the chapter with a discussion of how to use good engineering and good management to achieve a compelling and effective experience for your users and your customers when they operate your system, through what we call the *user experience.*

4.1 The Four Steps to Understanding Your Users and Your Other Stakeholders

In Chapters 1 and 2, we introduced the idea of our engineering project having users and other stakeholders. In this chapter, we go into the actual details. Who are these people? How can they affect the success of our engineering project? How are we to interact with them?

It is worth reminding ourselves *why* we consider users and stakeholders: our engineering project does not exist in a vacuum – it has a *purpose.* Most likely, this purpose involves doing something useful and important for those users, and something that is beneficial in some fashion for those other stakeholders, and through them, beneficial to society. In order to specify and design our system so that it truly serves the needs of those users and stakeholders (and society), we need to identify them, and to understand them.

In Chapter 2, we discussed the creation of the *requirements* for our system. In particular, in Figure 2.14 I presented a 10-step process for performing this creative task. In Figure 4.1, I repeat the first four steps from Figure 2.14.

Engineering Project Management, First Edition. Neil G. Siegel.

Companion website: www.wiley.com/go/siegel/engineering_project_management

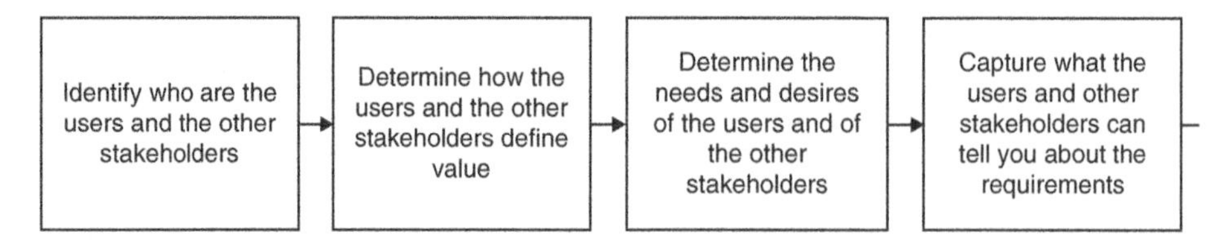

Figure 4.1 The first four steps of the requirements creation process (extract from Figure 2.14).

The first thing to notice about these steps is that each and every one of them contains both the word *users* and the word *stakeholders*. In this chapter, we will discuss the users and our other stakeholders in terms of these four steps.

If we are going to build a product or service – and in Chapter 1, we agreed that was in fact the purpose of a *project* – some set of people somewhere are the intended *users*: the people who will operate and/or employ the product or service, in furtherance of some mission or purpose.

The set of *users* may in fact be more encompassing than the set of people who actually operate our product or service. If our product is a new military fighter jet, the pilots who actually fly the system are certainly among our users, but the military commander who is deciding what mission those fighter jets are going to go out and try to accomplish tomorrow will also certainly consider himself/herself as a *user* of those jets as well! He/she is using those jets to accomplish his/her purposes – some combination of tactical, strategic, and political[1] – even though that commander is not going to get into an airplane and personally participate in the flying. That is why I say the *users* are the people who will operate and/or *employ* the product or service, in furtherance of some mission or purpose.

If those are the users, who are the remaining *stakeholders*? We use that term in this book to designate anyone who believes that they have a *vital interest in the outcome of the project*. Examples might include:

- **Those who are *paying* for the product or service** that your engineering project is going to produce (the *paying customer*).
- **Those who are acting as the *buying agent*** for the product or service that your engineering project is going to produce (the *buying customer*).
- **Those parties who are *close collaborators with your users*.** For example, that military fighter jet we are building may be flown by the Air Force, but if its mission includes providing what is called *close-air support* to ground troops (as it likely does), then the Army and the Marine Corps will have a legitimate interest in the planned capabilities and delivery schedule of our fighter jet.
- **Those parties who have a *mission or financial stake in the resulting product*.** For example, once our fighter jet is completed, there will be extensive testing by the Air Force agency that conducts testing and evaluation of new airplanes. There is also an entire infrastructure within the Air Force (and across the entire Department of Defense) that provides logistical support (diagnostic services, repair services, spare parts, and trained people to do all of these tasks) to military airplanes. Both of these organizations – the testers and the logisticians – certainly have a legitimate interest

1 "War is the continuation of politics by other means" – Karl von Clausewitz, *On War*, 1832.

in the design of our fighter jet too, especially with regard to what we might or might not be planning to provide in the way of diagnostic tools, and other matters that will directly affect their mission. They probably have a financial interest also: they will, for example, have strong opinions about what logistics-related work should be done by their organization, as opposed to what work should continue to be done by the prime contractor.

- **Those who feel that they are strongly affected by the *opportunity cost* of our project**. This group will include, for example, those who are involved in building and supporting the fighter jets that our project is designated to replace (and their political supporters); they may prefer that our project proceeds slower than planned (thereby allowing their older jets to remain in service for a longer period of time), or even oppose our project in its entirety. They perhaps would prefer for our project to get less money, so that more money is available for upgrades to the older jets. There might be others who believe that by receiving funding, our jet project is taking the opportunity to receive funding away from other projects, ones that they would prefer to be funded.
- **Those whose responsibility it is to in some way regulate our product or service**. If our project was to design a new car, there are agencies within both State and Federal Government that make regulations about car safety, pollution, gas mileage, tax treatment, and many other factors that can affect our project, and even stop our project. There are State and Local Government agencies that affect our plans and execution directly; for example, if we need to build or modify a factory, we certainly will need building permits, and probably additional State and Local Government approvals and permissions as well.
- **Our employees** are always very important stakeholders in our projects; we must investigate and determine what will make the project a success for them too. This will usually involve much more than just salaries and benefits. If some or all of these employees are represented by a labor union, that union is a stakeholder too.
- **The owners and/or stockholders of our company** (and of our important suppliers) are stakeholders too. They will influence the terms of our contract, and constrain our operating alternatives.
- **Members of the general public** (**and their political representatives**) who believe that they have an interest in our project and its outcome. In our military jet airplane example, this might include: (i) veteran's groups; (ii) industry-funded political action committees; (iii) people who live near where our factories and our suppliers' factories are located; (iv) people who live near the airbases where our jets will be based (some may like the idea of the jobs that the maintenance of our jets will bring into the area, but others may dislike the noise that the jets will cause); (v) people who live near the airbases where the *current* fighter jets are based (those bases may lose jobs when our jet is deployed); (vi) people and groups that would prefer that the money be spent on other projects and activities, rather than on our jet; and many others, too.

Notice that not all stakeholders are necessarily *supporters* of our project; the fact of their opposition (whether temporary or enduring) does not prevent them from still being important stakeholders. They have the potential to influence those making decisions that affect our project, regarding money, capabilities, schedule, the eventual size of a production run, and many other things.

Most of the examples above deal with *externally funded projects* (remember, from Chapter 1: a project where the entity that pays for the project is *outside* of our organization). But there are analogous players and tensions for *internally funded projects* (again, from Chapter 1: a project where the entity that pays for the project is within *our own* organization). Don't be misled into believing that the other people in your organization or company see every aspect of your project the same way that you do. If your project is to build the next version of the core product that your company currently sells, there are people who want it designed differently; people who want to extend the life of the existing product instead of starting a new product at this time; people who want to lead the new project themselves, instead of you; people who want to work on the project, and are annoyed that you did not select them for some important position on the project; people who think that your cost estimates are outrageously high; people who think that your estimated schedule is outrageously long; people who want to subcontract out the entire job, instead of doing it internally; people who want the company to concentrate on some completely different product or service, instead of what your project is intended to build. And that is just among the other employees! There are also the stockholders, the market analysts, regulators at both State and Federal levels (and in other countries, if you plan to sell your product outside your own country), the advertising agency that your company uses, your suppliers, your competitors, and the general public.

That is a lot of different people and groupings!

If we want to be allowed to build the product or service that our project is going to create (which includes getting someone to pay for it!), we have to understand these people, pay attention to them, and get them *aligned* (remember Figures 1.10 and 1.11?), just like we have to get our employees aligned. In this chapter, we discuss how to *understand our users and stakeholders*; in Chapter 13, we discuss how to get these people aligned.

We have now shown you how to go about identifying your users; that's the first box from Figure 4.1. Now, let's move on to the next box: determine how the users and the other stakeholders define value.

What do we mean by the word *value* in this context?

One of the defining characteristics of engineering is that we aspire to make decisions based on facts (often, based on a *quantitative* assessment of facts and measurements – we will talk about the pitfalls of measurement in Chapter 8). Since the product or service that our engineering project is going to produce is intended for use by our users, and will impact the lives of our other stakeholders, we will make better and more informed decisions about our project if we understand those people. We try to understand the users and the other stakeholders, their needs and their constraints. What is their mission? What are their products? How do they do their mission today? Why is that how they do it? What constrains the possibilities?

Behind all of these answers, there is a *why*: what is the attitude that *causes* those to be the answers to such questions? That *why* is the *value* which we seek to understand.

An example will help. Remember the air defense system used as an example in Chapter 2? I presented the statement:

Never shoot down a friendly aircraft.

This is a *value*, a goal or aspiration that the group desires to become inculcated in all of its members. This statement is considered fundamental to being an effective air defender in the US Army.[2] Statements similar to the above are found in their training manuals. Statements similar to the above are made by instructors in their training courses. This attitude is even the subject of jokes,[3] which is perhaps a sure sign that this statement is a core part of the US Army Air Defense Corps culture.

There may be practical considerations that affect what becomes a value for an organization. This is certainly true in this case. Under US military doctrine, in an area where both US ground forces (Army or Marines) and US air forces (which might be any or all of US Air Force, US Navy aviation, US Marine Corps aviation, or Army aviation[4]) are operating at the same time, there is a person from the aviation forces (usually, someone from the Air Force) who is designated as the *area air component commander*. This person sets policy (called the *rules of engagement*) which is designed to guide how air operations and ground operations are coordinated, so as to ensure the safety of both pilots and people on the ground. US ground troops are equipped with highly effective ground-to-air weapon systems, and those systems can shoot down US aircraft as easily as they can shoot down enemy aircraft. In the stress and confusion of battle, mistakes are easily made. So, the establishment of these rules of engagement is intended to organize the air portion of the battlespace. For example, the battlespace might be partitioned into zones, where in some of these zones, the designated plan is that enemy aircraft will be dealt with by shooting at them from the ground (e.g. the air defense mission that we described in Chapter 2 – these are called *missile engagement zones*), but in other of these zones, the designated plan is that enemy aircraft will be dealt with by US fighter aircraft (these are called *fighter engagement zones*). By doing this separation, the idea is that the US fighter aircraft will be operating in areas where the US air defenders have been told not to shoot at anything in the sky, and it is hoped thereby that the chance of a US air defender accidently shooting down a US aircraft is very small.

But even in the missile engagement zones, we have to be careful. In the heat of battle, a US aircraft might stray into a missile engagement zone, or a missile stray into a fighter engagement zone. This possibility leads to another statement that I presented in Chapter 2:

> You may not shoot at any aircraft, even if you think that it is an enemy aircraft, until after a trained operator has conducted a visual identification of the target through an appropriate magnifying optic, and determined by that visual examination that it is actually an enemy aircraft.

This is clearly intended as another method of preventing a US air defender from accidently shooting down a US aircraft. The actual *rules of engagement* are even more

2 Although not always in the armies or air defense units of other countries. Values are part of a *culture*, and therefore can vary from culture to culture.
3 For example, the US Army air defense school used to sell fly swatters emblazoned with the phrase "If it flies, it dies."
4 US Army aviation consists mostly of helicopters. When I was doing a lot of work for the US Army, the Army had more flying objects than any other US military service, including the Air Force, but most of them were helicopters, rather than fixed-wing aircraft.

restrictive: there is something called a *weapon control status* for the air defense weapons. This has three possible states:

1. *Weapons hold.* When the weapons control status is *hold*, a US air defender may fire at an aircraft only after that aircraft first fires at him. That is, he may only fire in self-defense.
2. *Weapons tight.* When the weapons control status is *tight*, a US air defender may fire without waiting for the aircraft to fire first, but he still must perform a visual identification of the aircraft that he is considering firing at prior to actually firing a weapon toward it.
3. *Weapons free.* When the weapons control status is *free*, a US air defender may fire at his discretion. He need not perform the visual identification.

In actual combat, the weapon control status is almost never set to *free* for the airspace over ground combat; the battlespace is just too complicated. It might be set to free for a ship at sea, operating in a situation where for other reasons it is known that no US aircraft are in the vicinity.

So, for our notional Army air defender, most likely he is either placed in *weapons hold* or *weapons tight*. He would much rather be in *weapons tight*; no one wants to wait for the other party to take the first shot! But our air defender does *not* get to select the weapon control status currently in effect; the air component commander makes that selection. Having a culture of never shooting at a friendly aircraft, and (more tangibly) having a procedure (e.g. the visual identification process) by which that culture can be translated into effective action, may give the air component commander confidence that he can safely allow the air defenders to operate in *weapons tight*. Since that is the strong preference of the air defenders, they (in part) adopted this *value* (e.g. never shoot down a friendly aircraft) into their culture so as to achieve the desired practical effect of having the air component commander allow them into *weapons tight*, rather than keeping them always in *weapons hold*.

But there is more to this particular value than its aid in achieving this practical effect. More broadly, the entire US Army, and in fact the entire US military, embraces a value of avoiding killing or wounding their own soldiers, airmen, sailors, and marines. Such *friendly fire* incidents are in fact quite common in the confusion of battle – some estimates say that half of all war-time casualties are caused by friendly fire. And all militaries desire to avoid such incidents, but there is (in my experience) a marked difference in the *intensity* of such desire between countries, and in the policies and procedures established and enforced to achieve that outcome. In the US military, this desire is very strong. So, the *value* of the expression

Never shoot down a friendly aircraft.

exists first and foremost because of this cultural characteristic of desiring to avoid friendly fire incidents.

We must arrive at such an understanding of what our users and stakeholders value for every engineering project. We do that by the methods mentioned above: reading manuals and training materials, asking questions of expert practitioners, and watching them perform their mission. Since what we are looking for in the search to understand what the users value is the *why* behind their answers, and the *why* behind their choice of methods, tools, and procedures, indirect methods (such as watching) to find their values are usually required. I do directly ask *why*, and so can you, but that is seldom sufficient; sometimes the values are ingrained, but not explicitly articulated. We must therefore engage in actions that will bring them to the surface.

This search for the users' value system cannot be exclusively qualitative. In engineering, we measure things and make decisions based on our analysis of those measurements. We therefore aspire to make *quantitative statements about the users' value system*. This requires us to consider how we can measure those values in a manner that the users will find credible and useful.

Remember the example in Chapter 2, where I described an insight we acquired about the US short-range air defense system: they were not shooting at most of the available targets. After discussion with the users, we collectively realized that using our new system to help them not miss so many of the potential shot opportunities could be considered a *user value*. But we went further: we created a *metric* regarding increased shot opportunities, and used that metric as an *operational performance measure* to assess the goodness of candidate designs. That is, we translated a *qualitative* statement about the users' system of values into a *quantitative* metric that we could use to guide decision-making about our project. You must do that for your project too.

We are now ready for the third box in Figure 4.1: determining the needs and desires of the users (and the other stakeholders).

Note that I say *needs and desires* (or *wants*), not just *needs*. We do this for two reasons:

- Most projects are awarded through a process that involves *competition*: your company must compete against other companies in order to obtain the contract to build the project. As we will discuss in the next chapter, in order to win such competitions, you must certainly provide them with what they *need*, but your chances of winning are far better if you also provide them with what they *want*. For most of us, *needs* and *wants* are not exactly the same thing! My friend Bran Ferren[5] says that people will sometimes buy what they *need*, but they will much more often buy what they *want*.
- As we discussed in Chapter 3, in the segment about testing, it is important that the users and other stakeholders *like* your system; in order to accomplish that, you need to start by understanding not only what they *need*, but also what they *want*.

We have already read training manuals and other written materials to help us learn about the users and their mission. We have talked to them, and even watched them at work. Now we have to separate what they (and their current systems) do that is *necessary* from that which is *not*; work procedures and systems often have lots of steps that are unnecessary, make-work, or carry-overs from older methods (and their limitations) used in the past.

Here's an example. I once learned from such discussions that the users of a particular system hated the fact that when they had important information to enter, they had *also* to enter the name of a person to whom to send that information. We do that every day for electronic mail, and don't usually think of it as a burden, because, in general, the information that we are sending is intended specifically for that person, by name.

---

5 Former President of Imagineering at the Walt Disney Company, winner of multiple Academy Awards for movie special effects, and now the Chairman of Applied Minds, LLC.

Whoever built their current system just assumed that such an approach was appropriate in the context of *that* system. But the users disagreed; they were inputting information that was valuable for multiple people, but the people for whom it was valuable changed over time; for example, between work shifts – there might be one person on each work shift who dealt with a certain type of information. The users felt that it was an unnecessary burden for them to have to figure out who (by name) was on duty at this time, so that they could send the information to that person. They also worried that when it came time for a change of work shift, information would not be available to the person on the next work shift responsible for that type of information.

What we determined is that the mental model (borrowed from electronic mail) of sending information to a person by *name* was a poor choice for this system; what they wanted instead was the ability just to input their information, and have the *system figure out who needed to receive it.* On each shift, someone will be the person responsible at that time for each work role; what the users wanted was for them just to input their data to the system, without having to name a specific recipient by name, and for all of the appropriate information to somehow be made available automatically to the correct person when they started their work shift.

The current system provided what was *needed* (e.g. people could input relevant information, and get it to the correct person), but the method used to do so (the users had to know the name of the correct person) created a lot of work considered by the users unnecessary (and error-prone), and therefore was not *liked.* After we figured all of this out, we created the idea that each user was associated with a particular *work role,* that the system could have knowledge coded into it about which kind of information was relevant to each of these work roles, and therefore that information could be routed automatically by the system to the person(s) who occupied the relevant work role at this time. This lessened errors, and reduced the workload of the users. The users *liked* that design; it would cost extra money to build, but once they understood what we proposed, they *wanted* that capability, and they then persuaded the buying customer to incorporate that feature into our contract.

So, we can use the *mission threads* (Chapter 2) and other materials, and thereby separate what is needed, what is wanted, and what is unnecessary. We can also (this is harder, but still important) determine what is *not* being accomplished using the current methods, or could be accomplished in a better, more suitable, fashion.

The insights gained from this analysis can lead to the creation of ideas for new and improved mission threads: how the mission *could* or *should* be accomplished, after we remove all that is *unnecessary,* sharpen up what is *needed,* add what is *missing,* and identify and add what is *wanted.* This creation of new mission threads, that depict what could or should be, is a key step in understanding your users, and convincing them that you do, in fact, understand them.

The world is full of always improving technology. Now that we have a clear understanding of what is needed, what is wanted, what is missing, and what is unnecessary, we are well situated to assess these technologies, and determine which of them can actually help our users and their mission, and which, despite apparent potential, will not actually help at all.

Remember the story from Chapter 2 about the radio vendor on a military command-and-control system who had figured out (for a price!) how to make the signaling rate of their radio become twice as fast as it was at present. That was a great bit of technology and to them, it seemed certain that this technology would be great for our users. But you will also remember that when we plugged that doubled signaling rate into our system-level model, the result was *no improvement at all* in the system's overall performance, as measured by the loss-exchange ratio. The lesson is that *not all technologies that appear relevant and valuable actually are valuable in the specific context of your system.* The analysis of what is needed, what is wanted, and what is unnecessary, along with the operational performance measures (OPMs) we discussed in Chapter 2, will help you determine which technologies will actually help your system and your users, and which (despite appearances) will not help your users.

Furthermore, many technological products fall short of their advertised performance. We must determine what they really can accomplish, probably through measurements made by us (and not by the manufacturer) in our own laboratory, where we can subject that technology to the conditions expected for our particular system.

Technologies have *limits*; a particular database management system may be great, until you get more than XXX total records, or until the users start making more than YYY simultaneous queries. We ought not to select technologies that cannot support the quantitative goals of our system. This is one of the reasons that we create quantitative versions of the OPMs.

Technologies also have *side-effects*; remember the concept of unplanned dynamic behavior that we introduced in Chapter 2. Since we do not want accidently to introduce unplanned dynamic behavior that would seriously interfere with our user's mission, we must investigate this matter before we select technologies for our system. We wish to avoid such undesirable unintended side-effects.

We must also think about the maturity of these candidate technologies: are they really proven products, ready for incorporation into mission-critical systems, or are they new and still largely unproven technologies, whose use may introduce *risks* (which we will discuss in Chapter 9) that might manifest themselves as poor reliability, consistency, safety, predictability, manufacturability, etc.)?

At this stage, we are trying to understand broad needs and potential enablers, but not yet trying to define specific requirements or pick specific technologies for our system.

These perceived needs and wants will be in tension; there are likely more than one opinion among the users, and many of the other stakeholders may have differing opinions at times too. It is our job as the manager of our engineering project to resolve those tensions so as to achieve a workable *balance*: success consists of having a reasonable portion of the users and stakeholders (and especially those users and stakeholders that we consider most critical) support our selected approach. We cannot realistically aspire to make everyone 100% happy.

Finally, we are ready for the fourth box in Figure 4.1: we must capture all of this information in writing. Writing something down is an intensely creative act; most of us actually do a lot of our *creating* and *thinking* during the process of *writing*, when we have to face the need to be specific, detailed, thorough, and clear. I find that it helps to have a model or template for capturing this information, and therefore I use a structure that I call the *social architecture*[6] (depicted in Figure 4.2) for the purpose of capturing the information we learned through the analysis processes described in Figure 4.1.

6 The term comes to me from my friend Bran Ferren.

Section	Description
Purpose	What is the mission of the users and their organization? What is this system intended to accomplish, and how will it aid the users in accomplishing their mission? Who are the other stakeholders, and what is their role in that mission?
Use-cases	Describe the ways in which the mission is accomplished. Present and analyze the mission threads. Describe the constraints –both written (e.g. laws, regulations, etc.) and unwritten (e.g. traditions, social conventions, etc.) that limit the range of acceptable actions
Who	Who are the users? What are all of the roles and types of users? Is there some type of self-imposed segmentation? Which particular subset of the users perform which tasks? What are the qualifications needed for each type of user? Are there legal and regulatory constraints over who can perform what?
Values and metrics	List what you learned about the coordinate system of value for the users and each stakeholder. What metric can be used to measure each? What are the minimum threshold values which the users and other stakeholders will require? What are the metric values which you aspire to achieve?
When	When do the users do each activity? What are the stimuli that initiate each type of action?
How	What are the *mission threads* (discussed in the text) that depict current operations? How is each step in each mission thread performed? What are the new mission threads that you have created to improve the mission capabilities of the user?
Problem conditions	What can go wrong during conduct of the mission? How are these problems detected? How is each type of problem dealt with? Who has the authority to take or approve off-nominal actions?
The system's role	What, in light of the above, will the new system do in support of the users and their mission?
The user experience	In light of the above, how will the new system interact with the users?
Lessons-learned	A tabulation and discussion of what we learned from the social architecture: (a) the goals of the system; (b) considerations that will determine user acceptance of the system; (c) considerations that will drive the design of the system

Figure 4.2 The social architecture.

In Chapter 2, I also introduced the idea of what I call the *two coordinate systems of value*. I repeat Figure 2.8 below, as Figure 4.3.

We have already discussed everything in this figure except for the last item: how we actually accomplish the linking of the technical performance measures (TPMs) to the OPMs. We will discuss that aspect now.

As engineers in charge of an engineering project, we will find that our degrees of freedom in making choices are most often in what I have called the *engineer's coordinate system of value*. For example, should we make a part out of steel, aluminum, or carbon-fiber composite material? Which frequency band should we select for our radio communications? What type of sensor should we use to find and locate those airplanes in the sky: acoustic, optical, or radar? What programming language(s) should we use for

The *customer's* coordinate system of value	The *engineer's* coordinate system of value
Usually non-technical in nature	Usually technical in nature
Characterizes something about the *mission*	Characterizes something about the *technology that implements the system*
Captured in what I call *operational performance measures*	Captured in what I call *technical performance measures*
Measures the *goodness* of the design for the users	Measures whether or not the design is *feasible*, and helps us estimate the *schedule* and *cost* needed to build the system
The two types of metrics should be *linked*; that is, we need to credibly and transparently show how changes in design (which change the *technical performance measures*) cause changes to the *operational performance measures* Our *degrees of design freedom* are located in the engineer's coordinate system of value, but we *measure the goodness of the resulting design* in the customer's coordinate system of value	

Figure 4.3 (repeat of Figure 2.8) The two coordinate systems of value, and the two types of objective measurements. The mapping between the two coordinate systems must be transparent and credible – even to our non-technical stakeholders.

our system's software? We create and use the technical performance measures to help us assess which of these alternatives seem to be effective and feasible, and also to help us estimate the schedule and cost of the resulting system.

The important lesson that I have learned is this: such use of the technical performance measures is *necessary,* but not *sufficient*; we must *also* ensure that each such technical choice has benefits (or at least no significant liabilities) for the *users*. We measure the effect of our engineering choices on the users by using the *operational performance measures*. Therefore, the technical performance measures *must be in some fashion linked* to the operational performance measures.

We could accomplish this linkage between the technical performance measures and the operational performance measures in an *ad hoc* fashion, but that is error-prone and inefficient; we will be making such engineering choices in great profusion, and throughout the life-cycle of our project. In order to make consistently good assessments of the effect of our engineering choices on our users, we therefore probably have to have some *formal mechanism* in place that interconnects the two types of metrics: that is, a predictive analytic process that, each time we consider a new engineering decision, allows us not only to assess the *feasibility* of the design (using the *technical performance measures*), but also to assess the *goodness* of the design for the users (using the *operational performance measures*). We want a pre-established, at least partially-automated interconnection between the two types of measures, technical and operational.

By having in place such a mechanism that interconnects these two types of measures, we can be consistent in our analyses. Also, by making the method through which we accomplish that interconnection both transparent and understandable to our customers, we can create *credibility* with our users and other stakeholders for our assessments regarding the goodness of the design for the users. This last aspect is vital; if the users and stakeholders (most of whom, remember, are *not* engineers) do not find our analysis credible, they will not support our design decisions, and may lose faith that we are actually going to build a system that will be of value to them. When that happens, projects are usually canceled.

How do we go about making this transparent and understandable interconnection between the two coordinate systems of value, so that our predictions of operational goodness are credible? I usually do this through the *nested models* that I described in Chapter 2. This allows me to show the users, in a stepwise fashion, the logic and analysis methods that I use to transform the *technical performance measures* (such as the selection of a sensor technology, or the selection of a radio frequency) into *operational performance measures* (such as the examples I used in Chapter 2: the number of shot opportunities, or the loss-exchange ratio); it allows me to make the transition *understandable to the users and other stakeholders.* In fact, I will likely actually *involve* the users and other stakeholders in the effort to *create* that interconnecting logic; having them involved in that process both allows me to draw upon their expertise, and also creates the psychological benefit of their having some *emotional ownership* of the resulting interconnecting logic. Having your users and your stakeholders have such emotional ownership of your design is very useful in creating the *alignment* that we talked about in Chapter 1.

4.2 Case Study About the Value of Using the Customer's Coordinate System of Value: Role-Based Processing

During my career as a working engineer, I built many systems for the US Army. Several of these were what are called *command and control* systems; these provide information to decision-makers, and allow them to make and then disseminate their decisions, and finally to monitor the progress of their team, so as to guide and control their actions. Closely associated with command and control systems are *situational awareness* systems, which is technology that enables a summary depiction of information about a situation transpiring in physical space. The summary depiction is typically graphical in nature, such as a map with symbols indicating items of interest, where the symbols move and change as the situation changes (see Figure 4.4 for an example). The symbols might include icons corresponding to entities of interest, as well as lines or polygons that denote key areas. For example, a situational awareness depiction of a battlefield might use icons to denote the location of individual military vehicles or of military units (e.g. a collection of organizationally related individual military vehicles, such as an armor platoon of four tanks), while lines or polygons might denote the location of a minefield or the boundary between two military units. Such graphical information is usually supplemented with context-specific textual or other amplifying information. Such amplifying information might include notes or text describing something on the map, photographs that depict details or views from multiple vantage points, video clips, sound recordings, or databases. The maps, symbols, and amplifying information are updated in near real time to provide a moving depiction of the changing situation. Often, the tools for sharing information, for issuing commands, and for monitoring and controlling operations are integrated into the situational awareness system, so as to enable rapid action based on the changing information about the situation. In summary, situational awareness tools provide users with an understanding of an evolving situation, and the command-and-control tools allow the commanders to take, command, and monitor actions in response to that evolving situation. These two – situational awareness on the one hand, and command and control on the other – are inseparable sides of a complete information system for military or civil government use.

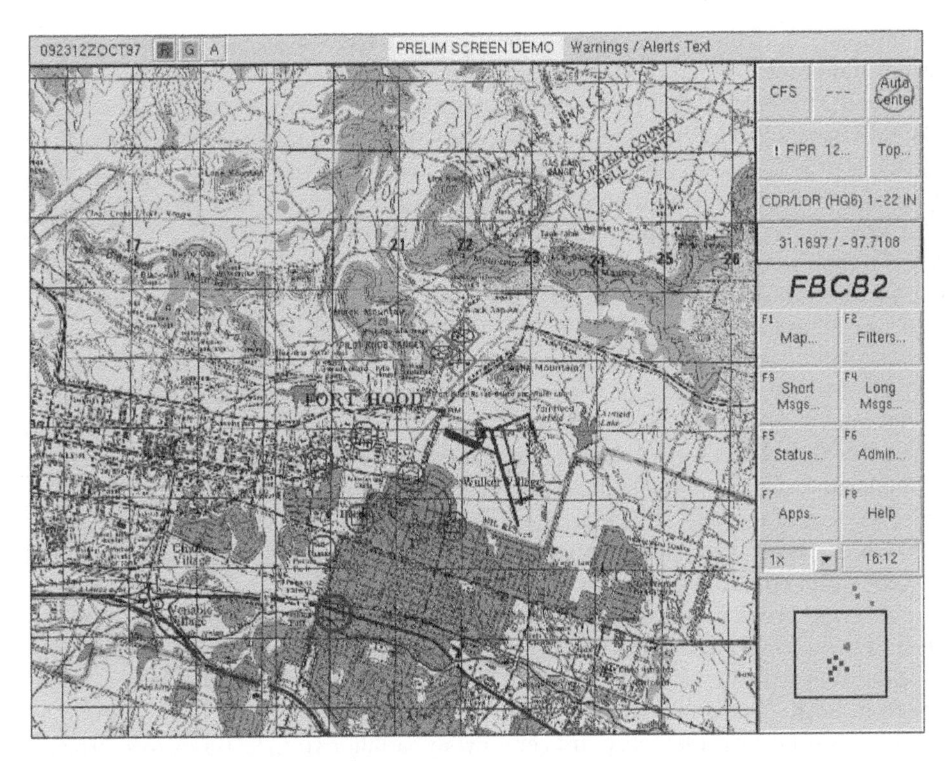

Figure 4.4 Example of a situational awareness graphic depiction. *Source:* From Northrop Grumman, 1999. Used with permission.

Let's consider a typical battlefield issue, and use it to illustrate the difference between the technical coordinate system of value and the customer's coordinate system of value.

In the engineer's coordinate system of value, we know that a computer cannot send information from one person to another without having some computer-readable form of addressing information: something that identifies the intended recipient. We all have everyday experience with things like electronic mail and text messaging, where the sender is responsible for entering the address of the intended recipient. In the engineer's coordinate system of value, this is a normal approach.

Our Army customers had used electronic mail too, and knew that this was the ordinary approach. When they wrote their requirements for the battlefield, they *assumed* that this was the *only possible way* to get information from one user to another.

We, however, were on the lookout for ways to revolutionize their mission, and we understood better than the customers how technology worked, or could work. We conducted what we called "user jury" exercises with groups of soldiers, where we had them use prototypes of computer screens to perform mission operations. We watched.

Every army in the world makes use of a type of guns called *artillery*. These are long-range, large-caliber guns that can fire their ammunition such long distances that you cannot see the destination from the location where you fire the gun; the destination might literally be beyond the horizon. What happens is that someone on the battlefield *calls for fire*, that is, asks that the artillery group have one or more of their number shoot at a particular location (e.g. a latitude and longitude) with a particular type of ammunition. The artillery's *fire officer* determines which of his gunners will service this mission, and orders them to perform that *fire mission*. Someone is in a location where he can see the intended target (a *forward observer*), and can provide feedback about the shots: the first shot might have been too long, going past the target, and so forth. This feedback is provided to the gunners, who adjust their guns and fire again. Eventually, the forward observer tells them that they can stop firing at this particular target.

This process has many opportunities for errors. The guns may not be aimed perfectly, the explosive charge that propels the ammunition will vary in size, the wind and other weather factors will vary, and of course the latitude and longitude provided for the gunner to aim at may not be perfect. At times, therefore, artillery fire might come down *too close to our own soldiers*. There is a message called *checkfire* that is intended to tell the artillery on an emergency basis: stop firing at this particular location, and stop right now!

One of the things that we did during our user jury activity was watch soldiers executing simulated artillery missions, including (deliberately introduced) errors that would require the use of the checkfire message. We watched soldiers under the (simulated) stress of being fired at by their own gunners (or seeing their buddies being fired at by their own gunners) trying quickly to enter and send the checkfire message. The slowest and most error-prone part was always the problem of determining the correct address of the intended recipient. Even if they knew the electronic address of the correct person who could stop the gun fire, under stress they might type it wrong, or select the wrong entry from a list. There was another complication too: on the battlefield, nothing is static. John may be the artillery fires officer for your battalion from 8:00 a.m. to 4:00 p.m., but Joe comes on duty at 4:00 p.m., and Jack at midnight. And of course, this is a battlefield. John may be dead or injured at noon, and James has to take over right away. This was a situation that was not going to ensure prompt receipt by the correct person of every single checkfire message! In fact, previous systems suffered from all of these problems; the users hated it, but just put up with it, because they had no inkling that a better approach was possible.

We saw the issues during our user jury exercises. But we were engineers, and we decided to go off and invent something better.

We came up with the idea of what we called *role-based processing*. People would not log into the system using only their name and password; they would also be assigned to a *role* in the combat team. Most of the information that was sent around the battlefield was in fact intended to go to one or more people who were each serving at the moment in one particular role. For example, every checkfire message needed to be sent to whomever was performing in the role of fire officer *at this instant*.

There were a lot of technical difficulties to be solved, but the invention of role-based processing eventually solved *all* of the problems associated with the checkfire message. The user *never again* had to enter either the name or address of the intended recipient; the system *always* knew that a checkfire message was to go to the fires officer, and the system *always* knew who was serving in the role of the fires officer at this exact instant.

The soldiers loved it! How did we arrive at this desirable outcome?

- My engineers took it upon themselves to acquire some significant domain knowledge about their customers. We had to learn about what the soldiers did, and why they did it that way.

- We had to talk to them, but much more important, we had to *watch them work* (I once spent a month in the field, sitting in the back seat of a brigade commander's vehicle).
- We had to make measurements. In this case, we had to gather data from our own user jury experiments, and also from the actual operation of preceding systems, about specific measures such as how long it took the average operator to create a checkfire message, what percentage of checkfire messages never got to the correct person, how long those that did get to the correct person took to get there, and so forth. This created a quantifiable characterization of the problem, identified the actual bottlenecks and pitfalls of the current approach, and allowed us to engage in discussions with the users about what would constitute success.
- We then put our engineer's hat back on, and brainstormed about how to fix the bottlenecks and pitfalls, while not inadvertently introducing other problems.
- We took the approaches that we invented to fix the bottlenecks and pitfalls; put them into our prototype software; reconvened the user jury; trained the users in our new, proposed methods; and then re-ran the scenarios. We (again) made measurements, watched them operate, and then discussed everything with the users. We iterated around this process probably three times before we were satisfied.
- We then had a concept that would revolutionize the creation and addressing of the checkfire message, and could incorporate that new, improved concept into our requirements, and into our design.

One of the key aspects of this example is that the users knew of the problem, but assumed that it was insurmountable, and therefore did not ask us to fix it, or even think to tell us about it. We engineers had to discover the problem by ourselves (using our domain knowledge of the customer), realize that this was a *pressure point* for our design, and recognize that the problem could in fact be feasibly corrected by the appropriate design. The *customer did not know* to ask in advance for role-based processing; we engineers had to use our insight gained from our unique combination in one brain of both *technology* and *domain knowledge about the problem* in order to *invent* that idea. But once we had invented it, the customer embraced it enthusiastically. As the old saying goes, the "customer is always right" but they are seldom *complete*; *we* must be the ones to *invent* the insights that really revolutionize the mission, and provide the technology that ensures that this revolutionary new approach is in fact feasible.

4.3 Special Topic: Designing the User Experience

We engineers are great at designing the internals of a complex system, but I have found that we are not always very effective at designing how our system ought to interact with the users: that is, what information should be displayed to the users at what times, what work we will perform within the automation of our system versus what tasks we reserve for the users, what mechanisms the users will use to provide their inputs and controls, how the users navigate through the capabilities of the system, how work is coordinated amongst the complete set of users, how the system is structured so as to be suitable for the intended operating environment (which might be inside a tank, or might be in the control room of a complex industrial process), how we bring items of likely interest to the attention of the users, how we minimize the degradation of the user's effectiveness over the course of a work shift, and so forth. These are important matters, and clearly of great interest to the user. They are also a chance to create those *want* and *like* factors in your users that we described earlier (and which we will discuss in more detail in Chapter 5).

What I have learned is that the design of the user experience is its own specialized field with its own cadre of expert practitioners, just as is the case for mechanical engineering, electrical engineering, chemical engineering, and so forth. We therefore need to have *people who are adept practitioners in designing and creating the user experience* on our team. The skills required by these people blend cognitive science, time and motion studies, art and esthetics, and many other fields. Fortunately, in the last few years, universities have actually started to teach user experience as a formal field of study, and even grant degrees in the subject. This means that you can find and hire people who are interested in this field for your project team. True expertise comes only with practice, but a set of basic skills and an interest in the subject can get those people started. Like everything else, you need a range of skills in your team; you need a small number of real adepts, and a larger number of people with more routine levels of skill. You will have to go and find those adepts, and then motivate them to spend some of their time training and mentoring the younger practitioners. (This, of course, is true in most of the skill areas across your project – you must *always* be growing better practitioners. Generally, the experts find such mentoring opportunities very satisfying, and the younger practitioners like receiving the attention from the experts, so this process is also very good for team motivation and alignment. We will say more about this in Chapter 13.)

But as we discussed in Chapter 1, where we pointed out that we as project managers have to know a little bit about each of the specialty skills on our project team – so that we can talk to those people, engage them, direct them, and understand what they have to say – we, the manager of our engineering project, also have to learn a little bit about how to design the user experience ourself. Let's teach you to do that.

The term *human factors engineering* has been in use for decades. It includes the analysis of detailed parameters, such as determining the space required for a person to perform a motion, how much light is required to work without excessive eyestrain, and so forth. These details are useful and important, but starting with them is to start in the middle, rather than at the beginning.

The *user experience*[7] is intended to cover the entire range of activities required in order to design our system based on the *needs of the user*, rather than on the *preference of the designer*. This idea is clearly related to the concepts that we presented earlier in this chapter.

Sometimes, the *user experience* is described as a focus on *esthetics*, that is, how things look. Dr. Norman, however, considers esthetics a *secondary* issue within the field of user experience. What Dr. Norman, and the other expert user experience practitioners that I have come to know through my work, consider as the primary factors for the user experience are these:

(a) Simplifying the structure of tasks.
(b) Making the appropriate things visible.

7 I believe that the term was coined by Dr. Donald Norman, during the time when he worked at Apple Computer.

(c) Devising an appropriate partitioning of tasks between the automation and the users.
(d) Allowing the system to continue to operate effectively even in the presence of errors, especially in the presence of human errors.

The user experience actually matters quite a lot. Consider the following little scenarios:

- *The car accident.* The driver took her eyes off the road for a moment to adjust the air conditioning. She had to look down because it was impossible to identify which was the correct control by touch alone (this actually happened to someone I know).
- *The alarm clock.* The alarm didn't go off because the time was wrong. The time was wrong because the buttons to change the time were on the top of the clock, and when the cat stepped on the clock in the middle of the night, she reset the time (this actually happened to me).
- *The slow checkout line.* The cash register was complex and confusing, and unless the clerk operated it slowly while paying it a lot of attention, he might get it wrong and have to start all over. Customers left in frustration without completing their purchases (me again, and more than once).

The consequences can, of course, be far more severe than in these little examples. I have read that both the Three Mile Island and Chernobyl nuclear power plant accidents were made far worse than they needed to be due to operator errors likely caused by overly complex and error-prone user controls.

Furthermore, remember what I said earlier:

> People can *sometimes* be counted on to buy what they *need*, but they can *always* be counted on to buy what they *want*.[8]

We can use the user experience to create some of that *want* factor! Think of all the attributes of your system or product that might create such a *want* factor, such as those listed in Figure 4.5.

Armed with the information we create through the social architecture process, adept user experience practitioners can contribute to all of the items on that list that are in **boldface type**. That is quite a significant potential contribution.

But how do we, the manager of our engineering project, interact with such practitioners? How can we convey to them a set of strategic guidance that we consider necessary for the approach we desire for our project, while allowing our expert practitioners the ability to employ their skills for the benefits of our users?

- ***Providing functionality, performance, or capacity that no competing product or system can provide***
- Value-for-money: a disruptive price–performance point
- A higher level of ***safety***, reliability, or other tangible benefit
- ***Unusually well-suited for the intended purpose***
- ***Can be operated by the intended users***
- ***Style, look, aura, and other intangible factors***

Figure 4.5 Obtaining that *want* factor.

8 Bran Ferren, again.

We do it the same way we interact with other specialists on our team: through the definition of a *hierarchy*.

Do you know about the International Standards Organization (ISO) seven-layer model for electronic communications?[9] It attempts to codify methods of electronic communication by dividing the responsibilities and functions into *layers* (that is, a hierarchy), designed so that you can perform the work inside one layer without having to have detailed insight and understanding of what takes place in the other layers; this concept is called *abstraction*.

In the seven-layer model for electronic communications, the first (lowest) level is called the *physical* layer, and embodies all the capabilities actually needed to send signaling over a physical communications mechanism, which might be a radio, a wire, a fiber-optic cable, or some other physical mechanism for electronic communications. The physical layer can then be designed and implemented without requiring detailed knowledge and understanding of how the other six layers are each implemented. The same principle applies to each of the other layers in the model too: they can be designed and implemented without requiring detailed knowledge and understanding of how the other six layers are each implemented. This allows for segmentation of skills, and for work to proceed in parallel.

We can create a hierarchy for describing and designing our user experience. For example:

- At the top of the hierarchy, we could place our *most abstract* considerations about how we want our system or product to interact with the users: what tasks the users will perform, and the sequencing of steps and actions (derived from the mission threads, of course); what data the users will need to see; what decisions the users will need or want to make (distinct from the decisions we want the automation in our system to make); and therefore what data the users will need to provide to the system.
- In the middle of the hierarchy, we could place *more concrete* considerations about how we want our system or product to interact with the users: how we will allow them to navigate through the capabilities provided by the system, and in general how they will interact with the system; how we will structure the data, so that it makes sense to the users; where each item of data will come from; how we detect and recover from errors; and so forth.
- At the bottom of the hierarchy, we could place the *most concrete* considerations about how we want our system or product to interact with the users: words and phrasing to be used on the computer displays; when to use buttons versus sliders versus voice inputs versus other options; text fonts, colors, and sizes; traditional human factors engineering (as described above); and esthetics.

This type of hierarchy – you can create more layers, if you prefer[10] – will allow you to interact with your user experience experts in a productive manner, and will create artifacts that will allow all of the other experts on your team, together with your users and your stakeholders, to understand the plan for the user experience, and to participate in a meaningful way in its formulation and review.

9 ISO/IEC 7498-1.

10 For example, Jesse James Garrett, in his book *The Elements of User Experience*, creates five layers for this purpose.

A warning, based on my experience, designing the user experience is an area where there is sometimes a temptation to just "ask what the user wants." I do *not* recommend that! The users are likely to have a strong understanding of their current mission, but are highly unlikely to be expert at designing systems and effective user experiences.

In my opinion, the correct way to involve the users is to develop *your own* understanding of what they do; as I have described earlier, you will do more *watching* than *asking.* There is often a significant experience-based and judgment-based component to what our users do in order to accomplish their missions; I like to think of what they do as being half technique, half art. I also think that about what *I* do as a systems engineer and project manager.

In addition to their not being experts at designing systems, it is my consistent experience that while experts (including our users) can do great things, they often do *not* actually know *how* they do a task (try asking my wife how she paints! Figure 4.6). Sometimes, they just can do it ... so *asking* them may not help. Watching is likely to be more useful, albeit slower.

Figure 4.6 A painter can paint, but probably cannot explain *how* they do it! Experts can *do* it, but don't always know *how* they do it. *Source:* Painting copyright © 1995 by Robyn Friend. Used with permission.

Also, it is often vital to understand the social constraints of the customer, as well as the nominal task – this can drive user acceptance. These constraints are often tacit, and hence *watching and listening* elicits different information than *reading and asking.*

A useful methodology is to use those *mission threads* to moderate your *watching* and your follow-up *asking.*

One last point about the user experience. Because of the visibility of the consumer electronics market, there is a lot of "buzz" about the desirability of making things *easy to use and/or easy to learn.* That is, in my view, *not always a desirable goal.* Making something easy to use or easy to learn is a *trade-off*; it is almost always accomplished by slowing the use of the system or product by *experts.*

Notice, however, that many important societal systems are intended to be operated *only* by *trained, expert users*:

- Nuclear power plants
- Military systems
- Airplanes and air traffic control systems.

There are many, many other examples, too – even *elevators* used to require trained operators!

The goal for a system intended to be used by this sort of expert user ought not to be to make it *easy to use* or *easy to learn,* but instead to make it *as effective as possible for those trained, expert users,* through emphasizing aspects such as:

- Speed and consistency of operation by an expert
- Flexibility of work style
- Avoidance of operator errors, and making it easy to notice when such errors have been committed
- Transparency of decision-making, and perhaps even supporting an audit trail of operations and decisions.

But if you are building your product for the consumer market, no one is willing to be trained! Most consumer products don't even include a user manual in the box anymore. The goal in these markets is to enable *moderately effective use* of the products by *untrained users,* which is very different from intending to enable *highly effective use* by *expert, trained users.* You need to know your users and their mission, and design accordingly.

The user experience process can provide an important contribution in both of these contexts (e.g. trained or untrained users). But the resulting design will be quite different, depending on that context.

4.4 Summary: Understanding Your Users and Your Other Stakeholders

Everything about engineering project management starts with understanding your users and your other stakeholders, and convincing them through tangible actions that you understand them.

As the project manager, you of course organize all of the tasks described in this chapter, select the people who will lead each task, and ensure that they then follow through on the execution of those tasks.

You must also *personally* participate in the efforts to acquire knowledge about the customers, their missions, and their values, and you also personally participate in the efforts to create the mission threads, the operational performance measures, and (most especially) their socialization with the customers. We must strive to get this knowledge of the customer not only into our own brain, but also into the brains of the key members of our project team, including the chief engineer for the project.

4.5 Next

The discussion in this chapter about understanding our users helps us understand why our engineering project has been created. In the next chapter, we learn about the actual process through which engineering projects come into existence.

4.6 This Week's Facilitated Lab Session

We are going to form you into teams, with four to seven people on each team. You will be working in these same teams for the remainder of the course, during these weekly facilitated lab sessions, and working together on the homework that is assigned during these sessions.

These facilitated lab sessions will each be about a specific topic. Each topic will involve your team having discussions, and then creating a written artifact, which will become a section in a *final report* that your team will prepare as graded homework; the complete report will be due during the final week of the term. Also, during the final week of the term, your team will also make an *in-class presentation* of a subset of the materials in your team report. Every member of the team is required to participate in every section of the written report, and every member of the team will personally present a portion of the final in-class presentation.

The first of these team sessions – today – will be about creating examples of the *two coordinate systems of value* that we use in engineering project management. We discussed these two coordinate systems of value in each of the first four chapters, and in detail in this chapter.

You should refer back to Figure 4.2, which defines these two coordinate systems of value that we use in project management:

- The *customer's* coordinate system of value, and
- The *engineer's* coordinate system of value.

As depicted in Figure 4.2, we use *operational performance measures* to quantify things in the customer's coordinate system of value and *technical performance measures* to quantify things in the engineer's coordinate system of value. The figure also defines what we mean by each type of measure.

Today's assignment

- Each team will pick a *real company*. The company you select should be large enough to have a fair amount of information available from its own website, and perhaps also from news sites and other on-line sources. It does not matter if more than one team picks the same company.

- Pick an example project that this company *might* perform someday, in order to create a new or improved product. This need not be an actual product that the company is already building, but it should be something that is close enough to what they do that they *might* build something like it someday.
- Determine who are the "customers" for such a product? What do *they* value?
- Create three examples of operational performance measures that are in those customers' coordinate system of value. Explain why you believe that this is what they would actually value.
- Create at least five examples of technical performance measures that relate to this product. Create a discussion about how you would relate the technical performance measures to each of the operational performance measures. Not every technical performance measure need map to an operational performance measure.
- Hold a group discussion of your findings and insights.
- Get together outside of class and write this all up (three to five single-spaced pages of text and figures; full sentences and paragraphs, rather than bullet points); this will become one section of your final team report that will be turned in as homework near the end of the term.
- You will also need to prepare a couple of briefing charts summarizing this topic, for the presentation by your team during the last week of class.

Hints

- You will need to start with a short textual description of your product. A drawing too, if you think that would help.
- You need numbers (actual pass–fail criteria) in each operational performance measure and in each technical performance measure.
- Think carefully about what you make into an operational performance measure versus a technical performance measure. In general, make those statements that relate to the *user's experience* into operational performance measures and make those statements that relate to the *feasibility of building your product* into technical performance measures.

5

How Do Engineering Projects Get Created?

When we get our first job, we are likely to be assigned to work on an existing engineering project; we are not troubled by the question of how this engineering project came into existence. Who created it? Why? How is it being paid for? How did it come to pass that it is our company that is doing the work? But as we progress in our careers, we come to realize that these aspects matter a lot. In fact, understanding them, so that *you* can help your company win new projects, is an important path for you to achieve attractive assignments and career success. In this chapter, I will therefore teach you the basics about winning engineering projects for your company, which centers around something called the "proposal."

5.1 Engineering Projects are Created in Response to a Need, or a Vision

Somewhere out there in the world is a group of users who need something; let us imagine a scenario wherein they decide that they can best obtain it by *buying* it from someone outside of their own organization, rather than making it themselves. Let's say that what they need is a loaf of bread.

First of all, they need to talk among themselves, and reach agreement on exactly what is it that they want to buy. White or whole wheat? Non-GMO flour? Gluten free? Sliced? What kind of packaging? As you can see, even to buy a simple loaf of bread, there are a set of decisions to be made.

Probably, they will decide at some point to commit the decisions that result from those discussions to writing; there is no point in getting it wrong later due to poor memory. If I go to buy a coffee for my wife, you can be sure that I write down "a small decaf peppermint mocha with extra whipped cream." What if I came home with a latte instead? She would not be happy.

If we write down simple things like the above (or a grocery shopping list), you can be sure that when someone plans to pay money for a complex engineered system, they will also try to write down something about what they want to buy. They will write down not only the characteristics of the item they wish to buy, but perhaps other things too: what is the date by which they need it, what kind of payment terms are they offering (e.g.

Engineering Project Management, First Edition. Neil G. Siegel.

Companion website: www.wiley.com/go/siegel/engineering_project_management

payment upon placing an order versus payment upon delivery, or some other arrangement), and so forth. We call such a document a *request for proposal.*

Engineered systems tend to be more expensive than a loaf of bread. When you buy a car, you probably *shop around*; even after you have decided on a make, model, accessory package, and color, you may still visit a few dealers and see who has the best price, or the quickest delivery.

Buyers of engineered systems shop around too. The way they do this is to send their written *request for proposal* to a set of companies that they believe might be able to build what they want (or these days, post their request for proposal onto an on-line forum where vendors know to look), and ask them to make an *offer* – in writing – about what they could provide, when they could provide it, how much it would cost, what kind of warranty they would offer, and many other relevant factors. We call such a written offer a *proposal.*

Companies can then look at the request for proposal, decide if they wish to respond (or not – it costs money to write a proposal; we call this decision a *bid / no-bid decision*), and if they decide to respond, they create a written proposal that they send back to the buyer. The buyer has likely specified a date by which such proposals are due.

The buyer will then read all of the proposals (and might invite some or all of the proposers to make an in-person presentation), and eventually will select the provider that appeals to them, based on whatever *selection factors* the buyer deems important: price, schedule, capability, reputation, warranty, other factors, or some combination of all of these. The buyer then negotiates additional details with the selected vendor, which allows the two parties to agree on some type of formal purchase agreement; this might be a *purchase order*, a *contract*, or some other form of written agreement.

For the typical complex engineered system, this formal purchase agreement will take the form of a legally binding *contract.*

And then the selected vendor is ready to start the engineering project, to build the item that the buyer wanted, in accordance with the terms of their proposal and that formal purchase agreement.

This sequence of events is depicted in Figure 5.1.

The vendor in general will not (and ought not) to start working on the project until the formal purchase agreement (of whatever type) is executed by both parties. Otherwise, they may get stuck with unreimbursed costs if the buyer ends up selecting some other vendor, changes their mind and decides not to buy anything at all at this time, changes their mind about what they want to buy or when they want to receive it, or any other material factor that would be defined in rigorous terms in the executed purchase agreement.

Let's define the term *proposal* in a little more detail. A proposal is a *written formal offer to perform a specified piece of work*, under specified conditions, to create a specified set of deliverables (products, services, and data), on a specified schedule, for a specified price, under a specified set of payment terms and timing, and under a specified set of contract terms (e.g. how to resolve disputes, etc.).

Since the proposal, as an offer to the buying organization, usually legally commits the offering organization to the terms specified in the proposal, the proposal must be *signed* by someone who is *authorized to commit the offering organization* (e.g. an officer of the corporation).

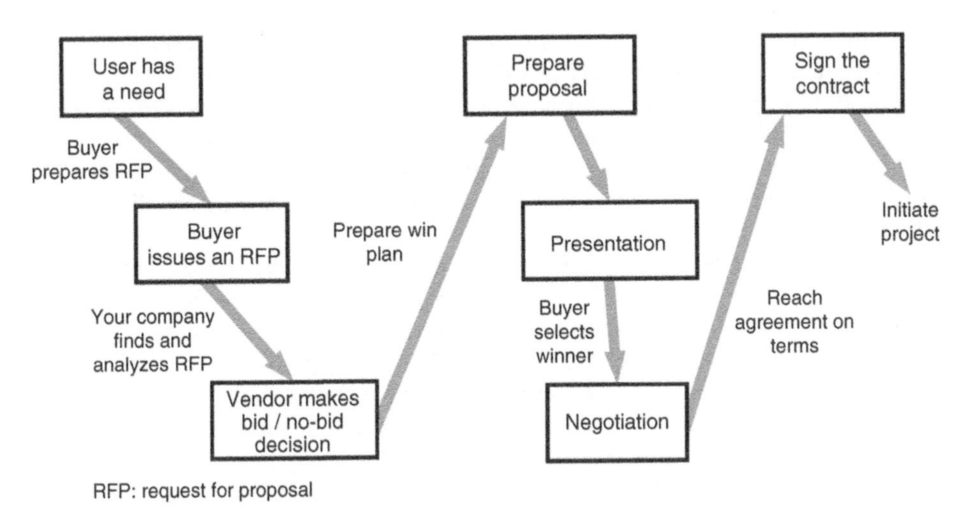

Figure 5.1 The steps leading to the commencement of an engineering project.

Once the buying organization has read your company's proposal, they may choose to *accept your proposal*, that is, tell you in writing that they wish to engage your company to do the work described in their request for proposal and your proposal. Alternatively, they may choose to *reject your proposal*, which is their way of saying "we have decided not to engage your firm for this piece of work."

Receiving written notice from the buying organization that they have chosen to accept your proposal is good news, but you still ought *not to start work* until the purchase order or contract is actually negotiated and signed by both parties.

Proposals are often prepared in the context of a *competition*, that is, someone who wants to buy a product or service will ask *more than one company* to prepare a written proposal, stating in advance that they will only be signing a contract with one of the bidders (*the winner*).

The buying organization then selects the winner and notifies the other bidders that they were not the winner.

On occasion, the buying organization might already know that they want your company to do the work, and therefore they do not plan to have a competition. Perhaps you built a related product for that buying organization already, and they were happy with your work. Running a competition takes time and money, and the buying organization may decide that running one for this particular product is not to their advantage, since they have confidence that your organization is the supplier they want.

We use the term *sole source* for such a situation, that is, the buying organization will solicit a proposal only from a sole organization. Notice that they *will still ask for a proposal*; in this case, not as a basis for a competitive selection process, but instead to make sure that there is agreement between them and you about what is to be done, how it will be done, when it will be done, how much it will cost, and so forth. Even for sole-source contracting, it is still necessary and appropriate to get all of the terms and conditions agreed to in writing before work starts ... and the proposal is a big part of that task.

Why do we care about proposals? Why am I giving it an entire chapter in this book? Remember, in Chapter 1, we discussed the fact that virtually 100% of company revenue comes from projects. And nearly every one of those projects (whether internally funded or externally funded) comes from a *proposal.* Proposals are what bring revenue – in the form of projects – to companies. Writing winning proposals is the *only* way to generate revenue, pay the bills, pay employees, and keep the company in business.

And if you are the buyer, proposals are what allow you to select the right team and organization to build your product or service. Whether you are buying or selling, *proposals are essential.*

Let's describe the process for externally funded and internally funded projects in more detail.

In many businesses (e.g. aerospace, construction and building contracting, public works, architecture, and many others), virtually 100% of company revenue comes from *competitive contract awards* (or, on occasion, sole-source contracts that follow successfully completed competitive contracts). *Every single one of those awards is the result of a proposal* in response to an externally sponsored competition.

In other types of businesses (e.g. those that build for the marketplace – like car companies and computer companies), there is an *internal* competition to decide what products or services to build and offer. Those too are usually resolved via proposals, albeit sometimes less formal than externally sponsored proposals.

Here is an example of how it works in such internal competitions. Let's say that you are the chairman of Apple. Your executive team probably has dozens of ideas (*proposals,* in various degrees of formality and completeness) about what the company should invest in and bring to market next. You, as chairman, will in essence run an *internal competition* (the degree of formality will vary) to decide which are the products and services in which you will invest as the "next thing."

As you can see, whether your company lives off external competitions or internal competitions, proposals are the *mechanism that creates engineering projects.*

There are additional (and more personal) reasons why you should care about proposals:

- I can assure you that becoming recognized as a good proposal manager is a sure path to enhanced career opportunities!
- Furthermore, one proven method to get the chance to be a project manager is *to be the person who wins the project,* by having led the proposal.
- And finally: working on proposals can be a very interesting work assignment. It is hard work with long hours, and sometimes a little stressful, but it is also intensely creative, interesting, and makes a difference to the company and to the customer.

5.2 How to Win

By now, I have probably convinced you that proposals are important. So, next I will tell you how to write proposals that can win those competitions.

We talked about *internal* competitions and *external* competitions. Here's a twist: there is *always* an internal competition, even when the explicit competition is external.

What do I mean by that? Let's consider a scenario:

- An organization wants to buy an engineered system, and has issued a request for proposal (actually, as we will discuss a bit later, you must start *long before* the customer issues the request for proposal, but for this little scenario, we will ignore that detail).

- Your company knows this buying organization, finds them reputable and reasonable, and furthermore, you think that your company has the skills and people appropriate to build the engineered system that they seem to want to buy.
- You spend some time and money gathering all of this information. You also have to estimate business-related items, such as:
  - How much revenue and profit could this project provide to your company?
  - How much will it cost to win? This includes preparing the proposal, conducting negotiations, related marketing costs, and so forth.
  - What are your chances of winning? Which other companies are likely to bid, and what are their skills, what is their reputation with this buying customer, and so forth?
  - What people will you need to make this bid? What facilities will you need to make this bid?
- Since it will require money to make the bid (we call this an *opportunity cost*), you need to ask the appropriate person in your company to authorize that expenditure. They will certainly ask for information like that indicated in the previous bullets. You therefore gather all of your information, and create some sort of written artifact, and probably also get to make an in-person presentation to the person within your company who has the authority to approve such an expenditure.

Now, put yourself in the shoes of the person who gets to make the decision: will your company bid or not bid on this job? In actual fact, this person not only has to consider the merits of the bid that you want the company to make, but this person *also* has to consider the merits of *every other course of action* open to the company at this time. For example, there are likely to be many other such *bid opportunities* available to the company; the company probably does not have enough money or people with the appropriate skills to bid on all of them; therefore, someone has to *choose* from among the options. Every possible course of action (e.g. bid this opportunity, bid this second opportunity, but do not bid this third opportunity) has its costs and potential benefits. The person with the decision authority considers all of these options – not just the merits of *your* potential bid – when making the decision. That is, you are in a *competition* against all of the other potential courses of action (and all of the other teams that have their own favorite bid opportunities) inside your company!

This is even more complicated than I just described, because not all of the opportunities competing for attention, funding, and people need to be bids; for example, someone in the company might want to spend money enlarging a factory instead of making your bid. The landscape of potential opportunities against which you have to compete in this internal competition can be very complicated.

Therefore, you must first win the *internal* competition for attention and funds, just so that you can gain permission to prepare a bid to the *external* customer.

In the little scenario above, I hinted at some of the information that you must develop in order to win this internal competition. The following is a more complete version of that list:

- *Understand the problem.* Who is buying it? Who will use it? By when will they need it? How many will they buy? What is it supposed to do? How is that different or better

than the way they do it now? These are many of the same questions we discussed in Chapter 4.
- *How to solve the problem.* What will our solution look like? What will we invent or make, and what will we buy from other companies? What exists already, and what, therefore, must yet be invented in order to create this solution? How long will it take to do this research and invention process? How much money will it cost?
- *How to measure the goodness of your solution.* This, of course, starts with the technical performance measures and operational performance measures that we discussed in Chapter 4. But it goes on to include something that we have not yet discussed: what will be our areas of positive differentiation (e.g. where will we try to be better than the other bidders, and where will we be content to only be as good as the other competitors)? We discuss achieving such *differentiation* later in this chapter.
- *How to prove that you can solve the problem and achieve differentiation.* What will we prove by building prototypes? What will we prove by benchmark measurements in our laboratories? What will we prove via computer-based models?
- *How to show that it can be a viable business proposition.* What are the business model, pricing/costing/margin, teaming, intellectual property rights and licensing, sales projections, cash-flow projections, etc.?

If you are working at Apple, and want to build a new gizmo, something like the above may be enough to get you funding and authorization to build your team, and to start designing your gizmo. Probably, however, you will not be given enough funding actually to finish it; instead, you will be asked to come in every few months and show your progress, and to update your market assessment. If your progress is satisfactory, and the market assessment still seems credible and favorable, you have a chance to get a next increment of funding in order to do a few more months of work toward finishing your new gizmo.

If, however, you are working at Lockheed Martin, and your goal is to get funding and permission to bid on a new fighter jet contract, you are just getting started. The good news is that the company has authorized you to start getting ready to make the bid, and authorized some funding for you to get to work on the proposal. However, that is just **Step 1**. You still have three more steps to get through:

- **Step 2**. Do what is needed so that you are *allowed* to submit the actual bid to the buying customer.
- **Step 3**. Win the external competition.
- **Step 4**. Close the deal.

We will talk about each of these next.

A proposal is a formal, *legally binding* commitment. If you submit a proposal to a buying customer, and they accept it and then you both sign a contract, your company is legally committed to providing the products and services specified in the contract (and in the proposal, if the contract incorporates that proposal, as is sometimes done), and to do so on schedule, for the price, and in accordance with the other terms defined in those documents.

Because of this, companies go through an analysis and decision process *before* you are allowed to submit a proposal, in order to determine if – in light of what we continue to learn – they actually want to submit this bid. Step 1 (above) got you authorized to start preparing the proposal, but the company will still want to review all of the terms and conditions that will be embodied in a potential contract *before* letting you submit the proposal; that review is Step 2. The following list describes the typical items that you company will want to review before they authorize you actually to submit the proposal to the buying customer:

- *Price to win*. Of course, this is a competition, and we do not get to see the other bidders' proposals, so we do not know the actual price that they will bid. But we may have some public information about what our competitors bid on similar jobs in the past,[1] and we may also have legal access to some information about how much the buying customer wishes to spend.[2] Your company will want you to use that information to estimate what price the company is likely to have to bid in order to win the competition. This amount can then be compared to your estimate for how much it will actually cost for the company to build the item, and therefore the company can estimate whether it will make or lose money on the project. This will be one of the factors used by the company to decide whether or not you are allowed to submit a bid; most companies will not bid if they believe that the price needed to win is so low that they cannot make a fair and reasonable profit margin on the project.
- *Time phasing of costs versus time phasing of revenues*. On most projects (as depicted in Figure 1.6), the company must expend money *before* receiving revenue. Therefore, the company will want to understand the cash flow involved: can they provide the cash to pay salaries, suppliers, and other bills while waiting for payment from the customer? There are also a number of financial measures that the company will calculate to quantify the quality of the financial deal: for example, something called the *return on assets employed*, and other similar financial measures.
- *Opportunity costs*. To perform on your project will entail a commitment by the company of money, facilities, equipment, and talented people. The company always has the option to use those assets for other activities; to assign them to your project is to *forego* the opportunity to use them on something else. We call this an *opportunity cost*. Your company will do an assessment of what else they could do in this same time period with those funds, facilities, and people. Would those other opportunities be better for the company, and/or better for those employees?
- *Risks to the company*. Obviously, any project could lose money, but many other things could go wrong too: accidents that damage facilities or injure people; legal liabilities if the company is late, or unable to perform on some portion of the contract; damage to the reputation of the company; damage to the environment through an accident or a poorly designed work process; or other things that could go wrong. The company

---

1 For example, if a competing company wins a contract for the US Federal Government, much of that contract becomes public information, and a copy of that contract can be obtained via the Freedom of Information Act. The information that you can legally obtain in this fashion will likely include the negotiated price for that contract, and the statement of work, which describes the work to be performed and the items to be delivered.

2 For example, if this buying customer is a US Government agency, their budget has to be approved by Congress, and the amount approved for this project is likely to be contained in the publicly available budget act for this agency. The amount shown for this project will include the budget for the Government project office, in addition to the funding for a contractor, but you might be able to estimate how much of that funding will go to the Government project office, and thereby estimate how much is left, which is the amount that the Government expects to spend on a contractor to build this system.

will want to assess all of these, as part of making a determination of whether they want to make this bid or not. Do we appear to understand all of the risks, and have appropriate strategies for mitigating those risks in place? (We will return to the subject of risks and risk management in Chapter 9.) Examples of such risks that must be identified, assessed, and plans for their mitigation created include:
  - *Capability.* Does the company believe that we can deliver all of the promised capabilities?
  - *Schedule.* Does the company believe that we can deliver all of the promised items by the promised dates?
  - *Price.* Does the company believe that we can complete all of the tasks required by the contract for an amount that will allow the company to make a reasonable profit on our efforts?
  - *Contractual.* Are there contractual terms in the request for proposal, or in our proposal, that the company cannot tolerate?
- *Probability of winning.* Writing a fully fledged proposal, and supporting the presentations and negotiations, can cost a lot of money; literally, at times, millions of dollars. So, if this is an externally-sponsored competition, the company will want to estimate their *chances of winning* the competition. If this is an internally-sponsored competition, the company will want to estimate the probability of achieving the projected levels of sales and profits.

If you get through all of this successfully, you have passed Step 2 and are allowed to submit your proposal to the buying customer. Now, you want to win! How do we maximize our chances of actually winning? This is Step 3 from the above list, and is the main topic for the remainder of this chapter. I will summarize the answer briefly now, then discuss Step 4, and finally return to a more detailed discussion of Step 3: *how to win.*

You will mostly likely win or lose based on what is *written in your proposal.* There may be some opportunities to make face-to-face presentations to the customer as well, but what is written is usually the main determinant of proposal success.

So, as we plan and prepare our proposal, we consider factors such as:

- Our value proposition, as expressed in the customer's coordinate system of value.
- The *differentiators* that we offer.
- Trust – have we earned this customer's trust?
- Contractual terms – have we offered attractive terms and conditions for the contract in our proposal?
- Schedule – are we proposing to deliver the product by the date that the customer desires? Are our claims about delivery date credible to the customer?
- Price – how much do we propose to charge? Does the customer have to pay us more than that amount if the job runs into complications, or is our company committing to absorb any such additional costs?
- Compliance – are we proposing to meet every single requirement in the request for proposal?
- *Want* factors – are there aspects of our offer that are attractive enough to the customer, so that they will prefer to select us over the other bidders?

Getting good answers to these items into your proposal comprises Step 3.

We will return to this aspect in a moment, but first we will talk briefly about Step 4: closing the deal.

Let's say that you win: you are formally notified by the customer that you have been selected as the company that they want to hire. You must still sign a contract. Creating a contract that both parties are willing to sign involves a lot of face-to-face discussion, which is termed *contract negotiations*. Sometimes, a customer will conduct such negotiations with all of the bidders, prior to announcing the winner; this allows them to negotiate with the contractor before they are selected, and customers think that this provides them leverage in the negotiations. It also, of course, allows the customer to be sure that the company they select is willing to sign up for contractual terms that the customer can tolerate, because they do not finalize the selection of a winner until after all of the bidders (including the eventual winner) have agreed to contract terms. Some customers, however, find negotiating with all of the bidders too expensive and too time-consuming, and defer some or all of the contract negotiations until after they have selected a winner.

Contracts are both very detailed and legally binding, so negotiating a contract is a complicated process. You will be directly involved, together with your contract specialists and your company's attorneys.

Congratulations! You finished all four steps, won the external competition, and were able to negotiate an acceptable contract. Now you can start work.

An important insight: to do all of the above really well takes a *lot* of time. For a big engineered system, your company probably needs to start *two or three years before* the request for proposal is expected to be issued. Your company will need to pay for the team working to get ready to write this proposal for a long time.

Here is another, equally important insight: starting early is really important. The data that I have seen indicates that the probability of winning is much higher for those teams that start early, as compared to those teams that wait to start until shortly before the request for proposal is issued. Companies that enter a competition late seldom win.

And still another important insight: there will be delays along the way.

- The customer may issue the request for proposal on a later date than that indicated in their original plan.
- The source-selection process (e.g. the process conducted by the customer to select a winner) may take longer than originally expected.
- If this is a government procurement, one or more of the losing contractors may file a protest, which holds up the awarding of the contract to the winner until an independent government agency assesses and adjudicates the merit of the claims by the losers.

What all of the potential sources of delay mean is that it will take a lot longer (and cost a lot more) than you might think to write a proposal and win a competition for an engineering project. You, your key personnel, and your company need to be prepared for the reality of how hard, how long, and how expensive this process really is. You and your company must be committed to this proposal! If not, it is better to decide early on that you are not going to make this bid, and find another, more suitable business opportunity.

We now return to Step 3: how to win. I will present two different approaches, one created by a former colleague and one created by me. You can pick and choose between them, based on what seems best for you, your company, and your particular bid opportunity.

5.2.1 Approach #1: The Heilmeier Questions

For about five years,[3] I served on a US Government advisory panel called the Defense Science Board.[4] One of the other (more senior) members of the board was Dr. George Heilmeier. Dr. Heilmeier was a distinguished engineer, best known for having been part of the small team that invented the *liquid crystal display,* which is used throughout the world on millions of electronic devices. He also served from 1975 to 1977 as the director of the US Department of Defense's Defense Advanced Project Research Agency (DARPA; originally called ARPA). During his tenure at DARPA, he started using a set of questions to help people within that agency who wanted to start a project sharpen their thinking about whether or not this was actually a good project for the agency to undertake. I have seen more than one version of these questions, but Figure 5.2 shows the version presented on the DARPA website.[5]

By the time I worked with Dr. Heilmeier, he was advocating that teams who were going to prepare competitive proposals for engineered systems use these questions to guide their planning and preparation process. That is, he advocated using them for what we are calling Step 3.

Note, however, that these questions are similar in many ways to the questions that I posed above, for Step 2.

Let's look at each question in a bit more detail:

- *What are you trying to do? Articulate your objectives using absolutely no jargon.* That is, what is the goal of this project? If you are completely successful, what will you have created?
- *How is it done today, and what are the limits of current practice?* The item that you wish to create is intended to help someone perform a mission; how do they do that mission today? This might be represented by the *mission thread* that represents current practice, as we described that item earlier.

- What are you trying to do? Articulate your objectives using absolutely no jargon
- How is it done today, and what are the limits of current practice?
- What is new in your approach and why do you think it will be successful?
- Who cares? If you are successful, what difference will it make?
- What are the risks?
- How much will it cost?
- How long will it take?
- What are the mid-term and final "exams" to check for success?

Figure 5.2 The Heilmeier questions.

3 1996 to 2001.

4 The Defense Science Board advises senior government officials about matters that affect national security. More information about the Defense Science Board is available at http://www.acq.osd.mil/dsb.

5 https://www.darpa.mil/work-with-us/heilmeier-catechism. During the time that I knew and worked with Dr. Heilmeier, he always referred to this set of questions as "The Heilmeier Questions." I note that the DARPA website refers to them as "The Heilmeier Catechism"; a *catechism* is a religious doctrine (the term is especially associated with Christianity) that takes the form of a set of questions and answers, used for instructional purposes. I do not know why there are two titles for this set of questions. As I said, Dr. Heilmeier in my presence always referred to them as "The Heilmeier Questions."

- *What is new in your approach and why do you think it will be successful?* The first portion of this question can be answered by the *new* mission thread diagrams. The second portion of the question asks two separate things: why is that approach likely to be an improvement for the users (e.g. what are the *operational performance measures* that indicate improvement) and why is your approach to implementing that likely to be feasible (e.g. what are the technologies that you will use to implement the new mission threads, and what are the *technical performance measures* that indicate that it can actually be achieved).
- *Who cares? If you are successful, what difference will it make?* The first portion can be answered by identifying the users and other stakeholders for this mission. The second portion again can be answered by the operational performance measures. (I would put this question much earlier in the list.)
- *What are the risks?* That is, what can go wrong? For each of those items, if one did go wrong, what would result? For each item, how can we lessen the *likelihood* of such an occurrence? For each item, how can we (separately) lessen the *negative impact*, if in fact one of these items did occur anyway? (We will discuss these items in greater detail in Chapter 9.)
- *How much will it cost?* Not only how much to build and test the new system, but how much to operate and maintain it over its entire anticipated lifetime. How much will it cost to dispose of the system when its useful lifetime is ended? What is the time phasing of these costs? What is the range of uncertainty for those estimates? We will discuss this subject in detail in Chapter 7.
- *How long will it take?* Not only how long to build and test the new system, but also the range of uncertainty for those estimates. We will discuss this subject in detail in Chapter 7 as well.
- *What are the mid-term and final "exams" to check for success?* That is, how do we objectively measure progress along the way? We will discuss this subject in detail in Chapters 10 and 11.

This is a pretty good list. As you can see, we have either discussed each subject already in this book, or are going to do so in a future chapter. Certainly, you need to address each of these items in a competitive proposal, whether the audience for that proposal is external to your company or internal to your company.

I do believe, however, that this list misses a couple of vital points, and so in order to address those particular items, I now present the method that I created and used during the time that I was writing (and winning) a lot of competitive proposals.

5.2.2 Approach #2: Neil's Approach: Achieve Positive Competitive Differentiation

My approach is slightly different. I focus on a systematic series of steps that lead to the creation of what I call *differentiators*, and what I call the *socialization* of these differentiators with the users and stakeholders, especially the buying customer. I created a sequence of steps aimed primarily at achieving that particular purpose.

What do I mean, in this context, by the term *differentiation*? By this, I mean creating a story line, supported by facts, so that the customer will conclude that my proposal offer is *different*[6] from the other, competing proposals in tangible and important ways that are beneficial to the customer. Of course, to be pedantic, I should have said *positive differentiation*; one can be differentiated in a negative sense too. That is not what we want; we want to be *better* than the competition.

My view is that in order to win a competition, you must be – in the eyes of those who are scoring and evaluating all of the proposals – the *best among those submitted*. That is, you win by achieving *relative* merit (i.e. merit relative to the other bidders), not by achieving any *absolute* level of quality. If you are manufacturing an item, you may aspire to an absolute measure of quality (e.g. 1 defect per 1 000 000 items manufactured, or whatever), but in order to win a competition, you must be the *best* of those who submit proposals, rather than meet some absolute standard of quality.

What I decided during my tenure as the person responsible for winning competitive proposals was to focus on *achieving this relative advantage*. In order to do that, for each proposal, I defined a set of specific topics in which I aimed to make our offer clearly the best: the *specific areas* where I desired to achieve positive differentiation. Since one usually cannot afford to be the best at everything, I also therefore accepted that I would only be just even with the other bidders in everything else; there might even be some areas where I accepted that some other company might be better than mine. But the idea was to *carefully pick the areas* where I believed that being the best would cause the buying customer and the source-selection process to choose my proposal offer over all of the others. That is, I pick the areas where I want to achieve *positive differentiation* and craft my proposal and marketing efforts so as to convey that message to the customer, and get them to believe it. If I choose the items where I want to achieve such differentiation properly (that is, I am able correctly to determine those items that are really most important to the customer), and I can actually convince them that I have credibly established that positive differentiation, then the customer will *want* my company to do the work ... and we will win the competition.

This approach seems to work. Using this approach, we won far more competitions than probability would indicate. In competitions that routinely had three or four other serious competitors (and therefore, if we won only an average percentage of these competitions, we would have won 20–25% of the time), we won nearly 75% of the time, over many competitions and over many years.

The written guidance that I created in order to accomplish this is pretty simple; it consists of just three pictures. The first is depicted in Figure 5.3.

First of all, you can see in that figure that quite a bit of the content is similar to that of the Heilmeier questions, although the sequence is quite different, and at times the focus. But there are also items on my list that do not appear on Dr. Heilmeier's.

As I do throughout this book, I start with the *users*, the *other stakeholders*, and their *mission*. We can use all of the techniques and methods that we discussed in Chapter 4

---

6 Here is a 2000-year-old quote about the importance of achieving differentiation:
מַה נִּשְׁתַּנָּה הַלַּיְלָה הַזֶּה מִכָּל הַלֵּילוֹת (Mah nishtanah ha-laylah hazeh mikol ha-laylot? *Why is this night different from all other nights?*) The point is that the night to which the reference is made is a *holiday* night, and is it therefore supposed to be markedly different from other (e.g. non-holiday) nights. This quote is from the Jewish book called the *Haggadah* that serves as a guide to the holiday festival called *Pesach* (Passover). The phrase originally appears in the *Mishnah* (a Jewish book of law and interpretations), Pesachim 10:4.

1. Customer measurement of value. Who are the users? What is their mission? How does this customer measure operational (not just technical) utility, and therefore how do they measure value from a contractor?

⇩

2. Differentiators. In light of the above, what positive differentiators will we try to obtain? Why are they the right ones? What items will we measure to show the customer that our approach will have operational value? What are the "threshold" values that we want to reach for each of these measures?

⇩

3. Story line. What is our "story" for what our system does that will make our system not only compliant, but desirable? What are the customer's current mission threads, and what are the new mission threads that represent our improved approach? Why, and in what ways, in that approach better? What have we invented or created that will enable us credibly to provide those differentiators?

⇩

4. Data. How will we collect the data and make the measurements to prove the achievement of the operational benefits? What scenarios and test data? What instrumentation and tools? Models? Prototypes? Benchmarks? What technical performance parameters need to be measured, in order credibly to predict operational performance? What calibration have we performed of our predictive methodology?

⇩

5. Design. What is our design? Is it reasonably complete and consistent? How do you prove that to the customer (models, benchmarks, etc.)? How do our technical accomplishments map back to deliver the operational benefits identified under Step 1?

⇩

6. Proposal artifacts and representations. How ought we best to represent the story of our design in a visual manner? What are the key pictures and graphs? How do we show that we deliver value in the customer's value system? How do we communicate this in a fashion that the customer will believe?

⇩

7. Risks and opportunities. What are the resulting risks of this design? The resulting opportunities? Which of those risks and opportunities are the most material, in terms of both likelihood and impact (a Pareto analysis). For the key risks, what are the mitigation approaches (for both likelihood and impact) and timelines?

⇩

8. System engineering management. How have we modeled the size of the technical effort? What metrics will we use to evaluate technical and other progress? How will we integrate subcontractor efforts? How will we verify and validate that the resulting system meets both the requirements and the customer use-cases? What is our projected schedule for delivery? What is the projected cost? Why are those estimates credible?

⇩

9. Socialization. What have we done to socialize our understanding of the operaitonal measures, technical measurements, measurement / predictive approaches, our "story" for delivering value, our improved mission threads, our design, our design validation, and the risks / opportunities with the customer? What feedback have we received? How are we adjusting our story in light of that feedback? What feedback have we received? How are iterations of this process? What tangible steps (e.g. intellectual property agreements, etc.) are we taking to transfer emotional ownership of these ideas to the customer?

Figure 5.3 Neil's approach to achieve competitive differentiation.

to understand the users and their coordinate system of value, and then use those insights in order to create the operational performance measures that they will recognize as measuring value that matters to them.

In the next step, I diverge significantly from the approach recommended by Dr. Heilmeier: I try to define the features of a potential design and offering that would make the customer *want* our company to do the work. This becomes my list of desired *differentiators*. I will describe below my technique for creating this list.

I must, however, *limit* the set of items that I place onto this list in two ways:

1. I must be able to afford to implement (and have the time to implement) all of the items on the list.
2. Each item on the list must not only be technically feasible, but I must also have a method for conveying that feasibility to the customer in a way that they will believe. Remember what we said previously: many of the customers are not engineers or scientists, so the explanations must be accessible to an adult of ordinary education and experience.

The idea of obtaining positive competitive differentiation from your competitors may sound obvious, but in my experience most proposal teams simply do not make such a list. It is not surprising, therefore, that the proposals that result do not convey strong positive differentiation. I believe that it is essential – absolutely the core of a winning proposal – to do this step *explicitly*, and to create a written list of candidate differentiators that are then explicitly discussed among the team, with your management, and (once you think that you have a pretty good list) with every element of the customer: the buying customer, the paying customer, the users, the other stakeholders, the nay-sayers, and so forth. This is not, in my experience, common practice. But having such a list provides guidance to every person participating in the proposal, and provides your potential customers with a clear view of what you are offering. In my view, every proposal ought to do this.

The next step is story-telling: how to tell the story about our differentiators, and their feasibility, in a clear, comprehensive, and easy-to-understand fashion. This is a separate step because I have found that it is often the case that the people who can identify the good candidate differentiators – and can do the technical analysis to demonstrate their feasibility – simply cannot tell the story in an effective and convincing manner. They tend to get caught up in confusing technical details. Therefore, once you have the list and the technical analysis, I have found it worthwhile to then recast the differentiators into an effective story line. This must often be done by people other than (or in addition to) those who actually create and validate the differentiators.

The next step is data-gathering: how will we actually make the measurements that prove our story? To do that, we need at least a preliminary design, so Steps 4 and 5 go hand in hand, and may proceed in parallel.

Good proposals have more than text; they have a small number of key depictions (figures, tables, and graphs) that concisely convey our key points. We next create these depictions. We also write theme sentences (or short paragraphs) to go with each depiction. Together, the depictions and the theme sentences are the artifacts that will tell our story, both in the proposal and in our face-to-face interactions that lead up to the formal submission of the proposal, and the eventual selection of a winning contractor.

Steps 7 and 8 are similar to several of the steps that are included in the Heilmeier questions. What could go wrong, and how do we mitigate those risks?[7] How will we evaluate progress?[8] How long will it take? How much will it cost? (Notice that I will

---

7 Which we will discuss in Chapter 9.

8 Which we will discuss in Chapters 10 and 11.

always discuss schedule before cost; I will tell you why in Chapter 7, and revisit the subject in Chapters 10 and 11.)

The last step – Step 9 – is another step that I have found is often done poorly. It is not enough for *us* to create great differentiators; we must get the customer actually to *believe* them. In my experience, explaining these great differentiators to your *customer* for the first time in the proposal is *not sufficient* in order to get customers to believe in your differentiators. In order to get them to believe in those differentiators, we must *socialize* those differentiators (and our other key ideas too) via face-to-face meetings with the customer *before* we submit our proposal. We do this for two reasons:

1. First, we are likely to learn something from such interactions with the customers that allows us to refine and improve our differentiators, and our other ideas.
2. Second, people are always reluctant truly to believe other people's ideas; they are far more likely to believe an idea if they feel that they have had a hand in creating it. You therefore need to allow your customer to have intellectual *skin in the game* of creating your differentiators. I like to say that we must *transfer some emotional ownership* of our ideas to our customers; if we can make them feel that these ideas are in part theirs, they are much more likely to believe in those ideas, to accept the arguments that led to them, and to accept the inferences that derive from them. That is what we want!

The value of your differentiators must be measured in the customer's coordinate system of value, not just in the engineer's coordinate system of value. We have talked about this before: customers are seldom engineers or scientists, and therefore have a different "coordinate system of value" than we do. Therefore, if we want them to like our ideas and select our proposal (especially if we want them to like us so much that they will pay a price premium to get us – always a great aspiration), then we must convey our value proposition in *their* coordinate system of value. We are engineers, so we measure things ... even when they are in the customer's coordinate system of value.

Here's a little complication: almost always, there is more than one "customer." There will be multiple influential people within the buying organization, multiple influential people within the user's organization, and multiple influential people in other places. We must therefore determine the "coordinate system of value" for *each* of these people and/or groups. Usually, the needs and desires of these various groups will be in conflict in some ways. *We* must resolve that, through balance; *we* must convince them that our proposed balance is the best possible outcome.

That seems like a lot of trouble. Can't we just ask the customer what would constitute a set of differentiators?

The conventional wisdom, of course, it that "the customer is always right." My view is different: my experience suggests that many customers are so busy operating their mission that their view of the future is "three minutes from now." They know well what they do now (and you should learn from that), but you usually cannot depend on them to create an effective vision for revolutionary future improvements.

In any case, if they did it for you, they would have to share that information with all of your potential competitors (in a competition, the customer will try to be sure that they are providing all of the bidders the same information, so that the competition is fair). And, since all of the potential bidders could avail themselves of that knowledge, this would deprive you of any potential differentiation from your competitors!

But if *you* create a vision for the future *for the customers* – and then convince them that it is valid, valuable, and feasible – they need *not* share that information with the other bidders, and therefore the differentiators that *we* create can provide you with competitive differentiation.

Now let's return to the meat of the discussion: how to choose those desired differentiators.

When the competition is finally over, the customer will have a set of reasons for why they selected a particular offer. In competitions for US Government contracts, the Government will actually tell you in advance what the "source-selection criteria" are going to be:

- It might be *price*. City governments do this when buying a commodity, or a low-tech service (such as repairing potholes in streets).
- It might be the *lowest price among those technically acceptable*. In this situation, bidders whose offer does not seem credible, who don't have qualified people, who don't seem to understand the problem, and so forth, are eliminated from the competition *before* the customer looks at the prices offered by each bidder. They will then select the lowest-price bidder *among those left*.
- For most engineered systems, however, the source-selection criteria are something more complicated: an approach called *best value*, which consists of what – in the judgment of the customer – seems to be the most advantageous combination of price, capability, risk, schedule, credibility, and contractual terms.

Since the customer has criteria for how they are going to select the winner, you must be able to credibly show that your proposal is *superior* in a set of factors (some of which are *quantifiable*; price is certainly quantifiable) that relate to those criteria. And remember the *want* factor too. You therefore need *differentiators* for each of the following:

- The explicit source-selection criteria
- The unstated source-selection criteria
- The "want factor."

The general process for selecting differentiators that I use is as follows:

- Understand the customer's point of view:
  - What they do now, and why they do it that way
  - What they have tried and rejected in the past
  - Their "unwritten rules."
- Use this to derive a statement of what they value.
- Use that to create actual quantifiable metrics (the operational performance metrics) that can express the goodness of a candidate solution in their coordinate system of value.
- Now that you have metrics that reflect the customer's coordinate system of value, you next (and this is the critical step) create ideas for *what the system ought to do* to score well against those metrics; these are the core of your potential differentiators. I do this by creating alternative versions of the mission threads, and by working through the process depicted in the matrix that constitutes Figure 5.5.
- Socialize the above with the customer, continuously.

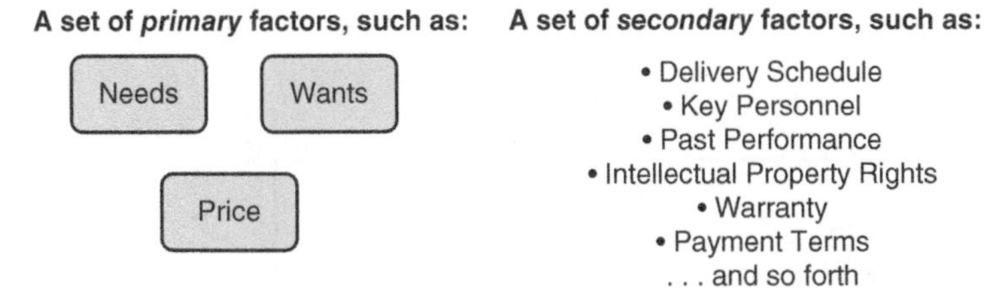

Figure 5.4 Factors that can help you create positive differentiation for your proposal.

You must plan where you want to achieve differentiation. You must figure out in which factors you want to be superior, and in which factors you are content to be even with the competitors (it might cost too much to be superior in every factor). Those factors where you intend and plan to be superior to all other competitors are your intended *differentiators*; you might prioritize your differentiators, so that you have primary differentiators and secondary differentiators (Figure 5.4).

Remember, if the customer has identified something as an explicit *need* (usually, by writing a mandatory requirement), every bidder will claim that they are doing that. Therefore, the mere fact that you are meeting a requirement *cannot be a differentiator*. It may be, however, that there is something superior about *how* you are implementing that requirement (e.g. something about your design that lowers risk, provides more design margin, or some other characteristic that is of benefit to the customer). That *how* may be an opportunity for differentiation. This is a frequent mistake that I see teams preparing competitive proposals make over and over again: they believe that the mere fact that they are meeting a mandatory requirement can constitute a differentiator. That is *never* true.

One way to test the effectiveness of your candidate list of differentiators is to write the briefing slide that you imagine the customer would use to brief their boss about why they chose *you* over the other competitors.

To guide the overall process of creating and proving differentiators, I use the matrix depicted in Figure 5.5.

The first column lists the key customer *wants* and *needs*. These, of course, are usually measured by *operational performance measures*, rather than by *technical performance measures*. The matrix works its way through a series of topics and questions; when you have completed every column for each of your desired differentiators (e.g. one row per differentiator), you are ready to start socializing them with your customers.

Examples of questions to ask the customers during the socialization process are:

- "If we could do this, would it help?"
- "What would constitute adequate proof that this idea is credible?"

Such socialization is a long-term process, *never* accomplished in a single meeting.

A word about where to look for potential differentiators. I find that most people today think that competitive differentiation comes from *improved technology*. I believe that this is generally wrong, on two counts:

1. Technology by itself tends to provide small, incremental improvements (this year's disk drive has 15% more capacity than last year's, and costs the same). But systems are expensive, and *no one buys big expensive systems to get small incremental improvements*.

Key customer wants and **needs**	**How this customer measures success** (operational performance measures)	For each operational performance measure, **What are the specific values that are** reachable today, and are desired by the customer for the future	**What, therefore, is our desired positive competitive differentiator(s)**	The **candidate implementation / mechanization approach** for this desired differentiator.	The **source of the candidate implementation**	How we **prove / measure the intended level of capability**	What **tools, infrastructure, scenarios,** etc. do we need in order to make those measurements	**Who is our expert?**	**Who** within our organization is responsible for achieving this differentiation?
Wants must be specific enough that we can name the parameters by which the customer will measure success			A statement of capability, with metrics and specific values to be achieved. Or a statement that we only want to achieve "table-stakes" against this want, rather than differentiation.	This is a design features piece of software, an algorithm, a material, etc., that can reach the level of differentiated performance described in the previous column.	Will we invent this within our company, or will we partner with an outside organization (another company, a university, etc.) in order to get what we need, or some mixture. We call this a *make–buy decision.*	Will we benchmark an actual component, use a system-level model, and so forth. Also, how do we convince the customer that those predictions are credible (e.g. model calibration, etc)?		What are the credentials of that person (this is always a person, not an organization or a product) that the customer would recognize?	

Figure 5.5 The matrix that I use to plan competitive differentiators.

2. And even if a specific set of commercially available technologies provides a big improvement, they are probably available to the other competitors, and therefore do *not provide competitive differentiation*. You must look *elsewhere* for the differentiation that you need in order to win.

Of course, if you have a *proprietary* technology, that will not be available to your competitors. But the first point still applies; my experience is that big improvements tend to come from *changing the business process* of a mission (which we represent in the mission threads), rather than solely from improved technology. This is not at all the "conventional wisdom," but that is fine by me; it gives me the opportunity for differentiation from my competitors!

Let me say that again: in my experience, instead of coming from technology, the differentiators seem to come from improvements to the *mission threads*. Only *after* you have created the specific ideas for what the system must do in order to create that differentiation (by creating the new mission threads) do you start to think about technology. The role of technology is seldom to create differentiation; instead, it is to make the implementation of your desired differentiation *feasible*. Your differentiation comes from the improved mission thread; that will usually relate directly to the operational performance measures, and therefore be credible to the customers.

Having differentiators is necessary, but not sufficient. The step of socialization with customers is also vital. As I have said previously, most of us can be counted on liking our own ideas, but are far less open to the ideas of others. Therefore, in order to get the customer to like your idea, you need to allow it (at least, in part) to become "their idea." My phrase for this is that I want to *transfer emotional ownership* of the idea to the customer. Then it is (at least in part) "their idea," and they are far more likely to fight for it!

There are many ways to accomplish the transfer of emotional ownership:

- Joint brainstorming
- Collaborative research, under some sort of formal agreement
- Grant generous intellectual property rights
- ... and many, many others.

A few examples of actual differentiators will probably help. Here are some that I have created over the years:

- *Example 1.* Prescribing drugs
  - Desired differentiator: Significantly decreased number of deaths due to adverse interactions among multiple prescription drugs taken by a patient.
  - Associated operational performance measure: Patient deaths per year in the United States, due to adverse interactions among prescription drugs.
  - Discussion: When we started, the number of deaths due to interactions among prescription drugs in the United States was estimated to exceed 100 000 per year. We were able to show that our improved system for managing and monitoring the prescription process would avoid a majority of of those deaths. As a result, our approach was eventually adopted, *despite* intense resistance by doctors.

- *Example 2.* The US Army's Blue-Force Tracker
  - Desired differentiator: Significantly improved land combat power, at a modest cost (e.g. a small percentage of the cost of the brigade's equipment, and far less than competing approaches to obtain a similar increase in combat power).
  - Associated operational performance measures:
    - The loss-exchange ratio, a direct measure of improvement in combat power.
    - The ratio of the cost of fielding the Blue-Force Tracker to the entire Army, versus the cost of upgrading every tank in the Army.
- Discussion: We were able to show that the Blue-Force Tracker would provide far more increase in combat power than a particular anticipated upgrade to the fleet of tanks. After conducting its own modeling efforts, and a series of large-scale, unscripted, force-on-force exercises, the Army accepted this estimate. The Army's own assessment showed that the anticipated cost of developing and fielding the Blue-Force Tracker to the entire Army was far less than the anticipated cost of fielding the upgrade to the tanks. As a result, the Army funded our Blue-Force Tracker, and elected to upgrade only a small number of the tanks an Army-wide upgrade of the tanks.
- *Example 3.* The US Army's short-range air defense
  - Desired differentiator: Significantly increased number of enemy aircraft shot down, while still never shooting down a US aircraft
  - Operational performance measures:
    - "90% of the time, the correct target shall be in the narrow field of view of the weapon's optic, and correctly enter auto-track" (remember Chapter 2!)
    - Ratio of achieved shot opportunities (that comply with the rules of engagement) *with* and *without* the system, against a set of standardized scenarios.
  - Discussion: We were able to show a better than 5× improvement in the number of shot opportunities achieved for a set of standardized, realistic scenarios, as proven through free-play field exercises, while also meeting the "90% in the narrow field of view" measure, which ensured that US aircraft would not be shot down.

This creation of differentiators all sounds pretty complicated. Why not instead just aim always to be the lowest-priced bidder?

First of all, you might be the lowest-priced bidder and still not win. If this is a best-value competition (as are most competitions for complex, engineered systems), and I have made the customer want my system through the creation of strong and credible differentiators, I will likely win even if your price is lower than mine. I once won a large contract for an engineered system, even though my bid was for twice as much money as any of the other bidders. Also notice that most of the cars sold in the United States are not the lowest-price model, and that Apple sells more phones that their lower-priced competitors. Lowest price is not a reliable path to winning!

Furthermore, life is not always pleasant for those who aim to win *only* by being the lowest-price bidder. You might be forced to bid such a low price that you can't make a profit; you can't pay your employees enough to attract and retain good talent; you can't provide benefits (vacation, retirement, sick-pay, medical insurance, suitable offices, etc.) to your employees sufficient to attract and retain good talent; and so forth. In my experience, you cannot build a great business around *only* being low cost.

But what about WalMart? Didn't they do exactly that? Actually, no. They do use their automated supply-chain management tools to try to achieve costs below their competitors. But they do many other things too; things that cost a lot of money, and thereby actually drive *up* their prices! For example, they have lots of locations, because they believe that having locations nearby is convenient for (and important to) their customers and potential customers. They also do many things that they believe improve the quality of the shopping experience: lots of free parking, greeters at the front of the stores, lots of checkout lines, no-hassle return policies, and many others.

They clearly believe that these items (even though they drive up their prices!) are an important part of their selling strategy – their *differentiation*! They pioneered many of these things, most especially the no-hassle return policies. I believe the most important single ingredient to WalMart's original success was their no-hassle return policies, not their prices.

Figure 5.6 summarizes my key suggested steps to winning.

In contrast, Figure 5.7 summarizes some of the typical mistakes that I see over and over again; these cause you to lose competitions for project contracts that you could have won.[9]

While we are talking about mistakes, I want to say something about the topic of *past performance*. The buying customer for a complex engineered system will certainly want to understand how your company has performed on previous contracts. Did you finish the contract on time? On price? Did you deliver everything that was promised in the contract? Did the users find the system suitable, as well as effective (remember that we defined these terms in Chapter 3).

The buying customer will also try to determine if some of these previous contracts on which your company has performed are *relevant* as indicators of how well you are likely

- Start early
- Qualified, committed capture and proposal team
- Work hard on building a strong relationship with all elements of the customer community
- Within legal limits, influence what they intend to buy
- Marshal strong and credible examples of good past performance, and create viable arguments for why they are relevant to this procurement
- Understand the competition
- Conduct an unbiased price-to-win analysis
- Create a complete offering package and win strategy (including desired differentiators)
- Select teaming partners to fill in your weaknesses, and to strengthen your desired differentiators
- Have the design and the business strategy complete and approved *before* the request for proposal comes out
- Follow an orderly and systematic process to prepare the actual proposal

Figure 5.6 The key steps to winning.

9 I wish to acknowledge my former colleague Douglas Pell, who worked tirelessly to gather lessons-learned from many actual competitions. His insights contributed to this list.

- Failure to identify the *customer's principal issues or problems in advance*
- The *customer contact plan* is inadequate to uncover the customer's needs, desires, and predispositions. We should determine who the customer views as the other strong bidders, and why, and find out what the customer wants and expects to happen.
- You fail to *pre-sell your solution to the customer*. We should take every opportunity to conduct one-on-one discussions with customers in advance of the request for proposal
- Failing to use the available and appropriate (e.g. ethical and legal) methods to achieve influence on the customer's acquisition strategy. We should: validate our understanding of the procurement approach through *frequent interaction with the customer*; identify issues and *solicit customer opinion and comment*; *help the customer* clearly communicate their needs toward developing an RFP package that will reflect an understanding of available technologies and solutions
- *Failing to identify key personnel* (especially the program manager) early enough to become known by the customer as someone they firmly believe can get the job done
- Failure to carry through frequent *customer-meeting plans*. We should meet frequently with the customer(s), and have a systematic process for bringing feedback from the customer contacts back into the capture activities
- There is not a complete and reviewed *win strategy* completed long before the RFP is issued
- *Teaming* is conducted on an *ad-hoc* or *sentimental* basis, rather than in accordance with the win strategies and competitive pricing strategy
- Fail to set and validate the target *price to win*. We should use both a top-down estimating model and generate a bottom-up price estimate; do the same for each competitor, so as to estimate their likely bid price
- Allowing the costs to "roll-up" as they will, rather than driving every portion of the bid to a *price-to-win* target
- There is not a regular tempo of *management reviews* and management involvement

Figure 5.7 Typical mistakes made during competitions.

to perform on this contract. Are some of those previous contracts of the same order of size and complexity as this new system? Do some of those previous contracts involve similar capabilities and/or make use of similar technologies as this new system? Are there portions of those previous systems that can credibly be repurposed by your company for this new system? Are some of the key technical personnel that contributed to the success of those previous projects committed by name to this new project? Your company is likely to be claiming all of these things, in order to show that you will be a reputable supplier, and that the low price you are bidding is credible.

In my experience, however, companies often seriously *hurt their proposals* through *overly broad and overly generic claims* about past performance.

How do proposal writers hurt themselves in this manner? First of all, customers are in fact not particularly interested in the fact that you solved some different problem for a different customer; they want only to know about your approach to solving *their* problem. They will also want to know why your approach for doing so is feasible and credible.

Therefore, you must tell them first *how you will solve their specific problem*; this is based, of course, around your design for their system, and your methodology for organizing and conducting the work: the tools you will use, and so forth. Past performance has *nothing whatsoever* to do with establishing this part of your proposal story.

Then, you must convince them that your *design and methodology are feasible and credible*. This is where past performance comes in: you successfully built the XYZ system for customer ABC last year, and it uses three of the same critical algorithms as you are proposing for this system. This provides a credible method for you to show the performance of those algorithms: you can run actual benchmark assessments on the XYZ system. This shows that your staff know how to properly code those algorithms in software. It might even back up your claim that you can reduce risk and cost a bit (but don't exaggerate how much) by reusing some of that software in the new system. And so forth.

What is to be avoided is the overly broad claim (e.g. "Our company built the XYZ system, so of course we can build your system"). The customer only cares about *their* system, and is far less willing to believe than you might expect that the experience of building the XYZ system for a different customer will be relevant to their system. You must prove, via tangible examples, *every single step* of relevancy, as described in the previous paragraph. If you are too broad in your claims, or too generic, you will create an air of charlatanism among the buying customers and the users, and end up not being believed, not just about past performance, but in other portions of your proposal too.

5.3 Your Role in All of This

Among all of the people involved in winning proposals for engineering projects, there are three key roles:

- The *capture manager*. This person is the *strategist*, the person who leads the effort from the earliest days of identifying or creating the opportunity, through the internal efforts that assess the opportunity, the shaping of the opportunity through continuous interactions with the various portions of the customer, the definition and validation of the desired differentiators, the qualification of the opportunity (that is, does our company, based on all of the factors described in this chapter, in fact want to make this bid), the selection of the designated project manager and their socialization with the customer, the selection of the proposal manager, and so forth.
- The *proposal manager*. This person is the one who actually leads the efforts to prepare the written documents that respond to the request for proposal.
- The *designated project manager*. This person is the one who will become the manager of our engineering project after we win.

On a proposal effort for a large project, these are likely to be three separate people. As is usual in systems engineering, there is *tension* between these people. For example, the capture manager wants to win, and will therefore be looking for ways to decrease the price that the company will eventually bid, in order to increase their chance of winning. But the designated project manager knows that he or she has actually to execute the eventual contract, and will press to ensure that the price bid is high enough to provide adequate resources to do the work. This tension (and there are many other examples of such tensions between these three people) is healthy!

For projects of a more moderate size, the designated project manager may also assume one of the other roles. But in this case, someone else must be appointed to focus on achieving a balance among the factors in tension; for example, if the designated project manager is also the capture manager (I have performed this double role at times), someone must be designated to be the person with the authority to ensure that the price bid is high enough to provide adequate resources to do the work.

As the capture manager, proposal manager, and/or designated project manager, you of course organize all of the tasks described in this chapter, select the people who will lead each task, and ensure that they then follow through on the execution of those tasks.

You must also *personally* be involved in the creation and validation of the desired differentiators. Since many of those differentiators describe desired characteristics of the design for your system or product, the creation of differentiators involves the use of engineering techniques and skills; this task *cannot* be performed by sales &/or marketing personnel.

5.4 Summary: How to Win

- Virtually 100% of company revenue comes as a result of *proposals*
  - Either *competitive contract awards*, or
  - *Internal competitions* to select our next product.
- Becoming a recognized good proposal manager is a sure path to career opportunities.
- There are recognizable steps to win:
  - *Some focused internally*. Reduce risk, don't sign up for a bad deal, etc.
  - *More focused externally*. What does the customer need and want, how do I obtain differentiation, etc.
- Obtaining credible differentiation is the key
  - We are engineers, and we measure this too
  - Persuasion! Transfer emotional ownership of your ideas to the customer.

And finally, a personal note. I *loved* running proposals. I felt that I was making a difference to the company and to the customers, it was interesting, and it was intensively creative. Maybe you will love working on proposals too!

5.5 Next

Now that we know how to win a competition, you are ready for day-1 of your new assignment as the manager of a new engineering project. What do you do first?

5.6 This Week's Facilitated Lab Session

- You will continue working in your teams. Today's topic is team exercises about proposals and how to win.
- Today's assignment for your team:
  - Refer back to the project that your team selected during the week-4 facilitated lab session.
  - Someone is going to have to make a decision about whether or not to select your team to build the product you defined; that might be either an internal or an external customer. You will therefore be facing the prospect of having to enter a *competition*, and having to win that competition.

- Create a short list of desired differentiators for your project. Explain how each relates to the operational performance measures that you created last week for your project.
- Answer the Heilmeier questions for your project.
- Hold a group discussion of your findings and insights.
- Get together outside of class and write this all up (three to five single-spaced pages of text and figures; full sentences and paragraphs, rather than bullet points). This will become one section of your final team report that will be turned in as homework near the end of the term
- You will also need to prepare a couple of briefing charts summarizing this topic, for the presentation by your team during the last week of class.

6

Organizing and Planning

Congratulations! You have been named the manager of our new engineering project. What do you do next? You decompose the work entailed in performing the project into smaller pieces, using a hierarchy. When this is done in a particular fashion, it is called a *work-breakdown structure*. Projects all over the world are managed using a work-breakdown structure. In this chapter, I both teach you the basics of creating and using a work-breakdown structure, and I also show you how to do it effectively within the specific context of engineering projects. Then, we move on to discuss the organizational structure of your project, and finally I show you how to use your work-breakdown structure as the basis to create a complete project plan for your engineering project.

6.1 The Work-Breakdown Structure

Congratulations! You have now been designated to become the manager of a new engineering project. You are still in the middle of the competition for the contract award, and will soon submit your proposal, as we discussed in Chapter 5. Figure 6.1 shows your project's organization chart so far.

OK, *what do you do next?*

You must *make a plan for how you will organize the work*. And, just like in so many other aspects of systems engineering and engineering project management, to do this we employ a *hierarchy*. That is, the first step in making a plan for how you will organize the work is to decompose the work into a *hierarchy*: you break the work into *smaller pieces*, and arrange those smaller pieces into a *nested tree diagram*. See Figure 6.2 for an example.

Why do we do this? Because:

- Smaller pieces of work allow you to allocate specific responsibilities to *specific individual people* (whether those people work for your own company, or work for other companies that your company has hired to perform a portion of the work).
- Smaller pieces of work can be the basis for a *plan*: who does what work, when, how, how do we tell when each piece of work is complete, what tools and methods will we use, and so forth.
- Smaller pieces of work form the basis for tracking progress in a meaningful fashion: we can check our progress against the plan.

Engineering Project Management, First Edition. Neil G. Siegel.

Companion website: www.wiley.com/go/siegel/engineering_project_management

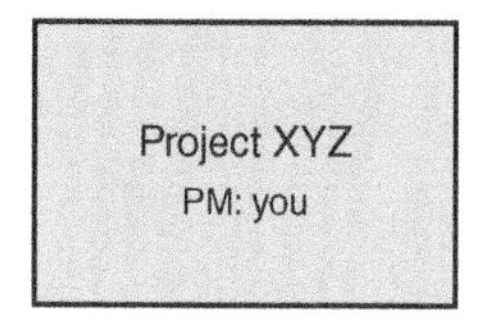

Figure 6.1 Your project.

Let's examine Figure 6.2. First of all, notice that the figure is arranged in *levels*, as indicated at the right-hand portion of the figure.

- The single box that represents the entire project (and which has your name, as the project manager, in the box together with the name of the project) is the *only* item at *level* 1 of this hierarchy.
- At the next level down (level 2), there are four boxes depicted (the dots . . at the right indicate that there could be more boxes at level 2 in your real version of the chart). Each box identifies a piece of work within the project. The work identified in each box is *disjoint* (that is, distinct and without overlap) from the work identified in any of the other boxes at the same level. Each box has a title that describes the work entailed by that box and (this is important), each box has the name of the *individual person* who is *responsible for accomplishing that portion of the work.* The figure also indicates that these four people – Jane, John, June, and Joe – report *directly* to you; there is no intermediate manager between you and them. We have now decomposed your project into four pieces of work.
- Each piece of work identified by a box at level 2 of the figure could itself be further decomposed into still smaller pieces of work. As an example, in Figure 6.2 we have decomposed the design task (at level 2) into two smaller pieces, one for hardware (labeled HW in the figure) and one for software (labeled SW). Each of these new boxes are at *level* 3 of the hierarchy, and each of those boxes also has a named

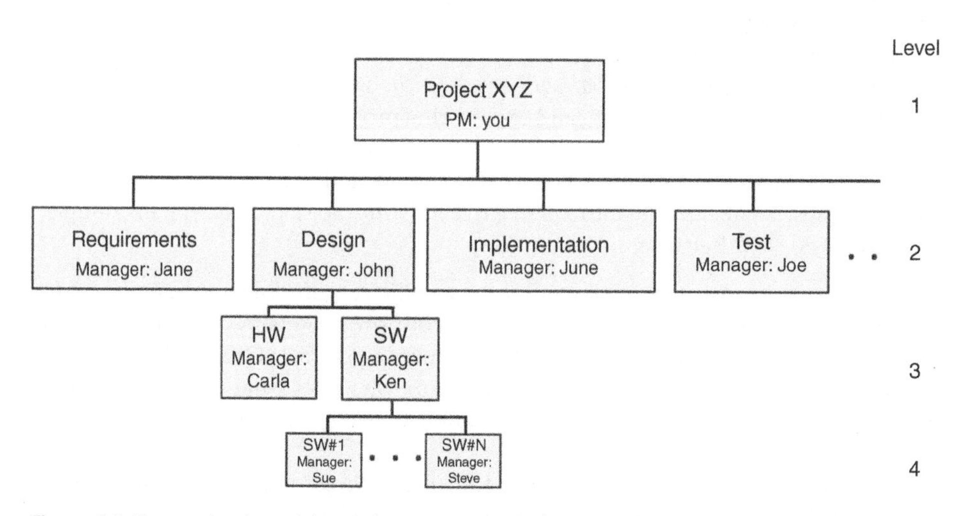

Figure 6.2 Your project's work-breakdown structure; in this example it is organized by *function*.

individual who is responsible for those smaller pieces of work. There is no intermediate manager between these two individuals at level 3 (Carla and Ken) and the manager (John) of the piece of work (*design*; at level 2) to which their pieces of work are subordinate.

- This decomposition can continue. In the figure, we have depicted a notional decomposition of the software task at level 3 into *N* tasks at level 4, the first headed by Sue and the *N*th headed by Steve. The other tasks at level 3 can also be similarly decomposed, and so can the tasks at level 4, thereby creating a level 5.

When we do a decomposition of the work for an engineering project in this particular way, we call it a *work-breakdown structure*. The term *work-breakdown structure* is used on projects almost everywhere, all over the world.

For a big project, the work-breakdown structure might be *eight levels deep*. Our example in Figure 6.2 is just four levels deep.

A work-breakdown structure, therefore, is a *hierarchical structure used to decompose the work that comprises a project into a nested set of smaller pieces*, which can then be used to *guide the planning, execution, and monitoring* of that work on the project. To create the work-breakdown structure, we break the work into smaller tasks, and to organize those tasks so that you can assign reporting relationships, assign someone by name as having responsibility for each task.

The work-breakdown structure also needs to be *complete*; that is, the work-breakdown structure must depict *every* piece of work that needs to be accomplished in order to complete the project.

The creation of a work-breakdown structure is a key element in project management.

The example in Figure 6.2 decomposes the work of the project by separating the *type* of work (e.g. requirements, design, implementation, etc.). We say that such a work-breakdown structure is organized by *function*.

You may elect to organize your work-breakdown structure in a different fashion. The other common way to organization a work-breakdown structure is by *component*, that is, the major parts of your system. For example, if your project is to build an airplane, you might have level-2 elements for: (i) fuselage; (ii) engines; (iii) avionics (e.g. the electronic devices needed to operate the airplane, such as flight computers, navigation equipment, and so forth); (iv) passenger equipment (seats, air pressurization, heating/air conditioning, lavatories, catering, entertainment, lighting); (v) cargo equipment (rails, containers, tie-downs), and so forth. See Figure 6.3.

How do you choose whether you will organize your work-breakdown structure by function, by component, or by some other principle? In fact, your customer may have a preference, or may even mandate that you organize your work-breakdown structure in a fashion that they dictate. Even if the customer does not require the work-breakdown structure to be organized in a particular way, your company might.

If the choice is left to you, you should think how you want to track progress for your project: what are the pieces of work that you – as the project manager – want to hear about? What are the pieces of work for which you want to have a single person responsible? For projects with a lot of invention, development, design, and (most especially) software development, I prefer organizing my work-breakdown structure by *function*. If the project has little development work, and is mostly about assembly,

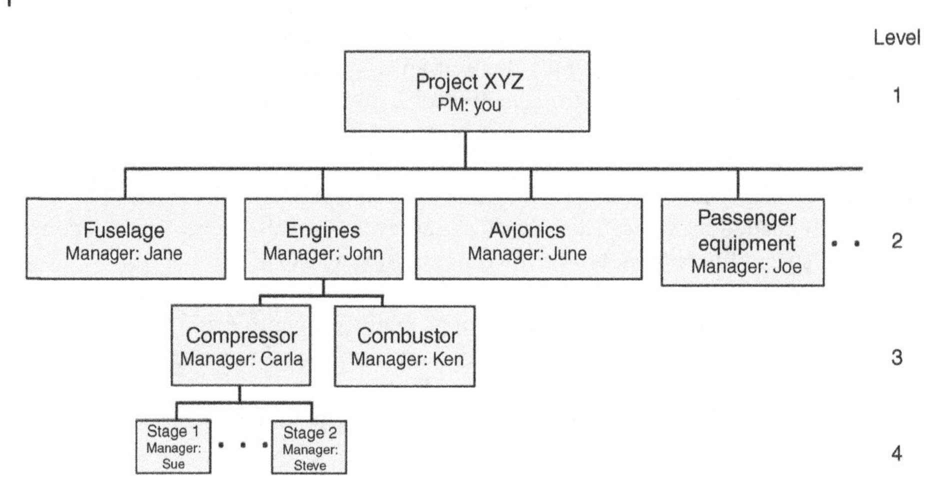

Figure 6.3 Example of a work-breakdown structure that is organized by *component*.

integration, and production, then I prefer to organize my work-breakdown structure by *component.*

On a big project, the graphical depiction utilized in Figures 6.2 and 6.3 can get unwieldy and hard to follow. You have another option: you can represent the work-breakdown structure in a more compact fashion as an *indented list*, such as that depicted in Figure 6.4. You can easily see that Figure 6.4 depicts exactly the same information content as Figure 6.3 (except that I snuck in the *cargo equipment*, for which I did not have room in Figure 6.3).

The indented list structure, being compact, also allows for the easy addition of additional information. For example, you could add columns to Figure 6.4 to show the amount of the approved budget for each row in the figure, the due date for each piece of work, and so forth.

	Levels			
Assigned manager	1	2	3	4
you	**Your project**			
Jane		Fuselage		
John		Engines		
Carla			Compressors	
Sue				Stage-1 compressors
Steve				Stage-2 compressors
Ken			Combustors	
June		Avionics		
Joe		Passenger equipment		
Jerry		Cargo equipment		

Figure 6.4 The work-breakdown structure as an indented list.

The work-breakdown structure depicts how the *work* is organized. We use several other similar depictions in managing an engineering project, including each of the following:

- The organization chart (how the *people* are organized). We will discuss the organization chart later in this chapter.
- The specifications (which define what the *system* is supposed to do, and how well it must do it), and the *specification tree* (how the requirements are organized into a set of nested documents). We discussed the specifications and the specification tree in Chapter 2.
- The product and system hierarchy (which show the *design*). We also discussed these in Chapter 2.
- The bill of materials (which lists the *parts* we need to buy). We will discuss the bill of materials in Chapter 14.

Don't confuse the work-breakdown structure with any of these other representations. They are all examples of hierarchies, and can all be represented by either tree diagrams or indented lists, just like the work-breakdown structure. But, as described above, they each have their own purpose, which is different from the purpose of the work-breakdown structure.

We use the term *element* for each item that appears somewhere on the work-breakdown structure, at any level. Elements are connected only by the *tree structure*; rules for such a tree structure are as follows:

- There are no *cross-links* (that is, there is never a connection between two elements at the same level of the work-breakdown structure).
- There is never a connection to more than one element at the level above a particular element.
- There is never a connection to any lower-level elements except those at the level immediately below a particular element.
- The elements depicted as linked to an element immediately above them in the work-breakdown structure must be *complete*; that is, as a group, they must encompass all of the work allocated to the element above them.

Notice that, under these rules, the *linkages between any two boxes correspond to a management reporting relationship between two named people.* Each link on the work-breakdown structure tree diagram therefore corresponds to a *specific agreement* between these two people. See Figure 6.5.

Experience shows that it is best to make this agreement *in writing.* The agreement between you as the project manager and the external customer (if this project has an external customer) is likely to be a legally binding *contract,* as we discussed in Chapter 5. The agreement between you and your boss (if this project has an internal customer) is most likely something less formal than a legally binding contract, but it is still likely to be in writing.

What I advocate is that *all* of the other one-on-one personal agreements that span the work-breakdown structure *also be in writing.* In essence, each person who has a link to a person above them in the work-breakdown structure treats that person above them as *their* customer; satisfying that person's needs and desires is their goal.

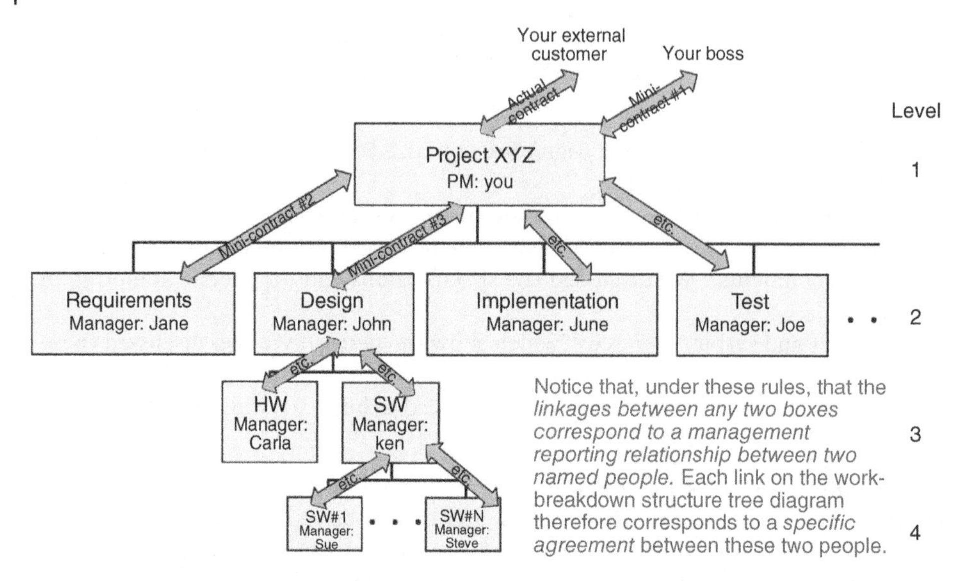

Figure 6.5 Each link in the work-breakdown structure is a management reporting relationship between two named people.

I call these agreements that are less formal than a legally binding contract a *mini-contract.*

By "tiering down" – that is, copying the relevant portion of the mini-contract that your boss has with *their* own boss into your own mini-contract with your boss, we obtain consistency in your mini-contract with your boss's mini-contract. By then continuing this process – copying the relevant portions of *your* mini-contract with your boss into the mini-contracts that you agree to with each of your subordinates – we maintain this consistency across the entire project. In addition, at each successive lower level of the work-breakdown structure, additional details also get added that pertain to the specific relationship and the specific tasks at hand.

Here are the items that I usually place into such a mini-contract:

- Who (by name) is accepting this responsibility? (e.g. who is the manager of this piece of the project?)
- What is to be accomplished? (we call this the *statement of work*)
  - The statement of work is a narrative description of products, services, or results to be supplied, and is described in detail below.
- What is to be delivered? How many? In what form? (e.g. for software, must source code be delivered, or is executable code sufficient? For hardware, packaging requirements, etc.) Where? (We call these items that must be provided to the customer before the conclusion of the contract "deliverables")
- What measurable criteria prove success? (We call these the "sell-off criteria")
- What could go wrong? (We call these "risks"). We will talk about risks in Chapter 9.
- By when must the project be started? By when must it be completed? Are there any mandatory mid-term milestones? (We call the collection of this type of information the "project schedule")
- How much money is allocated to this work? (We call this the "project budget")

- *Constraints*: (i) what must be done so that this piece of work fits into the rest of the system (complies with interfaces, complies with overall design, tools, and programming languages to be used, etc.); (ii) legal and regulatory requirements; (iii) corporate policies, standards, and requirements; (iv) customer-imposed standards; (v) any constraints of purchasing (e.g. required sources, additional approvals, and thresholds), etc.
- What equipment and/or information is being provided by the customer? (We call this "customer-furnished equipment", or if the customer is a government, we might call this "government-furnished equipment")
- Other obligations of the customer: facilities? people?
- ... and so forth.

As you can see, these mini-contracts contain a lot of the same items as we described in Chapter 5 for an actual, legally binding contract. That is why I call them *mini-contracts*. They constitute emotional and ethically-binding agreements, even if they are not legally enforceable.

And just like a contract is always the result of *mutual bi-lateral negotiation*, and *informed consent* by both parties, each such mini-contact must *also* be the result of *mutual bi-lateral negotiation* and *informed consent*. The mini-contract is signed by both parties too, so as to signify that the parties have voluntarily reached agreement.

The boss/customer *never* imposes the contents of a mini-contract onto a subordinate; if they did, the subordinate would not believe in the goals, dates, and capabilities, and would not feel emotionally responsible for success. As we will discuss in much more detail in Chapter 13, if we do not get our employees emotionally engaged through *mutual negotiation* and true *agreement*, our project is unlikely to succeed. We will devote an entire chapter – Chapter 13 – to such *social aspects* of project management. These social aspects are very important! One of the aspects that we will discuss in Chapter 13 is that any agreement between two people on a project must be mutually negotiated and mutually agreed to. In the United States, at least, agreements that are *imposed* seldom lead to productive outcomes. We will also learn that the most effective method for a boss to get a subordinate to perform a task is *persuasion*, and definitely not by saying *I have decided, just do it*.

Notice that some of these links may extend to organizations outside of your own; for example, your company might have hired another company to do a portion of the work (see Figure 6.6).

The agreements with these external organizations get some extra information in their agreement documents. In fact, as discussed previously, these agreements with external organizations are likely to be legally binding documents called *subcontracts* (used when there is development entailed in the work – legally, these are true, legally binding *contracts*; we use the term *subcontract* to indicate that this agreement is defining a subset of the work on a project) or *purchase orders* (used when the work consists primarily or exclusively of delivering material items that are offered for sale on a routine basis by the external company).

These external agreements get some extra information that is not usually in a mini-contract between two employees of the same company. For example, they will include:

- Payment terms (e.g. timing and conditions on payments)
- Contractual terms and conditions (e.g. penalties, dispute resolution, etc.)

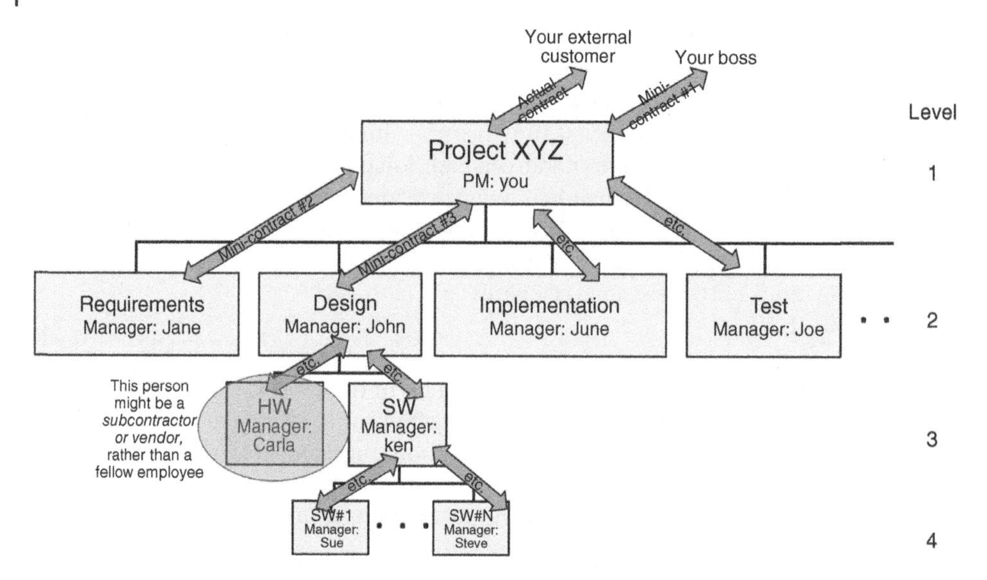

Figure 6.6 Some of these links may be with people who are outside your own company.

- Naming of key personnel
- Handling of intellectual property (who owns inventions made during the course of the project, what rights in those inventions accrue to each party, etc.)
- Requirements (if any) for co-location, where the work must be performed, the tools and methods to be used
- Agreements regarding sharing and protection of information
- Agreements regarding access to databases and computer systems (e.g. allowing a subcontractor's employees to have accounts on your company's email system, to have access to your software configuration management system, to your document production system, etc.).

Often, a project will create a written artifact that they call the *work-breakdown structure dictionary,* in which each work-breakdown structure element manager captures information about their element. As it happens, I don't like to use a work-breakdown structure dictionary; to me, what is important is not the element manager's view of their task, but the *bi-lateral agreement* between the element manger and his/her supervisor. That is why I insist on using the mini-contract format: it is *explicitly* a negotiated agreement, signed by both parties. The emphasis in the mini-contract is on the *agreement between the two parties* rather than (as in the work-breakdown structure dictionary) on the *description* of the element and its tasks. The mini-contract of course (as noted above) also contains the description of the element and its tasks, but its format emphasizes the agreement aspect, rather than the descriptive aspect. I like that.

6.2 The Statement of Work

In Section 6.1, I made reference to another artifact called the *statement of work.* We will talk about that artifact now.

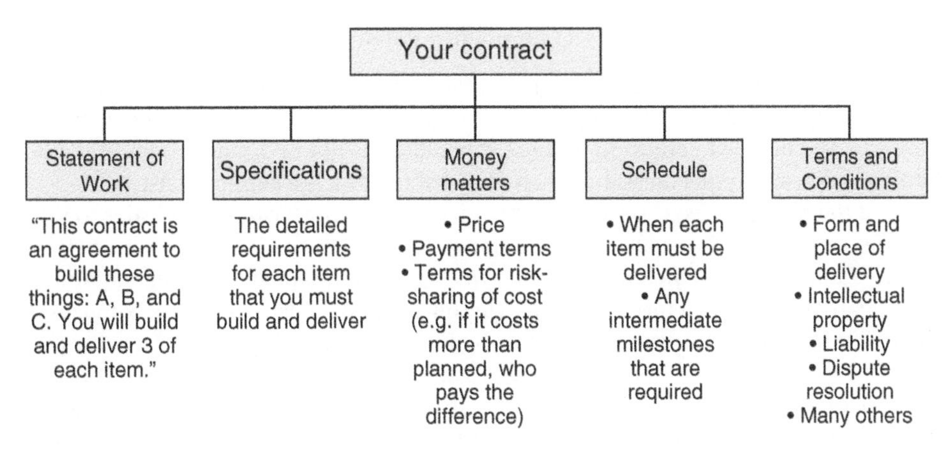

Figure 6.7 The principal portions of your contract for an engineered system.

The nomenclature *statement of work* comes from your *contract* with your external customer. We have already made reference to a *contract* as a legally binding agreement between two parties. Let us look at the contents of a contract for an engineered system in a little more detail.

Figure 6.7 depicts the principal portions of a contract for an engineered system. You can see that the statement of work is depicted on the left, right next to the specifications.

The purpose of the *statement of work* is to define *what work must be accomplished* (e.g. "design and build three each of items A, B, and C. When you are complete, deliver all of these items, along with training manuals, maintenance manuals, and this particular set of spare parts").

The *detailed definition* of these items – that is, what exactly are items A, B, and C supposed to do – is contained in the specifications. We already talked about specifications in Chapter 2.

Of course, a contract also talks about business matters, such as price, schedule, and various contractual terms and conditions.

In Chapter 5, we also pointed out that sometimes the contract incorporates your proposal. If your proposal is incorporated into your contract, this makes all of the promises that you made in the proposal legally binding, even if they call for more capability than is called for by the specifications. Be careful about what you write in your proposals!

Let me also introduce three new bits of nomenclature:

- We call the contract between you and your customer the *prime contract.*
- We call the contract between you and each of your subordinate external companies a *subcontract* (if you are buying custom-made products or services) or a *purchase order* (if you are buying commercially available products or services).

All contracts – whether the prime contract with your external customer, or the subcontracts and purchase orders with your external vendors – are *legally binding.*

In contrast, the mini-contracts between people who are all employees of the same company are *not* legally binding. They are emotional and moral commitments, and your status with the company, your next raise, and even your ability to remain an employee are on the line. A mini-contract is a serious obligation, even if it is not legally binding.

Back to the *statement of work*. As a minimum, a statement of work will contain:

- A list of the capabilities, products, or services that you must provide (that is, what your efforts on the project must create and/or acquire) and, if applicable, the quantity for each. The detailed definition of these capabilities is not in the statement of work, but instead is in the specifications; the statement of work is a summary list.
- A list of the deliverables (e.g. what must be delivered to the customer in order to satisfy the contract). This usually includes many more items than just the products that you are building; it also includes *ancillary items*, such as manuals, training materials, data, spare parts, and so forth. Every single thing that the customer expects must be listed. The details about these items may be in the specifications (this is usually the case for the required functional and performance capabilities of the products you will deliver), or may be defined directly in the statement of work itself (this is usually the case for quantities of items and the due date for each, as well as for ancillary items, such as data and spare parts that you must deliver).
- A list of the interfaces and constraints. Most products and services that are produced on an engineering contract are expected by the customer to interact with other systems that the customer already has and uses. This portion of the statement of work lists the characteristics that your products and services must have in order to perform those desired interactions (e.g. how it will fit in with the rest of the systems and procedures used by the customer). The technical details of these interfaces are not in the statement of work; instead, they are either in the specifications, or in separate *interface control documents*.
- If a particular person, organization, or company must be the one to produce a particular deliverable, they should be named. Sometimes the customer wants a part or subassembly to be provided by a particular vendor; if so, that must be named in the statement of work. For example, if you are buying a commercial aircraft from Boeing or Airbus, usually you will specify the particular company that you want to provide the engines. The statement of work in your contract would therefore specify that you want the engines to be provided by General Electric (or Pratt and Whitney, or Snecma, or whatever vendor you selected), and probably name the exact model number of the engine type you require.
- If particular tools, methods, or materials must be used, they should be named. For example, sometimes customers insist that a particular programming language be used for the software, or that a particular software language compiler be used, or that all parts must be designed and built using metric units of measure. All of these types of things must be defined in the statement of work.

There will be such a statement of work in your contract for the entire project with the customer (whether external or internal). Now, think back to the example work-breakdown structure depicted in Figure 6.2. When you negotiate a mini-contract with your subordinates (Jane for requirements; John for design; and so forth), you make sure that the relevant portions of the prime contract statement of work are included in the statement of work section of each of those mini-contracts. That way, nothing falls through the cracks and gets left undone. In order to be sure, you will probably build a *traceability matrix*, a diagram that maps every single item in the prime contract statement of work to one of your four subordinates (Figure 6.8).

Figure 6.8 A sample traceability matrix: who is responsible for implementing each portion of the prime contract statement of work.

SoW item	Assignee
1	Jane
2	Jane
3	John
4	June
5	June
6	John
7	John / June
8	Joe
9	Joe
10	June

Now you can be confident that the set of mini-contracts that you negotiate and sign with your subordinates is *complete*, in the sense that every item in the prime contract statement of work has been allocated to a subordinate manager for execution. Some items may have to be allocated to more than one manager; you must make sure that the mini-contracts specify the particular responsibility of each manager (e.g. what particular portion of that deliverable item is each person to provide).

Your subordinate managers will likely add details to the versions of the statement of work in their own mini-contracts with you.

And of course, when your subordinate managers negotiate their mini-contracts with their next-level-down subordinates, they ought to do the same thing: allocate every section of the statement of work in their mini-contract with you down to one or more of their subordinates, adding details as appropriate, and creating their own traceability matrix so as to make sure that their allocation is also complete.

The statement of work forms the basis for each contract, subcontract, or mini-contract, and at times it is useful to name things that you are *not* going to do (e.g. identify what will be *excluded* from your work scope), in order to be as clear as possible.[1]

6.3 The Organization Chart

Earlier, I made a distinction between the *work-breakdown structure* (which shows how the *work* is organized) and something called the *organization chart* (which shows how the *people* are organized). Let me say more about the organization and the organization chart at this time.

Why do we organize people? Mathematics tells us that inside any group, the potential communication links rise as a *factorial* with the number of people involved. There are six potential communication links in a three-person group. There are 3 628 800 potential communication links in a 10-person group. There are approximately 9.332 621 544

1 A famous example was the original specification for the EtherNet, developed jointly in the late 1970s by Xerox, Intel, and the Digital Equipment Corporation. It had a section near the beginning entitled "Goals and Non-Goals." The original EtherNet specification is available at http://ethernethistory.typepad.com/papers/EthernetSpec.pdf. I vividly remember reading that section title as a young engineer, and being struck by the utility of specifying both *goals* and *non-goals*.

394 42 × 10^{157} (the complete number is spelled out below[2]) potential communication links in a 100-person group. The scale of such communication complexity will overwhelm us. In addition, it seems likely that the opportunities for confusion will rise along with the number of such potential communication links.

Furthermore, the social scientists tell us that people can deal with only a limited number of interfaces at once; there has long been a rule of thumb that a supervisor ought not to have more than *seven direct reports*. To have more is confusing, and also dilutes the time that you can spend with each subordinate; you have only so much time per day. You will find that, as a manager, time is one of your most important resources.

It is also my experience that people do not *multiplex* (that is, perform multiple tasks simultaneously) very well. The social scientists agree with me on this point too.

But most real engineering projects have far more than three people working on them; how are we to cope? We do so by creating a *hierarchy* of people (yet another example of a hierarchy in systems engineering and project management!), so as to manage the complexity of scale inherent in large teams.

This hierarchy deals with scale, but also lessens ambiguity, through the assignment to specific named people of specific roles. The resulting hierarchy is the *organizational structure*; a depiction of this organizational structure is called an *organization chart.*

There are various principles that we can use to select an organizational structure for our engineering project. But *before* you select an organization, you need to have:

- Draft requirements (what the system is supposed to do, and how well)
- A draft work-breakdown structure (what the work to be accomplished consists of, and how you want to organize it)
- Draft schedule and cost estimates (which will determine how large an organization is suitable and affordable)
- Some of your key personnel identified by name (your direct reports and a few other key people, such as the project's chief engineer). People have skills, but they also have weaknesses, and I have found it useful to account for these as I design an organization.

My approach is to organize based on these four items. Only after you have these four prerequisite items are you ready to selection an organizational structure.

Note that since I included the strengths and weaknesses of my key personnel as a factor in designing the organization, this means that as people move off your project onto a new assignment – and are therefore replaced by other people, each of whom will have their own strengths and weaknesses – you may need to adjust the structure of the organization.

What are the goals of the organizational structure? I use the following list:

- Manage the stakeholders, both *internal* (e.g. your boss) and *external* (e.g. your customers)
  - Keep the project sold
  - Shape and manage expectations
  - Understand their goals, aspirations, and ways of measuring value

2 93 326 215 443 944 152 681 699 238 856 266 700 490 715 968 264 381 621 468 592 963 895 217 599 993 229 915 608 941 463 976 156 518 286 253 697 920 827 223 758 251 185 210 916 864 000 000 000 000 000 000 000 000.

- Continuously improve the plan for accomplishing the work
- Monitor progress against the work
  - Assess progress, risks, mitigation efforts, etc.
- Continuously assess how well individual people are performing
  - Does someone need help?
  - Does someone need to be replaced?
  - Is someone ready to take on additional responsibilities?
- Accomplish the actual work (notice that this is *only one* of the goals for the organization, not the *only* goal)
- Meet the terms of the contract
- Keep your people motivated, healthy, and safe
- Create institutional checks and balances, including against *your own weaknesses.*

Once you have the four prerequisite items, and understand the goals of an organizational structure, we are ready to select our organizational structure. There are two principal types of organizational structure (see Figure 6.9):

- *Projectized organizations.* This is a made-up term, used to signify that the people who work on the project *belong to the project,* rather than to a functional department within the company who are merely on loan to the project.
- *Functional organizations.* This signifies that the people have a permanent position within the company in a *functional department* (e.g. engineering, law, human relations, finance, etc.), and are simply on loan to the project.

Additional characteristics of each type of organizational structure are defined in the figure.

Of course, there are also hybrid approaches, that combine the characteristics of the two basic types in various ways:

- Some functions could be staffed by the project, but others by the functional departments. For example, engineering personnel could be recruited and managed by the

Projectized organizations:

- The people belong to you and your project
- You select them, and you must entice them to come and work on your project
- You must train them
- You determine their raises and bonuses
- You must discipline them, if there is a problem
- You must terminate them if they do not work out
- You must help them find their next assignment as the project nears its end!

Functional organizations:

- The people belong to a *functional department* (e.g. mechanical design department, software development department, etc.), *not* to you and your project
- You do not select people by name; instead, you write job descriptions, and the departments find the right people
- The departments train them, determine raises and bonuses, discipline them if and when required, terminate them if they do not work out; you get to provide input
- The departments find the people their next assignment as the project nears its end.

Figure 6.9 The two principal types of organizational structures.

project, but support disciplines, such as law, finance, human resources, and quality, could be recruited and managed by the functional departments.
- The project might transition over time:
  - A large project could be started depending on the functional departments; using the company's functional departments assists you in achieving the initial staffing at the start of the project.
  - This project could then be transitioned to use a projectized organization; this facilitates continuity for key positions, and is often less expensive.
  - This project could then be transitioned back to a functional organization as it nears its end; this facilitates the transitioning of the people to their next assignment.

There are many other combinations that are possible, of course.

Actually, in the real world, there is seldom a completely projectized engineering project. This is because of something that we have already pointed out: your company probably assigns some of the work to other companies.

- We use the term *subcontractor* if such an outside company's role includes development, and we therefore likely use a *subcontract*, as described above, to formalize our relationship with that company.
- We use the term *vendor* if such an outside company's role does not include development; that is, we are purchasing from them standardized products and services that require no significant adaptation in order to be suitable for use on our project. Our relationship with this outside company might be formalized by a document called a *purchase order*.

If you think about it, you will see that using such outside companies is similar in many ways to using your own company's functional departments:

- You write statements of work, and they bid against them
- They – not you – select the individual people
- They – not you – are responsible for finding the people their next assignment
- They might even be responsible for selecting the organizational model for their portion of the team.

In the company for which I used to work, about *half of our total expenditure* on a typical project came from such outside companies. As a result, every one of our projects had a significant functional aspect in its organization, even for those projects where our own people were projectized. My experience suggests that this is true in most companies; they acquire goods and services for their projects from other companies, and therefore almost every project is *actually* a hybrid organization with some functional elements.

When should you use which type of project organizational structure? Here are the strengths and weaknesses of these different project structures.

A projectized organizational structure, as compared to a functional organization structure:

- Is usually less expensive to establish and to operate.
- Entails more work and more risk for the project manager, as you try to get the project started; as you perform mid-project staffing actions as people leave the company, retire, or take other positions within the company; as you end the project.

- Can achieve stronger focus and higher levels of employee motivation.
- Can usually adjust faster to changes in requirements and market conditions.
- Can usually make faster decisions, because fewer people must be consulted. This seems good, but remember that *faster* decisions are not always *better* decisions.
- Can usually make faster decisions regarding poorly performing personnel and/or poorly performing vendors and subcontractors.
- Works well for long, complex projects.
- Maximizes the authority of the project manager, but also imposes the most responsibility (and the most work) on the project manager.

A functional organizational structure is more or less the opposite for each of the above (e.g. more expensive to establish and operate, less work for the project manager, often slightly slower at responding to change and making decisions). It achieves these effects by a sharing of authority and responsibility between the project manager and the functional department managers.

In hybrid organizational structures, we try to get the best of both worlds, picking and choosing a combination that we believe will be effective for our project and our company.

Therefore, you should design your organization based on where you believe:

- Your particular project needs emphasis
- You (as the project manager) have weaknesses
- There are particularly large risks
- Where the skills are located that are needed in order to do the work, starting with examining the question of whether those skills exist at all inside your company, or whether you must go outside the company for those skills. Not every company, even big companies, are good at everything. Be warned that companies often unreasonably pressure you to use internal sources.
- To take advantage of the skills of key personnel. For example, perhaps some key people work for a functional department, and are not willing to transfer to a projectized organization.
- How fast you expect changes to occur, and of what magnitude
- What will be desired or expected by your customer and your management. But don't let this factor unduly dominate; you must exercise your own judgement!
- ... and so forth.

Whichever type of organizational structure you select, you must make sure that it is clear which people have which responsibilities; the titles on the organization chart do not eliminate the potential for misunderstandings and disputes about authority and responsibility!

For example, your organization chart might have a box labeled *systems engineering* and a box labeled *software*; these are pretty typical entities on an organization chart for an engineering project. Which one of these managers is responsible for the development of the requirements specifications for the software? The box titles by themselves do not answer this question. I have seen projects assign this particular responsibility to one box or the other (although I prefer to assign this responsibility to the *system engineering* box). But the point is that the titles by themselves don't answer the question; you need some additional artifact.

Good projects therefore produce written *job descriptions* for each box on the organization chart, to address this problem. Of course, the mini-contract and its statement of work will answer this question too, but it is almost always socially useful to have written job descriptions corresponding to each box on the organization chart. People like to know their job responsibilities, and the responsibilities of the other people with whom they will interact.

Some projects also create a *responsibility assignment matrix*, which is a graphical depiction that maps project responsibilities along one axis and people/role names along the other. If, however, you have followed my approach for creating the work-breakdown structure – which includes assigning every element of the work-breakdown structure to a named person – you may not need to create a responsibility assignment matrix, because the names are already assigned to every element of the work-breakdown structure. But some projects follow a practice of *not* including names in the work-breakdown structure, in which case the responsibility assignment matrix is useful.

Since I prefer that you include names in the work-breakdown structure, you might wonder why we would also need to create an organization chart. First of all, the work-breakdown structure does not reflect our choice for the organizational structure (that is, projectized versus functional versus hybrid). We can depict that in an organization chart, by including annotations in a box indicating that the responsibility entailed by this box is assigned to a functional organization; the name of the individual selected by the functional organization to lead that work would also appear in the box.

Second of all, at times we will have the same person leading multiple items on the work-breakdown structure; we can also elect to group multiple elements of work from the work-breakdown structure into a single box on the organization chart.

6.4 The Project Plan

Projects require *written plans*; these plans are a key ingredient in achieving the alignment we talked about in Chapter 1. They allow the project to create a shared vision of what we are trying to do, how we will go about doing it, in what sequence we will perform activities, who will do what, what constitutes success, and so forth. Without project plans, we could not have a shared vision, and without that, we will *not* succeed.

The items that we have talked about so far in this chapter – the work-breakdown structure and the organization chart – become portions of the project plan. But there are many other things that require planning, in addition to those two items. Figure 6.10 shows a list of the subjects that might be covered in a project plan this table is long, and is therefore spread over several pages. The last column of the list tells you where in this book we discuss each item.

That is a long list! But you, as the project manager, do *not* prepare all of this material yourself. As we will discover in many other ways, your job is not to *do* the work, but instead *to ensure that someone else does it*, and that it is done according to standards and guidelines that you approve.

You will therefore organize a set of people to prepare the project plan. You will probably write a few of the earlier sections yourself, and you will read, review, make suggestions, and eventually approve the complete plan.

Figure 6-10: Typical contents of a project plan

Section title	Description	Reference
Overview	Summary of the project: What we are building. For whom. The objectives and scope of the project	Chapter 2
Business objectives	Our goals regarding profit, cash-flow management	Chapter 10
General approach	A short description of the approach to the work: What techniques and technologies will we use? Are there earlier products that we are enhancing? What work will be done by our company, and what work will be done by other companies? What are all of the work locations, and what work will be done at each of those locations?	Chapter 5
Identification of the customer and their value system	Who will be the eventual users of the products and services produced by our project? What do they need and want? What is their value system? Who are the other stakeholders for our project?	Chapter 4
Operation performance measures	How will we measure the goodness of our design and our products in a fashion that will be meaningful for those users?	Chapter 4
Desired differentiators	What are the areas in which we intend for our product to be better than other competing products? What are the actual parameters that we will measure? What are the actual values that we aspire to achieve?	Chapter 5
Project schedules	By what date must each item that we are producing be complete? Are there intermediate milestones that we must achieve? How do the various pieces of work depend on each other?	Chapter 7
Reporting requirements	To whom in our company will this project report? What information does that person expect, and at what intervals? Are there other corporate reporting requirements, (e.g. to the corporate finance department?) What is the organizational structure selected for our project? What is the organizational chart for our project?	Chapter 6
Deliverables plan	What must we deliver to the customer, and when? In what form? At what locations?	Chapter 6
Customer-furnished items	What material items, data, facilities, or services is the customer responsible for providing to us? When is each item due to be provided? In what form? Where?	Chapter 6

Figure 6.10 Typical contents of a project plan.

Section title	Figure 6-10: Typical contents of a project plan Description	Reference
Project financial resources	What is the customer-furnished budget for this project? What is the time-phasing of the availability of those funds? Are there additional financial resources (e.g. capital) being made available by the company, and if so, how much and when?	Chapter 7
Project personnel resources	Characterize the number, skills, and time-phasing of the personnel required. Are they being made available by one or more functional departments, or is the project undertaking its own staffing, or is some hybrid staff model being employed? Who, by name, is responsible for each portion of the staffing? Are there key personnel required, by name? At which locations will which people be located? Do any of these people require special certifications (e.g. accreditation or skill certifications), and if so, by when?	Chapter 6
Facilities	What are the facilities required by the project? Which are to be provided by our company? Which, if any, are to be provided by the customer? Which, if any, are to be provided by our subcontractors? By when is each facility required? Are there are any special certifications (facility security clearances, building permits, safety clearances, etc.) required for any of these facilities? How and by when will we achieve those certifications?	Chapter 6
Earned-value baseline	Who has the responsibility for the management of the initial creation of the project's earned-value baseline? How will this be accomplished? What are the milestones and deadlines that must be met? Who must approve the plans, and who must approve the final baseline?	Chapter 10
Technical plans and strategies	What are the top-level strategies and plans for each technical discipline on the project. For example: • Systems engineering • Software development • Integration • Testing • Fielding / deployment • Production • Quality	Chapters 2, 3, and 6

Figure 6.10 (*Continued*)

Section title	Figure 6-10: Typical contents of a project plan Description	Reference
Subcontract management plan	How will we manage the outside companies that are to provide us with products and services? Who has authority to approve each specific subcontractor and/or purchase order? Who is the manager assigned to each such outside company? What is the complete list of all such outside companies, what is their role, what are they providing, and what is the status (and plan for completion) of their subcontract and/or purchase order?	Chapter 14
Security plan	How will we protect the company- proprietary information developed by the project? How will we share or limit access to such data by the other companies participating on the project? Do any of our people require special security accreditations, such as government security clearances? If so, how will we go about obtaining such clearances, and by when?	Chapter 6
Communication plan	How will we communicate to all of the project personnel, both employees of our own company, and those employees of other companies who work on our project? Do we need specific permissions from those other companies to communicate with their employees? What mechanisms will we use, how often, and who is responsible for each? Do we need to obtain customer permission for communications that describe anything about the project, and if so (this is very likely), how do we obtain such permissions? How will we measure the effectiveness of communications? Who besides the project personnel do we need to communicate with? The buying customer? The users? The other stakeholders? The general public? How do we coordinate such communications with the buying customer and their requirements regarding such communications?	Chapter 13
Metrics plan	What items are we going to measure – technical, operational, and management measures –and how will we go about making those measurements, how will we process and analyze those data, how often will we do each measurement, and who is responsible for reviewing and approving each data item?	Chapter 8

Figure 6.10 (*Continued*)

Section title	Figure 6-10: Typical contents of a project plan Description	Reference
Preliminary bill of materials	What are the items that we anticipate purchasing outside of the company for use on the project, whether for internal use, or as part of a deliverable item? When do we anticipate purchasing each item? From whom? For what price? Do we have firm quotations? Who must approve each purchase? How do we perform receiving (and if appropriate) inspection of each received item?	Chapter 14
Project start-up plan	How will we start the project, (e.g. obtain the facilities, people, equipment, data, and other items needed to get the project going?) How is this time-phased? What are all of the project work processes that we will need? What form does each take, and by when must it be in place? Who is responsible for the creation of each, and who approves the installation of each? How are project personnel trained in the use of these work processes? How are they stored, and made available to project personnel? How and when do we create the various project systems (e.g. earned-value baseline, etc.) that we will use? Do we have time limitations from either the company or the customer on the creation of these project systems?	Chapter 12
Risk / opportunity management plan	What could go wrong on our project? How will we notice? What will we do in advance to lessen the likelihood and impact of such events? What are the opportunities to improve our project performance?How will we realize those opportunities?	Chapter 9
Tools plan	What tools and methods will the project use, for everything from routine (e.g. electronic mail, document archives, etc.) to more specific (e.g. software compilers and debug tools, software configuration management systems, etc.)? Who selects and acquires each tool or method? By when does each need to be in place? Who needs to approve each selection? Who confirms each installation as successful? How?	Chapters 2, 3, and 6
Safety plan	What are the risks to personnel, equipment, and facilities entailed in the project? How do we mitigate each risk? Who has responsibility for each? What outside authorities (e.g. city or corporate fire department, city building and safety department, Nuclear Regulatory Commission, etc.) must be involved, what is their role, and what approvals (if any) must they provide? By when must each such mitigation and approval be in place? Which facilities and locations are involved?	Chapter 6

Figure 6.10 *(Continued)*

Section title	Figure 6-10: Typical contents of a project plan Description	Reference
Team-building plan	How do we create alignment among the employees, corporate management, users, and other stakeholders for this project? Who is responsible for what portions of this activity? What is done at the beginning of the project, and what is done on a recurring basis? How do we measure the effectiveness of the team-building efforts? How do we incorporate the employees and management of the other companies who are working on the project?	Chapter 13

Figure 6.10 *(Continued)*

Then, you will provide that plan to all of your managers – everyone who appears by name in the work-breakdown structure and/or the organization chart, and direct that *each manager* at *each level* should create their *own* plan, that addresses just their portion of the project. Their "customer" is their direct supervisor, their deliverables are the items that their portion of the project must create, their schedule and budget reflect just their portion of the project, and so forth. Each such plan for a segment of the project must be approved in writing by the appropriate immediate supervisor. Each supervisor ensures that the set of plans prepared by their subordinates covers all of the responsibilities that are allocated to him/her by their supervisor, and so forth, all the way up and down the levels of the work-breakdown structure and the organization chart.

6.5 Your Role in All of This

The creation of the work-breakdown structure is usually done during the *business capture* and *proposal* (Chapter 5) effort, and in any case before you sign the contract and initiate work on your project. You will likely want to have your initial set of desired differentiators (Chapter 5), preliminary requirements and design (Chapter 2), and anticipated risks (Chapter 9) in hand before you prepare your work-breakdown structure.

As the project manager, you of course organize all of the tasks described in this chapter, select the people who will lead each task, and ensure that they then follow through on the execution of those tasks.

You will also likely personally make the decision about the structure of the work-breakdown structure (e.g. functional or component oriented) and the organizational structure (e.g. projectized, functional, or hybrid), although – as indicated earlier – your customer and your company may constrain your choices.

You will delegate much of the creation of these artifacts to your staff, but you should have a set of "hot-button" issues that you want to be able to see in the work-breakdown structure, most likely derived from the desired differentiators (chapter 5) and the risk register (chapter 9). You should insist that the work-breakdown structure be organized, even at the lower levels, in a manner that ensures that these items get attention and visibility.

Differentiators (Chapter 5), the design (Chapter 2), and the identified risks (Chapter 9) need to match up. For example, if one of your desired differentiators is for your computer system to achieve some fabulous input rate (e.g. it will be able to process millions of inputs per second), you certainly have some specific design features to achieve this, and you certainly have some entries in the risk register about what might go wrong, and (I hope – see Chapter 9) you have some specific mitigation plans to deal with those problems. You will then certainly also want to have *specific, visible tasks* in the work-breakdown structure that relate to these differentiators, design elements, and risks. This forces your staff to focus on these matters, and allows you, through the periodic management rhythm (Chapter 11), to see progress (or lack of progress) and to drive appropriate corrective actions. That is, the *work-breakdown structure must be informed and shaped by the technical aspects of your project*: your desired differentiators, the key aspects of the design that will provide those differentiators, and the risks that might prevent you from achieving them.

6.6 Summary: Organizing and Planning

Decomposition of an entity into a *hierarchy* is one of the core techniques of systems engineering, and for project management too. As we saw in Chapters 2 and 3, we perform systems engineering by decomposing the various aspects of the *system* into *hierarchies*: requirements, design, test, etc.

Similarly, we perform many aspects of engineering project management by decomposing the various aspects of the *project*; in this chapter, we saw that we create two important project artifacts, the *work-breakdown structure* and the *organization chart*, using the technique of decomposition into hierarchies.

Every project – large or small – needs a work-breakdown structure and an organization chart.

The work-breakdown structure depicts how the *work* is organized, and who by name is responsible for each piece of work. The work-breakdown structure is brought to a useful state by the negotiation and signing of *mini-contracts* between each pair of responsible people listed on the work-breakdown structure.

The organization chart depicts how the *people* are organized.

The work-breakdown structure and the organization chart are portions of a *project plan*. Every project needs an *overall project plan*, and furthermore, every manager at every level of the project needs a plan for their own segment of the project.

6.7 Next

Now that we have a work-breakdown structure and a project plan, we are ready to get into the details of creating useful project schedules, and from those schedules, project budgets.

6.8 This Week's Facilitated Lab Session

We have already done a couple of team sessions, one about the customer's coordinate system of value and the operational performance measures; another about proposals, desired differentiators, and how to win.

As we saw in this chapter, these are in fact *sections of the project plan*. During the remainder of the course, your teams will create additional sections of the project plan for your particular project.

If you are an *undergraduate* student, by the end of the course your team will create the following sections of a project plan:

- Overview
- The customer's coordinate system of value – operational performance metrics and technical performance metrics
- Proposals, proposal creation guidance (such as the "Heilmeier questions"), win themes, desired differentiators
- Work-breakdown structure
- Risk management. Special topics to include:
  - Create a risk register of at least six items for that project.
  - Create one mitigation plan for three of the items on your risk register, and two mitigation plans for the other three items on your risk register.
  - Create triggers for each of these mitigation plans.
- Project start-up
- The social aspects of the engineering project management role. Special topics to include:
  - *Staffing profile*. Create a staffing profile for your project. Discuss where it is feasible, and where it is not. Discuss what you might have to do to the activity network and other project artifacts to make the staffing profile feasible.
  - *People exercises*. Assign everyone on your team to a project role (e.g. project manager, software development manager, test manager, human relations manager, and so forth). Discuss how you will build effective interpersonal interactions within your team, and with your customer(s), boss, and other stakeholders.
  - *Customer/stakeholder exercises*. List and describe at least three different classes of customers/stakeholders for your selected project and its product. Create three operational performance measures for each that reflect their value system. For example, if you were building the Apple app store, the list of stakeholders/customers would include app developers.
  - *Boss exercises*. List and describe your boss's coordinate system of value. Create three operational performance measures that reflect his/her value system.
- Summary, conclusions, and recommendations.

If you are taking this course as a *graduate* student, by the end of the course your team will create the following sections of a project plan:

- All of the section listed for the undergraduate students
- Plus the following additional sections:
  - Deliverables plan, and a list of deliverables
  - Customer-furnished items
  - Facilities
  - *One* technical plan/strategy (e.g. systems engineering, software, etc.).

Your own instructor may, of course, modify this list.

Specific required products from each team:

1. *Written report.* Three to five pages per chapter (except for the overview and the summary, which can be shorter). Text and illustrations, not just bullet points. "Tell a complete story" (e.g. provide background, frame issues, present analysis, draw conclusions, and present lessons learned and recommendations for each of the main chapters). The written report will cover all of the topics listed.
2. *Briefing charts and oral presentation.* During the last week of class, your team will make oral presentations during class. 30 minutes per team. Every team member will present at least one chart. Each person is to introduce the person who comes after them: their name, and what they will present.

Your team must plan and organize all of this work. You will be provided additional facilitated lab sessions too, for each of the topics that are to be performed by both the undergraduate and graduate students. The four topics that are assigned only to the graduate students will not be the subject of a facilitated lab session; your team must plan and perform that work outside of class. Of course, you are welcome to talk to your professor about any questions you may have, or to request feedback on intermediate products as you progress.

The assignment for *today's* facilitated lab session is to create a work-breakdown structure for your team's project.

7

Creating Credible Predictions for Schedule and Cost: the Activity Network

In Chapters 2 and 3, I provided you with insight about some of the key factors regarding how we do engineering on projects. We will now use that knowledge as I start discussing the processes that we use for performing actual project management on our engineering project. In this chapter, I focus on the *activity network*, which allows us to make credible predictions regarding the schedule (that is, how long it will take us to do all of the work entailed in our project). We will also see how this same activity network is an essential first step toward estimating how much our project will cost. Predicting schedule and cost in a credible fashion are among the basic expectations for a good engineering project manager.

7.1 Setting the Stage

In this chapter, we will talk about how we go about creating credible predictions for two items: (i) how long it will take for our engineering project to complete all of the work entailed by the project's statement of work, specifications, and contract (we will call this the *schedule*); (ii) how much money it will take to complete all of the work entailed by the project's statement of work, specifications, and contract (we will call this the *cost*).

Notice that the information that people will want to know about schedule and cost are *predictions* about a future state. Yes, your boss and your customer will in fact want to know how much money you have actually spent thus far, but they will be far more interested in how much money you *predict* that you will need to spend in order to complete all of the work entailed in the project (a figure called the *estimated cost at completion*). When we say we need to estimate the cost of a project, this is what we almost always mean: our *prediction* for the *estimated cost at completion*. Similarly, when we say we need to estimate the schedule for our project, we almost always mean our *prediction of the date by which the project will be completed.*

First of all, let's discuss why we need these predictions.

The first reason is because, as the project manager, you will *constantly* be asked "When will the project be completed? How much will it cost when you are done?" Both your *internal* customers (e.g. your boss, his/her boss, and so forth all the way up to the president of your company) and your *external* customers (e.g. the buying customer, the users, the other stakeholders, etc.) will ask questions like these all the time.

Engineering Project Management, First Edition. Neil G. Siegel.

Companion website: www.wiley.com/go/siegel/engineering_project_management

In fact, you won't be allowed to start your project until you have answers to these questions that seem both *satisfactory* and *credible.* The answers you provide will be deemed *satisfactory* if the customers are willing to wait that long and to pay that much, and those same answers will be deemed *credible* if they are believed to be approximately correct.

There are other reasons (e.g. beyond the need to answer these constant questions) that inform why we have to create these predictions:

- We have to make efficient and appropriate use of all the people on our team; we can't have them sitting around on the payroll, just waiting for their particular task to start.
- We need to be clear to our partner companies and suppliers when we need the items that they are supposed to provide.
- We need to understand the inherent sequencing and dependencies of our project's activities – sometimes, something cannot start until something else is completed.
- We need to plan and organize people, equipment, and facilities:
  - Ensuring that we have access to the people that we need, at the time and place that we need them. We must also *load-level* the number of people that we will need (we will explain this term later in the chapter)
  - Ensuring that we have access to the facilities and equipment that we need, at the time and place that we need them
  - Timing the activities that need certain data or equipment so that those items are in fact available
  - ... and so forth.

One of the items that we are going to learn in this chapter is that our predictions about *cost* depend very significantly on our estimates for *schedule.* As a result, we *always estimate schedule before we estimate cost.* This is because, in order to estimate the project's schedule (e.g. the predicted completion date), you must define the sets of tasks to be accomplished, you must estimate how long you believe that each task will take, and you must decide in what order you will perform those tasks. We will then discover that you need all of that *same information,* plus some additional information, in order to estimate the cost (that is, the predicted cost of the project when it is finally completed). Said another way, you cannot credibly estimate the cost without having *first* credibly estimated the schedule.

Most books on project management talk about estimating *cost and schedule.* I prefer to talk about *schedule and cost,* providing a constant reminder about which item must come first.

The reasons given above for why we try to estimate schedule and cost seem pretty important, so you might expect that we already have methods in place that make schedule and cost predictions which consistently turn out to be pretty good. But that is *not true at all*; many – in fact, most – engineering projects have predictions for schedule and cost that turn out to be *wildly inaccurate.*

Why is that? My experience suggests that projects make the same mistakes over and over again:

- They plan around *fixed dates.*
- They fail to recognize the huge impact of statistical variation on the final predictions.

- They fail to calibrate their estimate for each task against actual past performance for similar tasks.
- They fail to recognize the extreme variation that can be introduced into their schedule and cost estimates by seemingly small changes or defects in the design; we introduced this subject already in Chapter 2.
- They make unjustifiable assumptions about the improvements in productivity that will result from the use of new, relatively untried methods and tools.
- They make the unreasonable assumption that everything will go well...

Let's look at the first item on this list: planning around fixed dates. We do this in our personal life all the time. If we have a task (say writing a report with a team of 10 people at work), we will say to ourself:

- On 1 April, we will start
- By 1 May, we will have finished this initial stage of the work
- By 1 June, we will have finished this next stage of the work
- By 1 July, we will have finished this further stage of the work
- And by 1 August, we will be done and ready to turn it in!

But what actually happens is something like this:

- Friday 1 April was a state holiday, celebrating April Fool's day, and the company gave everyone that day off. Our company in fact also shut down on the following Monday, using the resulting long weekend for some long-deferred facility improvements. Work did not actually start until Tuesday 5 April.
- 1 May arrived anyway. Would you even *notice* that you did not finish all of that initial stage of the work? Do you have a definition of what comprises that initial stage that is rigorous enough to allow you to tell if that work is completed or not?
- Did you understand all of the implications for when certain follow-on tasks can start (e.g. what tasks depend on the prior completion of other tasks)? Just because a stage of work was supposed to start on 1 May does not mean that it is actually possible to start that work on that date.
- Did you understand all of the implications about staffing? Will someone have to sit around for 10 days, waiting for someone else to finish something? Will a piece of work go unstarted for 15 days because the people scheduled to start it are still working on other tasks?
- Will a task have to be unexpectedly delayed for five days because a critical piece of test equipment and a special laboratory (of which your company has only one) is being used for longer than expected by someone else?
- If you even notice these problems, can you make a credible prediction about their impact, that is, make a credible prediction about when you can now expect to complete the work? Is there something that you can do to get back onto the original schedule, that is, get the work completed by the originally promised completion date, despite the problems cited above?
- And what is the impact upon your predictions for the cost at completion of the project, given all of the delays to the completion date? People and the rent on the building have to be paid every week, so these delays are almost certainly increasing the amount of money it will take to complete our little project.

Do you recognize that people plan activities against such fixed dates all the time? Here's another example: universities (and parents!) predict that the typical undergraduate engineering student will complete their degree in four years, and estimate the cost of their college education accordingly. But not nearly all of them actually do so; some studies indicate that fewer than 20% actually complete their undergraduate studies in four years. Making schedule predictions around fixed dates leads us to poor predictions.

7.2 Estimating the Schedule For Your Project

How can we learn to create these estimates for schedule and cost better?

- Instead of planning around *fixed dates*, we should instead plan around *tasks*, their *durations*, and *their interdependencies* ... and then *derive* the anticipated completion date.
  - This is the *first major shortcoming* of many project schedules: they plan around *fixed dates*.
- Furthermore, all statements about the future should be expressed statistically ... so we really need a way to make *statistically based predictions about the completion dates*.
  - This is the *second major shortcoming* of many project schedules: they do not account for the statistical variations.
  - The difference can be very significant, as we will see through our examples and our facilitated lab session.

Here is the basic recipe for a credible schedule prediction:

1. Define the tasks.
2. Identify the interdependencies between the tasks.
3. Estimate the duration of each task, in a *statistical* fashion.
4. Only the initial task in each chain is fixed to an actual calendar date; all other dates are derived from the task interdependencies and the statistically expressed task durations.

A schedule that is built in accordance with these four principles is called an *activity network*. Good project schedules are *always* built as activity networks.

Let's walk through each of these steps.

7.2.1 Step 1: Define the Tasks

Here is another example of the use of *decomposition* of an entity into a *hierarchy* as a technique: we want to break the work of our project into a set of smaller tasks. As it happens, we have already started to do that, through our creation of the *work-breakdown structure*. In the work-breakdown structure, we need only go far enough to define the work, and assign each resulting element of work to a responsible individual (and capture that agreement in a mini-contract). For the purpose of creating the project schedule, we may want to break the work into even smaller pieces, which I call *tasks*.

How small? My preference is to decompose the elements from the work-breakdown structure to a point where each task is *no longer than one or two months in duration*; we will see why I chose this duration when we get to Chapter 10. If your project is large and long (e.g. years in duration), breaking it down *before* the commencement of the project into tasks that are one to two months in duration may seem like too much work, and you may be concerned about expending so much planning work on tasks that are still years away when changes may occur that would render some of that planning work obsolete. In Chapter 10, I will also show you how to resolve this dilemma. For now, let's think in terms of tasks that are just one to two months in duration.

7.2.2 Step 2: Identify the Interdependencies Between Tasks

What do we mean by the term *interdependencies between tasks*?

Consider two tasks, A and B. One situation is that one of these tasks might be dependent upon the other; for example, task B requires something as input that is to be created by task A. Under that circumstance, you could not start task B until task A has completed. We say that task B is *dependent* upon task A. We can create a graphic depiction of this situation, representing each task with a little box and using an arrow drawn between the tasks to indicate the dependency relationship (see Figure 7.1). This depiction implies a *time dimension*, flowing from left to right.

Another situation, however, is that these two tasks might have nothing to do with each other, and neither might require as their own input any products from the other; we then say that these two tasks are *independent*. In which case, the date for which each task is expected to be performed can be established without reference to the other. Task A could be performed before task B, after task B, or at the same time as (we say *in parallel with*) task B. If we were to create a graphic representation of this situation, with each task again represented by a little box, we would show this *independence* by *not* having an arrow connecting task A to task B, and not having an arrow connecting task B to task A (see Figure 7.2). This depiction has a time dimension too: it is possible that these tasks could be performed at the same time.

These task-to-task dependencies will take various forms, but we will start with the simplest: the one we described above, *finish-to-start*, that is, task B cannot start until task A is completed ("finished"). There need be no time lag between the completion of task A and the commencement of task B; task B can start immediately upon the completion of task A. We say that task A is a *predecessor* of task B, and that task B is a *successor* of task A.

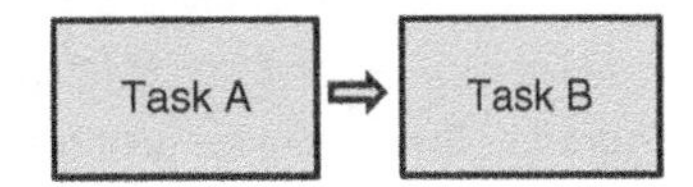

Figure 7.1 Depiction of two tasks, where the second is dependent upon the first.

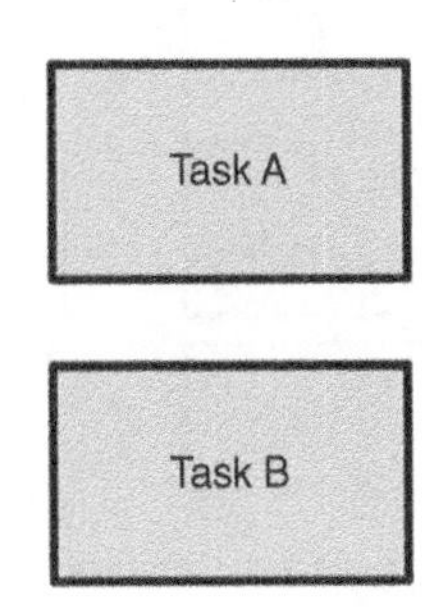

Figure 7.2 Depiction of two tasks, where the tasks are independent of each other.

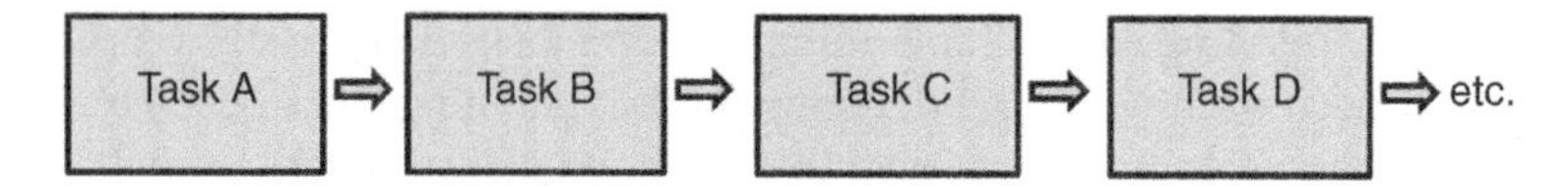

Figure 7.3 Graphical depiction of a chain of tasks.

This leads to the ability to create a *chain* of tasks, with the commencement of each successor being dependent upon the completion of its predecessor (see Figure 7.3).

Now, let's introduce a next level of complexity: instead of there being a *single* task that is awaiting the completion of a predecessor task, there might be *two or more such tasks* that are waiting for task B to complete before they can start (see Figure 7.4). In this situation, we say that task B is the predecessor of all three tasks, X, Y, and Z.

Now we are ready for still another level of complexity: similarly, there might be two or more tasks that must *finish* before a given task can *commence.* We say that tasks L, M, and N are all *predecessors* of task C, and that task C is a *successor* of all three tasks, L, M, and N (see Figure 7.5). Topologically, this is all of the complexity that we need in order to represent *any* schedule.

You can create more complicated dependency relationships, other than the *finish-to-start* relationship that we have been using. For example, you can have dependency relationships such as these:

- Start-plus-N-days → start
- Finish → finish
- ... and so forth.

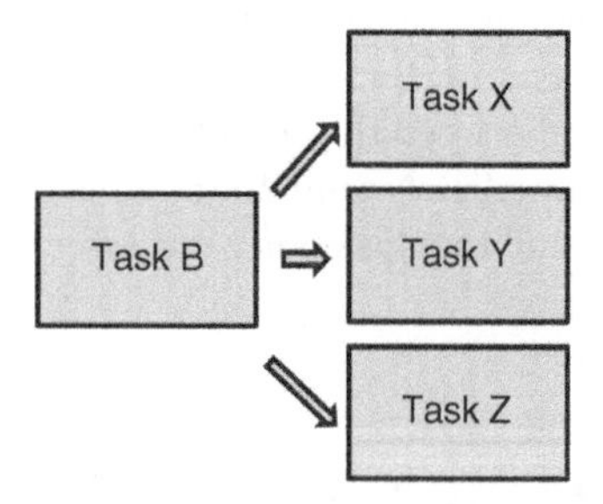

Figure 7.4 The commencement of multiple tasks might be awaiting the completion of task B.

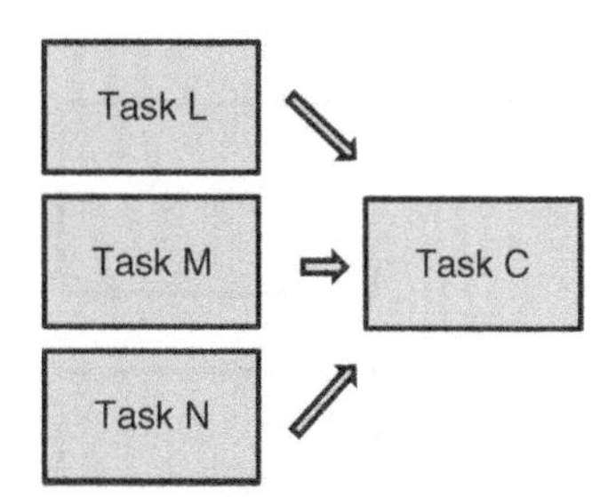

Figure 7.5 One task might be waiting upon the completion of multiple tasks.

But you actually *do not need* these other types of relationships ... these additional dependency relationships can in fact all be represented by finish-to-start relationships, by breaking one or more of the tasks involved in these more complicated relationships into their component parts. We will examine this aspect in Chapter 10.

With the above tools, you can build the *activity network* for any project. Even only using these simple finish-to-start relationships, one can build quite a complex sequence and relationship of tasks.

Let's look at a simple example. Consider Figure 7.3 again for a second, and assume that each of the four tasks (A, B, C, and D) are estimated to take 10 work days. Then we can calculate that the duration of the entire chain is 40 work days. If we then pick an actual calendar date for the commencement of task A, and make proper allowance for there being only five work days per week (and also accounting, of course, for company holidays and so forth), we could calculate the actual calendar date on which we predict that we will finish the work depicted in Figure 7.3.

This example is simple enough that we could do the calculations ourselves. But real project activity networks are usually too complicated for us to perform these calculations without making a lot of mistakes. So, instead, we use a *computer program* to make these calculations for us. In my class, we use a computer program called Microsoft© Project for this purpose, but there are literally dozens of such *scheduling programs* available on the marketplace.

7.2.3 Step 3: Estimate, in a Statistical Fashion, the Duration of Each Task

In order to create a credible schedule, we not only need to estimate the duration of each task, but we need to do so in a *statistical* fashion. What does this mean?

In our simple example above, we said that task A in Figure 7.3 is estimated to take 10 work days. But that estimate might be *wrong*! When we actually go and perform the work entailed in completing task A, it might turn out that it only takes 6 work days to finish the work. Or it might take 26 work days. Or any of an infinite variety of other potential durations.

It is a well-established principle that when there is uncertainty involved in an activity, better predictions can be made if *that uncertainty is taken into account*, rather than ignored.

It may be hard to believe, but in actual fact, most people truly do *ignore the uncertainty*! This is one of the main contributors to the poor predictive abilities of most project schedules. They assume that the estimate of 10 work days to complete task A is perfect. In probabilistic terms, we would say that they assume the probability of completing task A in 10 days is 100%.

But, of course, that probability is actually less than 100%. So, we can improve the quality of our predictions by explicitly representing this uncertainty in the estimation process. Fortunately, there is a reasonably simple way to do that, which we call *three-point estimation*. Instead of creating a single estimate of the duration of task A, we create three:

1. An *optimistic* estimate of the task's duration
2. A *nominal* (often called *most likely*) estimate of the task's duration, and
3. A *pessimistic* estimate of the task's duration.

	estimated duration, work days	probability of occurrence	duration * probability
optimistic	6	20%	1.2
nominal	10	30%	3
pessimistic	26	50%	13

17.2 work days
Expected value (e.g. a weighted average)

Figure 7.6 What we learn from our example three-point estimate.

We also estimate the probability of occurrence for each of these three durations, so that these probabilities add up to 100%. Here's an example:

1. Optimistic estimate of duration for task A: 6 work days, 20% probability
2. Nominal estimate of duration for task A: 10 work days, 30% probability
3. Pessimistic estimate of duration for task A: 26 work days, 50% probability.

A statistician might say that we need many more than three points to make a good estimate, but let's look at the effect on our overall schedule of just this simple three-point estimate. Look at Figure 7.6; I have used the values from the above example to calculate a weighted average of our three estimates for task A.

Notice that the weighted average turned out to be just over 17 work days, more than 1½ times as long as our original single-point (or nominal) estimate. And when we have a complicated schedule, with lots and lots of task interactions, delays in earlier tasks cause delays in the start dates for later tasks, which forms an additional source of overall delay to the project's actual completion date. Project schedule predictions that use three-point estimating are usually *two to three times longer in duration* than project schedule predictions for the same project that use only a single estimated duration for each task.

Technical Note

Some project scheduling computer programs simply calculate the expected value of the task duration (as in Figure 7.6), using the weighted average that derives from the three-point estimates, and then calculate the predicted end date by using that expected value as the task duration. Better project scheduling computer programs do something different: they perform what are called *Monte Carlo simulations*, using each of the three values for the task duration in different iterations of the simulation, with frequencies determined by the assigned probabilities. The Monte Carlo simulation approach is significantly better. Be sure that you take the time to understand how your computer-based tools work! Figure 7.6 was intended only to show an example of the impact when we account for the uncertainty, and not intended to be a methodology for incorporating that uncertainty into a predicted schedule. The proper way to incorporate the uncertainty into a schedule prediction is through the use of a Monte Carlo simulation.

Therefore, we can see what results from ignoring the uncertainty: your estimated project schedule is completely wrong, and wildly optimistic! Remember, we are going to use this schedule estimate to estimate the cost of the project too, and if the schedule estimate is wrong and wildly optimistic, the cost estimate will share those characteristics.

There is another important point to learn from this simple example: asymmetry. Notice that the difference between the estimated optimistic and the nominal durations is 4 days, but the difference between the nominal and the pessimistic durations is 16 days. Why is this? In real life, distributions are seldom *symmetric*; that is, the *probability of being early is seldom as large as the probability of being late.* Think of airline flights: there are hundreds of reasons why they might arrive late, but very few reasons why they might arrive early. The plane will *never* take off early (it will not leave the gate until its posted departure time), but may often take off *late* (mechanical troubles, pilots or other crew members get sick and we have to wait for a replacement to arrive, weather problems, long lines queuing up for the runway, and so forth). The duration of the flight in the air may be more or less than planned, mostly depending on wind conditions, but mechanical problems can contribute only to a longer duration of travel. And once you land, even if you are early, the arrival gate may not be available, and so forth. I once had a 3-hour flight arrive 26 hours (that's 1560 minutes) *late* (really!), but I have never had a 3-hour flight arrive and deplane more than about 15 minutes early; the *asymmetry* is more than a factor of 100! Note that there can be asymmetry in both the *deviation* from the nominal duration and the *probability* of the variance: the pessimistic condition is both a *bigger deviation* from the nominal condition than in the optimistic condition, and *more likely to occur* than the optimistic condition. This is *Siegel's golden rule of creating three-point estimates*: they almost always should be weighted toward the pessimistic, in both *deviation* and *probability.*

Once we add in the nuance of using three-point estimation, you will definitely want to use a computer program to calculate the estimated duration of your project; it is just too easy to make a mistake doing it manually.

Using a computer program also makes it relatively easy to *adjust* the schedule; no project ever gets all of the tasks, their estimated three-point durations, and their interdependencies correct the first time ... all of these items must be continuously reassessed, and the activity network updated as the project progresses. After all, you will learn a lot more about the project as you progress, and so your activity network (and therefore the predictions that it makes) get better as the project progresses too.

7.2.4 Step 4: Fixed Dates vs. Derived Dates

A vital aspect of an activity network is that *only* the very first task start date in each *chain* is fixed to an actual calendar date; *all* of the other dates in that chain are *derived* from the sequencing and the durations of each task. This is the essence of credible schedule prediction; you actually let the *data make the prediction*, rather than trying to force things to a desired end date.

We now know how to create an activity network! We will talk about the process of *updating* the activity network in Chapter 10.

7.2.5 Examples

I have created a very simple activity network – just 10 tasks, arranged in two chains with some links between the chains. The network is depicted using single-point estimation (that is, based only on the nominal task durations) in Figure 7.7, and with three-point estimation in Figure 7.8. This is still a pretty simple example, and yet the three-point estimation version is about 50% longer in overall duration. Both start on 2 March; the single-point estimation version predicts completion by 9 May, but the three-point estimation version predicts completion not until 2 June. If you ignored the uncertainty, and used only single-point estimates, you would likely have seriously underestimated the

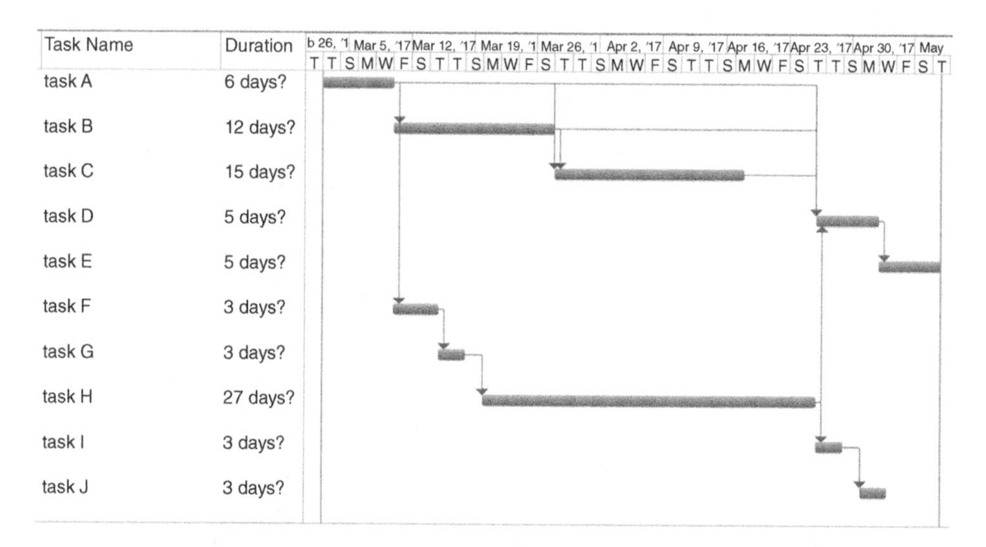

Figure 7.7 The activity network for a simple project, estimated with single-point task durations.

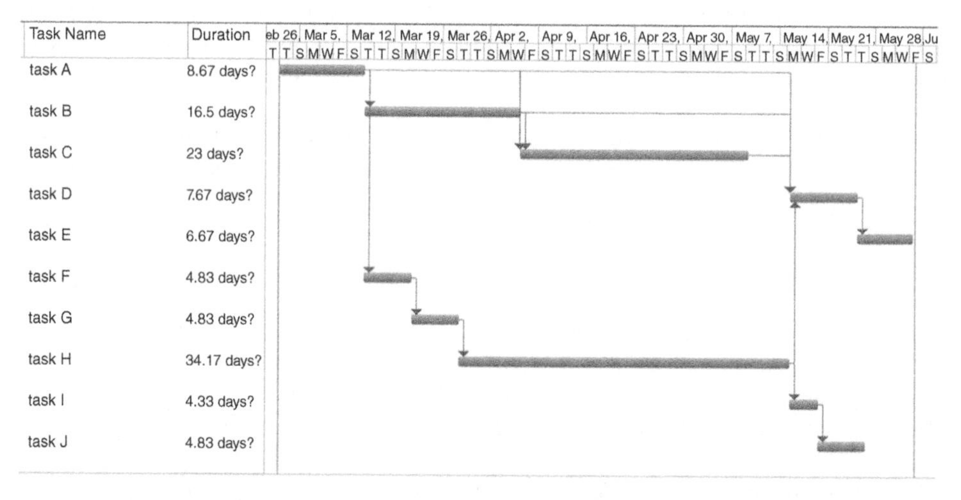

Figure 7.8 The activity network for the same simple project, but this time estimated with three-point task durations.

duration of your project. The larger and more complex your project, the larger the difference is likely to be!

To summarize what we have discussed so far: here (again) is the basic recipe for creating credible project schedules:

1. Define the tasks (ideally, each task is no more than one to two months in duration).
2. Identify the interdependencies between the tasks (e.g. the finish-to-start relationships).
3. Estimate the duration of each task, in a *statistical* fashion (using the three-point estimation methodology).
4. Only the initial task in each chain is fixed to an actual calendar date; all other dates are *derived* from the task interdependencies, and the three-point task durations.

I called a schedule that is built in accordance with these four principles an *activity network*. Good schedules for engineering projects are *always* built as activity networks.

7.3 Estimating the Cost of Your Project

Employing the approach described above allows you to create credible estimates for your engineering project's *schedule*. Remember how we said that cost is in large part derived from the schedule. Therefore, from these schedules, we can create credible *cost estimates*. Here's how:

Our project schedule consists of *tasks*, each with a *predicted start date* and a *predicted duration*. To create a credible cost estimate for your project from this information, we do the following:

- This task is somewhere on the work-breakdown structure, and the work-breakdown structure names the *assigned manager* for that piece of the work. Remember though, we said that a task on the schedule may be a *subset* of an element on the work-breakdown structure – because we may have broken that work-breakdown structure element into smaller pieces in order to have tasks that are no more than one to two months in duration (we did not impose a time limit on the duration of work-breakdown structure elements). So, the responsible manager for this particular task may be someone who reports to the person named in the work-breakdown structure. Every task must be assigned to a single responsible manager.
- That responsible task manager is, of course, the person who actually generated the information about this task for the project schedule: what tasks must be predecessors to this task (because this task needs their inputs), what artifacts will this task generate (so that other task managers can determine if this task must precede their task), and what is the estimated duration of this task.
- Now, that same responsible task manager generates some additional information, starting with an estimate of *how many people* will be needed to do the work entailed in the performance of this task. This is a count of people by time period (usually week or month[1]), segmented by *discipline* (e.g. mechanical engineer, software programmer,

1 Which leads to a unit usually termed the *man-month* (e.g. one person working for one month). Dr. Fredrick P. Brooks wrote a wonderful essay on the pitfalls of the man-month as a measurement method, which is contained in his excellent and worthwhile book *The Mythical Man-Month*, published by Addison-Wesley in 1975. Because of the age of the book, some of the examples are dated, but the main points in the book are still timely and important. I always recommend this book to my students.

etc.), *skill level* (e.g. expert, beginner, etc.), and *special experiences required* (e.g. must know the C++ programming language, must know how to machine titanium, etc.). If, at the time that the task manager is preparing this information, he/she already knows who some or all of these people will be by name, those names are provided too.

- The project will have a *business manager*, someone who handles the business matters of the project, including the financial accounting. On a large project, this will be an entire team of business specialists. Each task manager will have a business specialist assigned to help them. That business specialist will have access to information about how much per hour each type of person (by discipline, skill, and so forth) is likely to cost, and of course also has access to the actual salaries of the people who were identified by name (in most companies, the task manager will not have access to actual salary information; this is considered private information, and only a few specialists have access to this type of information). So, this business specialist can turn the information about the number and types of people into a *dollar value.*
- Companies and projects have lots of other costs, in addition to the salaries of the people who do the work. Those people, in addition to salary, likely receive *benefits* (e.g. some combination of vacation, sick pay, holidays, pension, medical insurance, etc.), and the cost of those benefits must be accounted for as we turn the schedule estimate for this task into a cost estimate for this task. Furthermore, they probably have an office or laboratory space, a computer, a telephone, and so forth – those must all be paid for too. The business manager is the person who then adds in all of these costs. In most companies, these types of costs are called *indirect costs*, and the company's accounting structure calculates them as a percentage of salary, averaged over the entire work force. For example, benefits might cost an additional 50% of the amount of salary. And the building, rent, desk, computer, phone, air conditioner, and so forth might be an additional 40% of the amount of salary. The business manager will know how to do this for your project, and for your work locations (some of these costs are likely to vary by work location).
- The task may also entail the purchase of materials and parts, may involve travel, or may in other ways incur additional costs that are not directly relatable to salary. Such costs are often called *other direct costs*. The task manager must identify the materials, parts, and travel (start and end cities, and the approximate duration of each trip), but it is the business specialist (and not the task manager) who turns these estimates into dollars.
- Companies have people (e.g. the president of the company, his/her secretary, and so forth) and business functions (human resources, the law department, etc.) that do not work directly on the project. But these costs must be paid for too; every project in the company pays a pro-rated share of these costs. The business specialist figures out the appropriate pro-rated share of these costs for your task.
- The sum of all these items forms the *cost* of the task; that is, what it will cost the company to do the work entailed in the task definition.

We have now converted the schedule for one single task on our project to a cost estimate for that single task.

The project's business manager can now add up the cost estimates for every task on the project, add in an additional amount of money in order to create a pool of resources that can be applied to correct problems that arise (we will call this pool of resources *management reserve*, and we will introduce this concept later in this chapter, and discuss it in detail in Chapters 9 and 11), and then we have a *cost estimate* for our entire project.

The above only covers the *costs* of the task; it does not leave anything over that would allow the company to make a *profit*. When we add the *profit* to the *cost*, we obtain the *price* – that is, what we actually need to charge the customer for this work. We will discuss profit and price separately at the end of this chapter.

7.4 Injecting Realism Into Your Estimates

7.4.1 The S-Curve

Of course, just like for the schedule estimate, the cost estimate has *uncertainty*; it might be wrong. So, we create *three-point estimates for cost*, similar to the way we created three-point estimates for schedule. In fact, since we determine a portion of our cost estimate by the *duration* of the task (duration × number of people × hourly cost of each person), we use the three estimates of duration (remember: these are *optimistic*, *nominal*, and *pessimistic*) to develop three cost estimates. But remember that (duration × number of people × hourly cost of each person) is only a *portion* of our cost estimate; there are all of the other elements of cost that we just discussed too. There is uncertainty regarding all of these other elements of our cost estimate, such as the number of trips to various locations required to perform the work, the quantity of materials and the unit cost of each material, and so forth. The responsible task manager must create *three-point estimates for these other items* too (e.g. the number of trips between Los Angeles and Washington D.C. entailed in the performance of the task), and the business manager puts all of these items together into a *Monte Carlo simulation* of the cost. The result is a range of potential project cost estimates, each with a probability of occurrence. This might be represented in a graphical form called the *S-curve*, as in Figure 7.9.

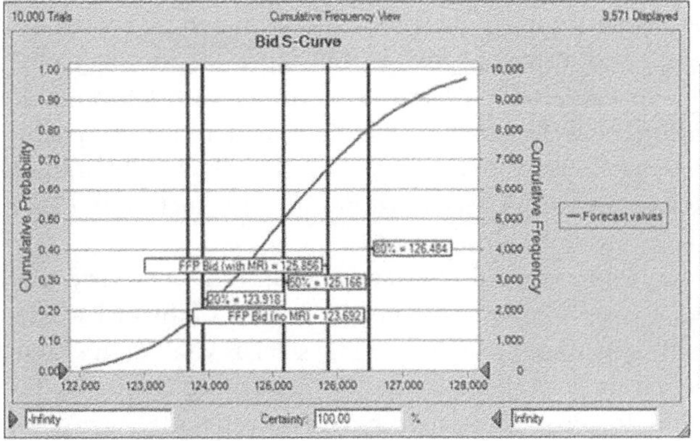

Forecast:	Bid S-Curve
Percentile	Forecast values
0%	120,336
10%	123,272
20%	123,918
30%	124,400
40%	124,812
50%	125,166
60%	125,573
70%	125,985
80%	126,484
90%	127,196
100%	130,866

Figure 7.9 Example of an S-curve.

How do we read the S-curve? The curve shows something called *cumulative probability* (Y, or vertical axis) plotted against *likely project cost* (X, or horizontal axis). Let's start by looking in the lower-left-hand region of the graph: find the point on the X-axis that shows "123 000." Look upwards to the curve immediately above that point on the X-axis; you see that the curve at that point on the X-axis is around .10 (e.g. 10%) on the Y-axis. The Y-axis is labeled "cumulative probability." So, what the curve is telling us is that the chance that the work of this project can be done for $123 000 *or less* (that is the *cumulative* aspect) is only 10%; there is a *90% chance that the project will cost more than $123 000 to perform.*

You can see by looking toward the middle region of the graph that we don't reach a 50% probability until the estimated project cost is just a little more than $125 000.

If you want to be confident in your cost estimate, you look for the place where the shape of the curve starts to *flatten out* (this is in the upper-right-hand region of the graph). In this example, that occurs somewhere between $126 000 and $127 000, and you can see that there is an 80% chance that the project will cost $126 500 or less to perform.

The table on the right-hand side of the figure lists the project cost at every 10% increment of increasing cumulative probability.

Of course, your business manager can make you an S-curve about the schedule too, and show you the probability that you will complete the project within various durations.

You can think of the methodology I have described in the above as a *bottom-up* cost-estimation methodology; we call it *bottom-up* because we start from the bottom, the lowest levels of the work-breakdown structure and the definitions of the individual tasks, and then aggregate our way up to an estimate for the entire project. *Every* engineering project makes such bottom-up cost estimates; in fact, as we will see in Chapter 10, you will be *updating* this bottom-up cost estimate (and the bottom-up schedule estimate too) on a periodic basis (most likely, once every month) throughout the duration of the project.

7.4.2 Another Aspect of Realism in Schedules: Margin and Slack

In Chapter 2, we discussed the importance of margin in the design: not designing our system so that it must operate at the very edge of its technical parameters. We used the example of a spacecraft: if the rocket booster is capable of lifting 1000 pounds to the correct orbit, we will design the spacecraft for a lesser weight, perhaps 950 pounds. That way, if things go wrong during the design and some articles end up weighing a bit more than planned, we can still deliver a spacecraft that weighs less than 1000 pounds and can therefore be lifted to the correct orbit. We must design with such *margin* for every technical and operational performance parameter.

In the same way, we must design our schedule to have margin too: not every task should abut against its neighbors; there should be gaps between many of the tasks. This schedule margin is called *slack*; the existence of such schedule slack is vital, in the same way that the existence of design margin is vital: it allows us to recover from the inevitable small delays in task completion. It is one of our cardinal roles, as the manager of an engineering project, to ensure that our schedules incorporate an appropriate amount of slack along each separate chain of tasks that comprise the schedule. Many of the software programs that are available to lay out your project schedule can calculate the slack along each chain within the schedule.

7.4.3 Calibrate Against Top-Down Estimation Methods

But even with the three-point estimates and the Monte Carlo simulations, most companies, and many customers, do *not* consider the bottom-up methodology for schedule and cost adequate by itself. This is because these bottom-up estimates *may not be appropriately grounded in past experience*, and therefore they may not be *realistic*. My experience is that people who are creating schedule and cost estimates for their task tend to be *optimistic*, and therefore they *underestimate* the time and cost of their task. How can we introduce some realism into the estimating process?

Fortunately, methods to accomplish this do exist. These methods are based on the idea that even if your project is creating something entirely new, most of the tasks you will be doing are in fact something that is *similar to work that has been performed on a previously completed project.*

For example, your project might be planning to use a radical new sensor; perhaps some new improved type of radar or global positioning system (GPS) receiver. Your system will then probably incorporate an algorithm called a *Kalman filter*, which processes the measurements from sensors. Your project involves the use of a new sensor, so your Kalman filter is a little different from all previous Kalman filters, but it is *not entirely* different. And since lots of people (including, most likely, your company) have built Kalman filters (these days, usually as a software program) in the past, you can start your estimating process from the baseline of real, past experience: how many lines of software code are there in the typical Kalman filter? How many man-months of effort did each previous implementation of a Kalman filter take? What requirements and features seem to make some Kalman filter implementations cost more than others? And so forth.

This is a *top-down* estimating approach; we are starting from the *actual data* from completed, previous *entire* projects, and using segments of those data to tell us information about actual productivity for both the entire project and parts of that project.

Good companies actually keep archives – organized as libraries – of information from previous projects (both successful and unsuccessful!). What was the original estimate for schedule and cost? How long did it actually take to finish? What did it cost when we actually finished? How many people did we bid? How many people did we actually use? These archives ought to be at the task level (or at least, at the lowest level of the work-breakdown structure), so that you can see the details for individual pieces of work. You can then go into that archive and find six previous projects that built Kalman filter software, find out what each bid for schedule and cost, and find out what each actually entailed, in terms of schedule and cost. There ought to be lots of other information (e.g. what went wrong, etc.) in these archives; we will return to the subject of historical project archives in Chapter 9.

What we actually do then is to compare the duration and amount of effort by labor category for each task that we have estimated in our bottom-up estimate against this library of historical actuals (man-months, calendar-months, and cost) for real previous tasks on real previous projects. If our bottom-up estimates for this task are materially different from the historical data, we need to *adjust those bottom-up estimates to bring them more in line with the historic data* (this is what brings realism into the estimating process!), and/or justify those differences. My advice is to beware of these justifications! Historical actual experience is almost always more accurate than uncalibrated bottom-up estimates, which tend to be wildly optimistic.

The historical actual results in these project libraries will also include what are called *cost estimating relationships* (e.g. what percentage of a task typically goes to systems engineering, to software development, to business management, to quality assurance, and so forth). The aggregated bottom-up estimate for your entire project should be compared to these historical experiences of what percentage of your project budget should be spent on each discipline too, and any significant deviations either corrected or justified. This will help you catch areas that you forgot in your work-breakdown structure and in your estimation process.

We also do a second type of top-down estimation as well: we build and operate *parametric models of schedule and cost* for the work entailed by our project. The use of such parametric cost models for engineering projects was invented by Dr. Barry Boehm,[2] and is now used on almost every engineering project in the world.

Parametric, top-down models of schedule and cost work like this: you create and input estimates of the size of the task (software lines of codes, number of requirements, number of parts, etc.) and choose values for various parameters (this is the *parametric* in the model title) about difficulty, development environment, personnel skills, and so forth, and the model produces *estimates of the schedule and the cost* (the cost will be in man-months, not in dollars). The creators of the model have gathered data from hundreds, or even thousands, of past projects, and built and calibrated extensive *data tables drawn from that real experience.* The model uses those data tables to estimate the schedule and effort involved in *your* project. Your business specialist then uses salary data and other company cost data to turn those estimates for labor effort into estimates of dollar cost.

The use of these two types of top-down modeling – both *libraries of historical data* from actual, previous projects (and it is vital, by the way, that such libraries include *every* project undertaken by an organization, not just the successful projects! We will discuss why this is so in the next chapter) and *parametric models of schedule and cost* – to calibrate and adjust our bottom-up estimates is an essential step toward creating credible estimates for schedule and cost.

It is important for you to know that many customers will *not even accept* a proposal that does not include the calibration of the bottom-up estimates for schedule and cost against these two types of top-down estimates; they see these calibrations as *essential for credibility*. Many customers will also create their *own* top-down estimates of the schedule and cost entailed in your proposal; if their estimates are materially different from your own, they are highly likely either to reject your proposal, or to lower the score of your proposal because of a lack of credibility.

2 This is the same Dr. Barry Boehm cited in Chapter 2 for his invention of the spiral development model. Dr. Boehm invented his first parametric cost model (for estimating the work involved in a software-development project) while he was employed at TRW in the 1970s. As of this writing (2019), Dr. Boehm is still actively teaching and conducting research.

Don't be unrealistic about how accurately we can estimate the cost of a large, complex engineering project that entails lots of new development. If you are in charge of creating a new type of engineered system, and you manage to finish it for no more than 25% over the original price estimate, you have done well. There are plenty of examples of such projects that have ended up at three to five times the original price, and finishing at 10 times the original price is not rare.

If you end up finishing your project at three times or more of your original bid price, this outcome will be bad for both you and your company. How do we avoid this? By *basing the estimate for the new work in actual previous practice*: as explained in this chapter, you break the work into small pieces, figure out which small pieces of previously completed successful projects these new pieces are most similar to, and use the *actual cost from those experiences* as the basis for the bid for each small piece on the new project. That is what the top-down use of historical actual costs from the company's library, or from a parametric cost model, provides. There is *no other method* that seems to work. Bottom-up estimates are usually hopelessly optimistic.

Furthermore, I strongly recommend that you do *not* believe claims for improved productivity based on new tools and methods; history says that they almost always far fall short of the promised improvements.

There is another, related but vital lesson to be learned here: if your customer insists that you bid such a contract with a lot of new development on a firm-fixed bid (I define this term below), *you and the company will lose your shirt.* Do not accept firm-fixed price contracts for development work. You can accept firm-fixed price contracts for *production* of articles, once the development and testing are complete.

We will cover two more aspects of estimation before we conclude this chapter: *resource leveling* and *cost* vs. *price*.

7.4.4 Resource Leveling

Let's say that your engineering project is going to require about 400 people. You probably can't get 400 people hired and assigned the day after you sign the contract; it will take time (most likely, several months) to accomplish all of that. Nor will you need exactly the same number of people throughout the duration of the project. In fact, not only will the quantity of people vary as the project progresses, but so will the mixture of skills that you need. For example, during some months you will need more software programmers; at other times you will need fewer software programmers but more testers.

Your team created an estimate of the people required for each task, by skill discipline and other factors. Your staff can turn that into a month-by-month staffing profile for your project (e.g. a simple *histogram* of the number of people who will work on your project each month). You can create specialized subsets of this staffing profile too; for example, histograms for the software development portion of the project, and so forth.

What you will likely discover is that the original staffing profile has big jumps up and down from month to month. This is a problem: you cannot actually have the size and constituents of your staff vary so rapidly. If you need a particular person for four months, but not for the fifth month, and then you need him again thereafter, you may find it impossible to find another project that will be able to employ that person for exactly that one month for which you do not need him. If you want this person to be available to you after that gap, you most likely have to keep that person, pay that person, and find something useful for that person to do during that one-month gap. That costs money – you must pay that person's salary during that month! So, you have to *smooth* the staffing profile – make do with fewer people at times (and therefore extend the duration of the tasks on which they are working); keep people at times between gaps – and through these techniques eliminate the worst jumps up and down in staffing, thereby keeping the short-term variation in staffing to a level that your human resources organization believes it can handle (which will not be very much!).

Such *resource leveling* (sometimes called *load leveling)* is an essential step in the creation of a *credible* project schedule and cost estimate. Your management and your customer will consider the big jumps up and down to be unexecutable, and therefore a gigantic risk to your project. But performing such leveling will likely drive up your price and extend your schedule. Sometimes this effect is modest, but sometimes it is quite significant. You cannot, however, omit this step: the staffing profile must be leveled, so variations are small enough that your human resource organization can handle the rate of change.

7.5 Cost vs. Price

The last aspect of estimation that we will cover in this chapter is *cost* vs. *price*. Earlier in this chapter, we defined the *cost* of your project as the sum of all the elements that will cause you and your company to spend money – *directly* (such as salary and items that you purchase) and *indirectly* (such as employee benefits, facilities, corporate management, and so forth) in the performance of the work entailed in your project. If our company were to offer to do the work for that amount of money, however, we would (i) have no reserve against uncertainties (in a large project, we may not know in advance *exactly* what will go wrong, but we can be quite certain that *some things* will go wrong) and (ii) make no profit for the company and its shareholders. In exchange for doing a useful piece of work, the company is entitled to earn a reasonable rate of profit on its work. So, these two items – *management reserve* (a fund of money set aside to deal with problems as they arise on the project) and *profit* – must be added to the estimate for cost; it is the combined sum of *cost + management reserve + desired profit* that forms the amount we bid to the customer. This sum of cost + management reserve + desired profit is called the *price*. The price will be *specified in the contract*.

As we have seen, every project entails uncertainty in cost, and therefore uncertainty in price too. In Chapter 9, we will show you how to mitigate some of these uncertainties. But *some level of uncertainty will always remain*. An important question then arises: which party holds the financial risk for this uncertainty? That is, if the project costs more to complete than planned, where does that extra money come from? There are many possible answers to this question, but the question will *always* be answered in the *contract*. Here are some possible answers:

- **It might be that the company doing the work holds all of the risk associated with the uncertainty about the cost**. This is often called a *fixed-price contract*. The contract defines a fixed dollar value that the customer will pay to the contractor, no matter how much or how little it actually costs the contractor to do the work. This makes sense if there is relatively little uncertainty involved in the work. Note that this is seldom, however, the case for *complex engineering projects* (such as those we are considering in this book); such projects usually entail a lot of invention, and that implies significant uncertainty.

- **It might be that the customer buying the system holds all of the risk associated with the uncertainty about the cost.** This is often called a *cost-reimbursable contract.* The contract states that the customer will pay *all proven costs* (no matter what the amount) that relate to the work defined in the contract, plus an additional amount as profit (sometimes called *fee*). The amount of the profit might be calculated in any of several different ways: as a percentage of the final cost, as a fixed dollar amount, on some sort of sliding scale that rewards the contractor for finishing the work at a lower cost, or many other methods. This is appropriate for many complex engineering projects, because of the uncertainty entailed by their invention and complexity. Customers obviously prefer that the company holds this risk, rather than them, but it might be the case that no reputable company will sign a contract for the project unless the customer holds some or all of this risk.
- **It might be that the customer and the company share the risk in some fashion.** There are a large number of hybrid approaches wherein the parties to the contract share the risk associated with the cost of doing the work entailed in the contract. By *sharing the risk,* we mean that both the customer and the company doing the work will be responsible for some designated share of any extra money required to do the work. The formulas for computing the relevant shares can vary a lot, and at times can be very complicated.

There is an important psychological aspect of risk-sharing too: if the customer forces the company doing the work to bear all of the financial risk, it creates an inherently adversarial aspect to the working relationship. This is because, inevitably, if the company is holding all of the risk, they will look for pieces of work that can be avoided so as to lessen the chance of the cost exceeding that which is the basis for the price in the contract. A *smart customer will realize that this is not to their advantage*; things are better for the customer when the interests between them and the company doing the work are more in alignment (just like we talked about in Chapter 1, see Figures 1.10 and 1.11). This motivates the use of *cost-reimbursable contracts* for complex engineering projects, usually with some type of hybrid approach to determining who pays any extra monies required, and the amount of profit, so as to create some *sharing of the risk.*

7.6 Your Role in All of This

The manager of an engineering project requires that his/her team creates estimates for the schedule and cost of the project. This is one of your most important and most watched tasks. But *you* do not actually *do* it, your subordinate managers do – you *coordinate* it.

Your employees must create these estimates, rather than you, so they *believe* in them. You must allow them to make real estimates, and not dictate the answer. You can mandate that they use particular estimating methodologies:

- Including structures and instructions for how the bottom-up estimation should be prepared, including three-point estimates for schedule and cost and (unless your project is really small) the use of Monte Carlo simulations for schedule and cost.
- Requirements for calibration of their bottom-up estimates against both (i) previous actual project results and (ii) parametric models.

You can also provide the task managers who are preparing these estimates with *target values* (e.g. your aspirations for the schedule and cost of their tasks). And you should negotiate with them about final values for their estimates. But in the end, you ought not to dictate the answer; you must reach a consensual, bi-lateral agreement with each task manager – they must be allowed to work to estimates which they psychologically "own." We will say more about the sociology of teams in Chapter 13. As a preview of that chapter: the task managers must psychologically "own" the estimates for their own tasks in order for them and their employees to feel *motivated to succeed in the work*. We want them to be motivated, because the data indicates that *motivated employees are far more productive than unmotivated employees*.

What do we mean when we say that motivated employees are far more productive than unmotivated employees? Just this: that motivated employees literally do a lot more work per day than unmotivated employees. Furthermore, the difference is not trivial; in software development, studies indicate that the difference in productivity might be a factor of 3! You therefore *want* motivated employees (there are other good reasons for wanting motivated employees, which we will discuss in Chapter 13), and the first step in achieving motivation of your employees is allowing them psychologically to "own" the estimates for their own work.

You must ensure that the resulting schedule incorporates an appropriate amount of schedule slack along every chain of tasks.

You will also be asked to make regular reports of your updated predictions for schedule and cost. We will discuss this more in Chapter 10.

7.7 The Intersection With Engineering

Much of what we discussed in this chapter could apply in some ways to *any* project: a construction project, an artistic project, and so forth. But there are vital intersections with *engineering* that transform the processes described in this chapter into processes suitable for engineering projects. There are three principal such intersections:

1. **Determining the actual interrelationship and dependencies among the tasks.** That is, what inputs from other tasks does this particular task require? For engineering projects, this is largely a matter of engineering and technology considerations. Your company's engineering processes, by defining typical sequences of activities, and the artifacts produced by each step on those sequences, will get you started, but the interrelationships and dependencies can only credibly be defined by people with the appropriate engineering background and skills.
2. **The creation of those three-point estimates for schedule and cost.** It is the details of the engineering and technology that will allow you to determine appropriate factors for the level of asymmetry that you should use in both magnitude of *deviation* and *probability* of occurrence.
3. **The calibration of the bottom-up estimates against the two types of top-down estimates** (historic data from previously completed projects and parametric models). On an engineering project, these comparisons and assessments cannot be made without understanding the actual engineering issues involved.

7.8 Next

Having established an estimate for the schedule and cost of our project, we now have some quantifiable measures against which we can assess progress for our project. But before we start telling you how to assess such progress (which we will do in Chapters 9 to 11), we will find it very useful to establish some background knowledge in the art of making measurements, and drawing valid conclusions from those measurements. This is a subject where many people, even highly skilled and highly trained people (such as scientists and physicians), make massive mistakes. So, I am going to show you how to avoid many of these mistakes and, therefore, how to draw better conclusions from numeric information.

7.9 This Week's Facilitated Lab Session

- Activity network, part A (part B will be during week 11). You will turn this in as an individual (not team) homework assignment, but not until after we complete part B during week 11.
- Build a schedule of at least 30 separate tasks, of various durations (but for the initial version, specify only a single most likely duration for each task). Give each task a unique name.
- All tasks are to use only finish-to-start sequencing.
- Include at least four chains of at least six items each.
- Also create at least two examples where a single task is a required predecessor to more than one task, and at least two examples where a single task itself has multiple predecessors.
- Include at least five cross-chain dependencies, all finish-to-start.
- Do this using the Microsoft © Project computer program.
- We will then have a discussion of findings and insights in small ad-hoc groups.

8

Drawing Valid Conclusions From Numbers

Invalid data and poor statistical methods can lead to bad decisions! There are many ways for an engineering project manager to make mistakes, but one of the most common and most insidious is through making logical and procedural mistakes that cause us to draw erroneous and invalid conclusions from quantifiable data, and as a result, making poor data-based decisions. As engineers, we measure things, and then we often make decisions based on those numbers. For example, we predict when our project will be done, how much it will cost when it is done, and what the technical capabilities of our product will be (e.g. how far will our new airplane be able to fly safely without refueling). And we use those data to make decisions for our project. Whenever we use numbers, however, there is a chance for error: our measurements always involve uncertainties, a particular assumption is only true under certain circumstances, we may not have collected appropriate samples, and so forth. In this chapter, I show you the most common ways that we undermine our own credibility through poor data collection, errors in logic, procedural mistakes, weak statistics, and other errors, and how you can instead use valid methods and strong statistics so as to create credible predictions for all of our project management roles and measures.

8.1 In Engineering, We Must Make Measurements

We are engineers ... and, as part of our job, we routinely *measure* things. *Qualitative* assessments are seldom adequate for our purposes; we most often instead must make use of *quantitative* data in our analysis and decision-making.

Most people, when they need to make measurements, don't give the matter much thought. They just step on the scale, pull out a tape measure, or get out the air-pressure gauge. In ordinary life, that might (or might not) be good enough. When we are making decisions about complex engineering processes, however, making measurements that can be the basis for useful decisions is much more complicated. Look at Figure 8.1, which summarizes the *measurement process.*

Engineering Project Management, First Edition. Neil G. Siegel.

Companion website: www.wiley.com/go/siegel/engineering_project_management

Step	Discussion
Decide what to measure	What measurements do we need to take in order to provide the data that we will need to make a certain decision?
Determine how accurately we need each item of data to be measured	Do we need to know the weight to the nearest pound, or to some other degree of accuracy?
How to achieve that accuracy of measurement	Once we know how accurate we need to be, what tools and methods do we need to employ in order to achieve that accuracy?
Determine how much data must be collected	Do we have to measure this item just once, or 10 times, or 1 000 000 times?
Determine how, where, and when to collect the data	Under what conditions do we have to measure this item? What tools do we need to do so? What are the operational states of our system when it is valid to collect these data? Furthermore, at times it can be very difficult to separate the item of interest from other factors. For example, we might want to know how long something takes, but the item of interest is contained inside another process, and it is hard to see the actual start and stop of the item in which we are interested separate from that other process
Understand the range of validity of the data	Are the measurements we collected valid only at certain times? Over certain temperatures? Under certain combinations of input load to our system?
Validate and calibrate the data	Is our weighing scale accurate? Is our voltmeter accurate? How do we know?
Analyze the data	What do the data indicate? Do the data suggest issues that we do not understand, and that must be tracked down before we can use the data to make decisions? Are there various alternate explanations for what the data appear to indicate?
Draw conclusions from the data	What are the correct logic sequences that can lead to decisions? Where are we on solid ground, and where are we making judgments? What is the level of uncertainty involved, and what are the risks that could result from drawing the wrong conclusions?
Document the entire process, so that it could be reconstructed	Engineering is an iterative process; many months from now, we may learn something that will cause us to want to make this measurement again, and compare the new result to the previous result. We therefore need to capture the process, tools, and data that we used to make this measurement

Figure 8.1 The measurement process.

8.2 The Data and/or the Conclusions are Often Wrong

People make awful mistakes in *every one* of these steps and, as a result, their conclusions are often *completely wrong.* Here's a real example that I experienced many years ago.

My wife has a PhD in Iranian linguistics. When she was still attending university, and considering faculty members to be on her doctoral committee,[1] she wanted at least one of the committee members to be a woman. She received a recommendation from one of her faculty advisors that person X might be suitable. I suggested that my wife get a couple of person X's publications, and we could read them together, and see if we thought her expertise was such that she could help guide my wife's research. I am not a linguist, but many linguists – and people in many other fields of academic study – use *quantitative methods* in their research; my wife could understand the linguistic portion of person X's papers and I (with my training as a mathematician, the academic field of study for my first two college degrees) could understand the quantitative methods.

I was astonished to find that the quantitative methods used in person X's work were *completely wrong*; even if her linguistic data and conclusions were perfect (which I was not qualified to assess), her research conclusions – which depended in an essential fashion on the quantitative assessments that she made via logic and statistics – were completely unjustified by her data, because her statistical method and analysis logic were simply *wrong*.

These were peer-reviewed papers, meaning that other, supposedly-qualified academics in the same field had reviewed these papers for correctness of method and rigor. In these cases, however, the peer-review process had completely failed; anyone trained in elementary statistics and logic could spot these errors in a few minutes. Social scientists are in fact supposed to be trained in basic logic and statistics.

This was a life-changing event for me; I suddenly realized that published, peer-reviewed work by experts might be *completely wrong*.[2]

Nor did this turn out to be an isolated example; I have since read hundreds of works by social scientists, physical scientists, public officials, and other experts … and discovered that *most of them make similar mistakes in handling their quantitative data*.

When I started working as an engineer and project manager, I also discovered that those domains are full of errors in quantitative method too.

It is not my role in this book to fix the field of linguistics; my role is to help you become effective managers for engineering projects. In engineering and engineering project management, we *inevitably and routinely make decisions based in part on quantitative data*. It is therefore vitally important that the methods, tools, logic, and procedures that we use to collect, process, and assess those data – which are going to influence our decisions – be logically sound and correct. *That is what this chapter is all about.* Once we have established this baseline of *mistakes to avoid*, we will in future chapters dive into the specific quantitative methods and analyses that we must perform in our role as managers of engineering projects.

The above provided an example of an error, but a second example that includes a discussion of the *exact mechanisms of the error* may be useful. Here, therefore, is an

---

1 The small group of professors who guide the research of a candidate aiming to complete a PhD. It is this committee who make the final judgment about whether the candidate passes or fails.

2 To finish the story, my wife selected a different professor for her committee.

example of decision-making based on quantitative data that includes a discussion of exactly what went wrong:[3]

- You are a physician. A patient has come to you and asked to be tested for a particular disease. This disease is fairly rare – on average, only about 1 person out of 1000 in the United States has it.
- After taking the patient's medical history, you and the patient decide to order the test.
- A week later, you get the results from the laboratory that ran the test: the results come back *positive*. That is, the test result says "yes, your patient has this disease." You look up the information about the accuracy of the test, and it says that the test has a *false-positive rate* of 5%. This means that 5% of the time when the test says "yes, the patient has this disease," that answer is *not* correct; the patient does *not* have the disease.
- You call the patient to come in for her follow-up appointment. After she arrives for this appointment, the patient asks you "What is the probability that I have the disease, given that the test came back with a positive result?" *What do you tell her?*

  Write down your answer in the box below, before you read on.

Your answer to the patient's question: What is the probability that I actually have this disease?

According to the sources cited by Taleb in his book, most doctors respond "*The probability that you have the disease is 95%.*"

Did you give that answer?

Unfortunately, that answer is *wrong.*

The correct answer is just under 2%; that is, it is *highly unlikely* that the patient has the disease ... *despite* the positive test result.

We need only some elementary arithmetic to figure this out:

- Imagine that you, over the course of your career, have ordered this test 1000 times on 1000 separate people. Over those 1000 tests, how many likely came back with a positive test result? Let's figure that out:
  - 1 person who actually had the disease (remember, we said that the *occurrence rate* is 1 person per 1000 of population in the United States)
  - 50 people who had a *false-positive test result* (we said that the false-positive rate is 5%, and 5% of 1000 samples is 50)
  - So, there ought to be *51 positive test results* in your group of 1000 tests.

3 I first read of this particular example in the excellent book *Fooled by Randomness* by Nassim Nicolas Taleb (Random House, 2005). I always recommend this book to my students.

- Your current patient is one of those 51. Is she/he the one who actually has the disease, or one of the 50 with the false positive? Within the framework of the question as asked herein, you have *no way of knowing.*[4] *So, the probability that this patient has the disease is 1 out of 51*, or just under 2%.

Technical Note

The problem statement did not include anything about *false negatives* (instances where the test said that the patient did ***not*** have the disease, but in fact she actually ***did*** have it). In real life, you would include that too. We will say more about false-positive and false-negative results later in this chapter.

Think about those two answers. Since the answer can only be between 0% and 100%, 95% and 2% are about as far apart as they can literally be. *The doctor's answers could hardly have been worse.*

Doctors are *supposed* to be trained to answer exactly this sort of question. What went wrong?

The reason for the error is easy to explain. The doctor was provided with *three* pieces of information: (i) that the disease was rare (only 1 person out of 1000 have it); (ii) that the test had a 5% false-positive rate; and (iii) that the test result came back saying "yes, the patient has this disease." To reach the correct answer, as we saw above, *all three of these pieces of information had to be used.*

Notice, however, that in deciding that the answer was 95%, the doctor *only used the fact of the 5% false-positive rate* for the test (and the fact that the test said "yes"); he/she *ignored* the other piece of information – the fact that only 1 person out of 1000 has the disease.

This is a very common method by which people analyze a problem: they look at the set of all the provided information, they then select the *subset of those items of information* that they *deem to be the most important*, and make their decision based *solely on those items, ignoring* all of the other information that may be available (presumably, because they deem this information to be less important than the information upon which they based their decision). Unfortunately, this method is often completely invalid, and leads (as in our example) to incorrect conclusions and decisions.

If this method is so easily proven to be wrong, why is it still very common? Human beings apparently hate the complexity and ambiguity of a situation with many pieces of information, and therefore we are evolutionarily conditioned to make an *informal assessment regarding which piece(s) of information seems to be the important one(s)*, and prefer to make our decision based *solely* on *those* piece(s) of information that we deem to be important, ignoring the remaining pieces of information.

Unfortunately, there is something in mathematics that statisticians call a *conditional probability*; in such a situation, *multiple pieces of information interact with each other*,

4 In real life, the doctor may have access to confirmatory tests before making a diagnosis. For example, if a breast mammogram comes back indicating a region that might be cancerous, we no longer make a diagnosis on the basis of the mammogram, but perform a confirmatory test (which is usually a test called a *biopsy*) before making a diagnosis. Not all medical conditions, however, have such confirmatory tests available.

and a reasonable answer can *only* be arrived at if *all the pieces of information are used* and *combined in the correct fashion*, as we did above in deriving the correct answer. This was the doctor's mistake; he/she was given *three* pieces of information, but elected to use only *two* of them in deriving the answer. Unfortunately, because of the conditional probability involved in the problem, the answer arrived at in this fashion – by using only two of the three available pieces of information – was completely *wrong*.

Notice that if the doctor had chosen to ignore the two pieces of data – the one about the 5% false-positive rate and the fact that the test came back positive – and used *only* the piece of data about occurrence rate (1 person out of 1000 has this disease), he/she would have concluded that the chance the patient had the disease was 1 in 1000; that is, 0.1%, which is *much closer to the correct answer*. But you could perform this (still incorrect) analysis without even running the medical test! In this case, running the medical test and using its results while ignoring conditional probabilities gave you a *worse* answer than if you had not even run the medical test!

Terms like *statistics* and *conditional probability* sound complex and arcane, and people often therefore anticipate that doing the analysis in an appropriately rigorous fashion is too hard. But, as we saw when we actually worked out the answer to our doctor problem, nothing more profound than some very basic logic and some simple arithmetic was actually needed to derive the correct answer. A textbook on statistics will provide a formula for something called Bayes' Law:

For events A and B, provided that $P(B) \neq 0$, then

$$\mathbf{P(A \mid B) = [P(B \mid A) * P(A)] / P(B)}$$

(the notation "P(A)" is read as "the probability of A" and the notation "P(A|B)" is read as "the probability of A given B")

Or it might be presented in another equivalent form:

$$\mathbf{P(A \mid B) = P(AB) / P(B)}$$

(the notation P(AB) is read as "the probability that *both* A and B are true")

This notation seems opaque to many people, but in fact the formulas just encapsulate the methods of the calculations that we just performed; I will leave it to you to work out exactly how. Think of P(A|B) as "the probability that the patient has the disease, given that the test said *yes, she has the disease*" and you should be able to lay the problem out using the first of the formulas above.

But dealing with the notation and the contents of statistics textbooks is *not really necessary* for us as managers of engineering projects; all we have to do is *recognize a situation that may involve a conditional probability* and request that the calculation be done based on that insight. As the manager of the engineering project, you are not likely to be doing the calculations yourself.

You will, however, find that conditional probabilities are *everywhere*! Therefore, the type of mistake made by our example doctor is very common.

But this is only one type of error that people routinely make in dealing with numbers. There are many, many others. Many of these are just inadvertent errors in data collection and/or data analysis, but some are actually willful: people often define their terms in bizarre ways, so as to improve the "look" of the answer. An easy to understand example of this is the *unemployment rate*, which is intended to measure the number of people out of work, as a percentage of the population of people who could work. The government naturally wants the general public to believe that their economic policies are effective, and so they define what they call the "employment rate" as something that is somewhere between one-half and one-quarter of what you and I, as reasonable citizens, would define as the unemployment rate.[5] See Figure 8.2.

Between the combination of the *mistakes* and the *willful distortions*, we are surrounded by numbers that are *incorrect.*

This leads to what I (jokingly) call *Siegel's Outrageous Simplification*:

> Most of the numbers in public discourse are wrong.

In Figure 8.3, I list some of the common mistakes in handling quantifiable data. It is a rather long list!

Below, I discuss a few of these items in a bit more detail.

8.2.1 The Fallacy of the Silent Evidence

If you only hear about *successes*, you will be misled – the failures may exist, but if you do not hear about them, you will misinterpret the likelihood of success (and of failure); what you need is a properly representative sample of the evidence, which will include *both* successes *and* failures.

Here's an example. Take an event that has two outcomes (a stock goes up or down, a sports team wins or loses, and so forth). Now send confident predictions to 2000 different people, in advance of the event – 1000 with one answer, and 1000 with the other answer. After the event has actually occurred, 1000 individual people will notice that you got that prediction correct. But they will *not know* about the 1000 people who simultaneously saw that you got it *wrong.* Now, send another two-outcome prediction to the 1000 people to whom you previously sent the answer *that by coincidence* turned out to be *right*; 500 with one answer, 500 with the other answer. After that second event has occurred, there are now 500 people who saw you being correct two times in a row! But that result is still just a coincidence. Now do it eight more times. There will be a person who has seen you be correct 10 times in a row! If you ask for a testimonial from that person about your predictive ability, you may well get a very flattering statement attributing you with deep insight and amazing predictive skills. But that person has that impression *only because he or she knows nothing about all of the other letters.* In fact, of course, you were correct no more often than would be predicted by random chance. But *all* of that other evidence, that

5 This under-reporting is so marked, and so noticeable, that a new metric – the *labor under-utilization rate* – has been created. This new metric addresses what I called fudging effect #1. But this is not the metric that the government uses in public discourse!

Fudging effect #1: In the United States, when you become unemployed, you have the opportunity to register with your state or local government for assistance, which usually takes the form of both *financial payments* (called *unemployment insurance*) and *assistance in finding a new job* (sometimes called *job placement*). If, after several months, your financial payments cease, and you find that the government office is not effective at helping find a new job, you may well elect to stop coming in to that government office every week; each visit takes several hours, and after all, it is not helping you. At that point, the *government no longer counts you as unemployed*! You are, of course, still just as involuntarily unemployed as you were before, but by no longer counting you in their statistics (justified by the feeble claim that you must be "discouraged" and therefore not actually in the market for a job), the government can lower the reported unemployment rate. I have seen reports which indicate that the number of such "discouraged" adults thereby no longer counted as unemployed is at times *nearly as large* as the pool of those officially considered unemployed; this implies that the officially reported unemployment rate could thereby be artificially lowered by as much as a factor of two from the actual figure. There is a similar fudging effect for people on various government disability programs; not all disabilities prevent employment, but governments almost always exclude anyone who is on a government disability program from the ranks of the unemployed, even if there is work that they could do, and they would in fact like to be employed.

Fudging effect #2: We should be measuring the unemployment rate as *per household*, rather than per individual. This is because the unit that is disrupted by unemployment is the *household*. In most of the two-adult households in the United States today, *both* adults work. Yet a household with two earners is highly likely to have their living expenses set to depend on *both* incomes; few two-earner households live exclusively on only one of those incomes. Because of this, about of unemployment by *either* earner is a catastrophe for the *entire* household. A two-earner household that requires all of the income from both earners is essentially *twice as likely* to face a catastrophic unemployment event as a one-earner household. This is similar to airplane design: if you design your two-engine airplane to be able to fly *only when both engines are operating properly*, the reliability of your airplane is *lower* than if you had designed the airplane to be able to fly safety with only one engine, and even lower in reliability than a similar single-engine airplane! Since the average number of earners per household in the United States today is nearly 2, the percentage of households affected by an unemployment event is nearly twice as high as the percentage of individual earners so affected.

Put these two effects together, and you can see that the *actual* unemployment rate in the United States bears little resemblance to the numbers reported by the US (or state) governments.

This is not a partisan issue; both of the major political parties in the United States do this.

That these errors both significantly *reduce* the reported unemployment rate cannot reasonably be judged to be a coincidence; there are plenty of smart people in the US Government who understand the above two points.

Figure 8.2 An example of fudging the numbers: the US unemployment rate.

negative evidence, is hidden from that one person. This is the fallacy of the *silent evidence*. Many investment scams and consumer frauds work on this basis.

There are many reasons why, in designing and analyzing our engineered systems, we might not see all of the negative evidence. One reason is that people do not like to report bad news. There are many others. So, in making design decisions for our engineered systems, we must take steps so as to be sure that we are in fact seeing all, or at least a properly representative sample, of the evidence, else we will fall victim to the fallacy of the *silent evidence* too.

We can now see that the fallacy of the silent evidence is why, in the previous chapter, I said that it is important that a company's archive of past project performance includes *all* projects undertaken by the company, *not* just the successful projects.

The fallacy of the *silent evidence* is, by the way, the fatal flaw in political polling (or any other sort of polling or opinion survey). In the United States, at least, participation in a poll is *voluntary* (nor, of course, are we required to tell the pollsters the truth about what we really think either). This voluntary participation means that the poll measures the opinions of those who are *willing to participate*, rather than the opinions of the voting population at large. *No one can foretell in advance* whether the opinions of those two different groups (e.g. those who willingly provide answers to the pollsters, and those who will actually vote) coincide. My impression is that precisely the most emotionally laden issues and elections – exactly when you would most desire your poll to be accurate – is the time when most people will decline to participate (or not tell the truth). Because of this, polling can *never* be accurate; this is an inherent, unrecoverable, and fatal flaw.[6]

8.2.2 Logical Flaws in the Organization of System Testing

The fallacy of the *silent evidence* (and the *induction problem and round-trip error*, which are also described in Figure 8.3) affects the test programs of engineering projects in a very significant fashion. Recall the procedure that we described for a test program (Chapter 3): you create test procedures that are designed to exercise the system in a manner that allows you to see if, under those particular circumstances, the system appears to operate in accordance with each requirement contained in the system's specification. We appear to be making the assumption that, just because the system passed each requirement under some scripted and controlled circumstance, there are no flaws remaining in the system. This is obviously a case of both a *round-trip error* and a "we see only white swans, so we therefore conclude that there is no such thing as a black swan" type of situation: the flaw listed in Figure 8.3 as the problem of *induction*.

But this, of course, is not actually the case: our system still contains flaws (all human creations do), and some combination of conditions and stimuli will cause your system to fail, and perhaps to fail in a very serious manner. It is only the case that the evidence of such failures is thus far *silent*, because we have not looked for it; we designed our tests deliberately so that the system would pass! That is, we deliberately looked only for positive evidence. You must therefore find out where in your system those boundaries in operating conditions are whose crossing will cause these failures, and do so *before* you deliver the system, and then either correct those failure mechanisms, or establish limits on the use of your system. Just because you ran some tests and the system worked, does *not* imply that the system will not fail; it only demonstrates that it has not failed *yet*. You can see now why I strongly advocated (in Chapter 3) pushing the system through your test conditions until it actually fails (as part of your verification program), and also why I advocated for unscripted operation of the system by the users (as part of your validation program). These are ways of finding some of those boundaries in operating conditions, and avoiding the fallacy of the *silent evidence* (and the problem of *induction*) in our test program.

---

6 And the accompanying statements of alleged accuracy (e.g. "this poll has a 3% margin of error") are wrong as well.

Figure 8-3 (multiple pages). Common Mistakes in handling quantifiable data.	
Type of mistake	Discussion
What you are measuring may not be exactly what you think you are measuring	Your measurement method may omit a portion of the item you seek to quantify, or include a bit of some other item
You underestimate the amount of error and noise in the data	*Every* measurement includes errors, both *systematic* and *random* ("noise"). You must separate these errors from the item you wish to characterize ("signal"). If you do not, you are likely to mistake the change in the *noise* (especially random variations) for a change in the *signal of interest*
Failing to understand that the answer from the measurement might just be wrong	The answer provided by a measurement or a test might just be wrong; there are false positives (the test says "yes" when the correct answer is "no") and false negatives (the test says "no" when the correct answer is "yes")
Missing that the answer depends on a *conditional probability*	You base your analysis and eventual decision on too few parameters, failing to notice the essential interconnection between parameters. This may come about through discarding some parameters, or failing to measure them
Assuming independence between measurements ignoring sequential effects	The expected outcome for 1000 people who each bet once in a casino is very different from that for a single person who bets 1000 times, because the *initial conditions* for the person who bets 1000 times *change with each successive instance*
Using ineffective or weak statistics as a basis for evaluation	The most common statistic in use is comparing a *current measurement* to a *single prior measurement*, and inferring a trend (and a cause for that trend) from that change. For example, comparing last month's budget prediction to this month's budget prediction, this quarter's company profits to last quarter's, etc. This statistic, however,has almost *no predictive power*; there are better, stronger statistics that should be used instead
Using data outside its range of applicability	We often collect real data, and then we *extrapolate* that result to different conditions (the data collected might have been about *red* cars, but we are drawing a conclusion about *blue* cars). This works at times, but is disastrous at other times. Based on an experiment where you cooled water from 70 °F to 60 °F to 50 °F to 40 °F, what would you predict about continuing to cool the water to 30 °F?
Not collecting repeated data on a meaningful time frame	There is no point in making measurements at a rate significantly faster than the underlying phenomena can actually change. If you were monitoring your weight in order to see if your new diet and exercise regime was effective, you would not weigh yourself every five minutes.If you did so, almost all of the change that you detected would just be *noise*, not an actual change in the *signal of interest*

Figure 8.3 Common mistakes in handling quantifiable data.

Figure 8-3 (multiple pages). Common Mistakes in handling quantifiable data.	
Type of mistake	Discussion
Poor selection of data	Selecting data that are not truly representative of your operating conditions
Changing the data or the measurement approach during collection	If you change the way you collect the data mid-stream, there may be no valid way to compare data collected by the first method from data collected by the second method. And, of course, doing things that allow the data actually to change (e.g. the temperature at which we collect samples is allowed to vary, etc.) invalidates comparison too
The limit of the utility of examples (the problem of *induction*)	In the real world, no number of observations of a positive phenomenon constitutes a proof that that phenomenon is always true. Yet a single negative observation constitutes a proof that it is not always true. You may observe thousands of swans, and they may all be white, but this *does not and cannot* prove that *all* swans are white. In contrast, a *single instance* of an observation of a black swan constitutes proof that the statement "all swans are white" is *wrong*.[1] In the world of formal mathematics, there *is* a rigorous principle called *induction* which can be used to establish a general principle from a set of examples, but this technique is valid *only* within the formal (and artificial) constraints of an appropriate mathematical construct; the real world *never* complies with such formal mathematical constraints. The existence of a valid principle of induction in mathematics causes people to (invalidly) attempt to apply induction to the real world
Attribution bias	We attribute our successes to skills and knowledge, rather than to random chance. We attribute our failures to rare-but-random events, rather than to a lack of skills or knowledge
Path dependence	We "fall in love" with the path of reasoning that we used to arrive at an answer, and refuse to adjust that answer even in the face of additional, contradictory evidence
The fallacy of the silent evidence	We see what appears to be a compelling set of evidence in favor of a proposition, but we fail to notice that all of the *contradictory evidence* has been *omitted* from our sample
Round-trip error	The tendency to confuse the condition "no evidence of flaws in our system" with the condition "there is evidence that there are no flaws in oursystem." These are not interchangeable statements! The condition of "no evidence of flaws" is *often* observable in real life (e.g. a patient no longer presents any visible indications of having cancer), whereas the condition "there is evidence that there are no flaws in our system" is very rare (essentially existing only in formal mathematical proofs). This latter condition would correspond to a patient truly being "cancer-free," a very different condition than that of merely no longer presenting any visible indications of having cancer. Some doctors (but not all!) have learned not to confuse these two conditions

Figure 8.3 (*Continued*)

Figure 8-3 (multiple pages). Common Mistakes in handling quantifiable data.	
Type of mistake	Discussion
The narrative fallacy	Correlation does *not* prove causation. Just because one event *precedes* another does not in any way establish that the first event *caused* (or contributed in any way to) the second event. But we humans like to create *stories* that explain observations, including data; in fact, this story-telling seems to be necessary in order for our brain to cope with the volume of information that is made available to us. People hate ambiguity, and therefore, we often do such story creation far *too soon* – before we have enough data actually to draw valid conclusions. This tendency to create stories for everything, even when not justified, is called the *narrative fallacy*. Just because a story is *appealing* does nothing to establish its *correctness*; an appealing story may be completely wrong. Then *path dependence* (above) causes us to be unwilling to change that story, even in the face of counter-examples
Failing to recognize the existence and significance of *outliers*: the problem of *scale*	Engineered systems exhibit *non-Gaussian behavior* (that is, the distribution of unusual events does not follow the well-known "bell curve"); our unusual events may both be more frequent than the bell curve would predict and asymmetric (that is, strongly biased toward events that degrade our system's performance, rather than events that improve performance). In addition, many engineered systems experience sampling or input rates that are orders of magnitude beyond normal human experience; I have built engineered systems that process 1 000 000 data instances per second (that is *31 trillion* data instances per year), continuously. People have *no useful intuitions* about such large sample sets. Such a system will experience, on average, more than three 6-sigma events every second (a 6-sigma event is an event that a statistician considers so rare that it never occurs in real life). Engineered systems, because of these high data rates, experience a significant number of unusual events (statisticians call these "outliers"). Many statistical techniques call for the *discarding* of outliers; but for us, that is exactly the *wrong* strategy; we must *focus* on the potential of outliers and (as discussed in Chapter 2), through design strategies, try to *prevent them from disrupting our system*
The tendency to believe what you want; the tendency to ignore evidence, and to explain away evidence that tells the "wrong story"	Humans have a strong tendency to interpret all new evidence in a way that supports their selected explanation. They are also quick to discount and eliminate evidence that appears to contradict their selected explanation. This is similar to *path dependence* (above), but deals with the tendency to be biased in our selection of data, whereas path dependence deals with the tendency to be biased in favor of an original line of reasoning
. . . and many more	

Figure 8.3 (*Continued*)

This is so important that I am going to say it all again:

- Projects design their test programs to *succeed*; that is, the test is considered done when the system performs properly on *all* of the test procedures. That is, however, an insufficient test program! Your test program should *not* be content looking only for confirmation that – under a carefully scripted set of conditions – the system works. You must also find the situations where your system fails!

- This is because you *cannot prove that your system is defect-free by gathering instances of it working*; that would be the logical error of *induction* (also called *confirmation*). Confusing the condition "no evidence of flaws in our system" with the condition "there is evidence that there are no flaws in our system" would be the logical *round-trip* error.
- A famous example of the error of induction is the *black swan*: you may observe a lot of swans, but even if *all* of the swans that you observe are white, that does *not prove* that there are no black swans, and therefore that it would be *invalid* to conclude from your observations that all swans are white. On the contrary, it only takes a *single instance* of observing a black swan to *prove* that not all swans are white, whereas *no* number of observations of white swans is sufficient to *prove* that all swans are white.
- Your system *will* have defects; every human creation does. *You* need to discover them, and to discover the boundaries of effective and safe operation, and not leave this discovery to your customers.
- What you *can* do with your test program is characterize what makes your system fail. I call these *boundaries*.
- The point of gathering information about failures is *not* to confirm our design decisions, but to give us the *earliest possible indication of the nature of their flaws*. So, don't sweep bad news "under the rug," or interpret it away. I find it helpful always to adopt the *worst possible interpretation* of the data, and the worst possible outcome/impact, and react accordingly: how can I improve the design to prevent this result?

8.2.3 The Problem of Scale

Most of our complex engineered systems today include computers and software. The most characteristic aspect of such systems is that they can process *lots* of data; I have built systems that process more than 1 000 000 new inputs every second, and do so 24 hours a day, 7 days a week. There are even systems in the world today that process more data than this, but even our more mundane systems process amounts of data so vast that humans have no reasonable intuitive grasp of the *scale* involved.

In such large sets of input samples, there will be a shocking number of *outliers*; that is, data or processing that *deviates materially from the nominal conditions, values, and/or sequences expected*. Since your system cannot avoid having to process these *outliers*, there are two aspects of design that become magnified in importance:

1. Getting the system to do what you want under approximately normal conditions.
2. Preventing the system from doing horrible things under severely off-nominal conditions.

This is pretty much analogous – but in a slightly more general version – to what I said in Chapter 2: implement the dynamic behavior *you want*, but take active steps to prevent (or limit the damage caused by) the dynamic behavior you *do not* want.

Consider the following example. Let's say there is a set of actions that are a key portion of the mission of your system that must be performed within a time constraint. This is very common whenever computers intersect with real physical objects, whether that object be a part of the processing system (e.g. a disk drive whose data platters rotate at a given fixed rate) or an external mechanical process (e.g. some machine whose motion is being commanded by a computer). Let's say that in the nominal case, you want the computing process to complete in 100 milliseconds (1/10 second). A statistician will tell you to collect a lot of samples, and see if most of them are near (or below) 100 milliseconds in duration. They will use terms like mean, median, variance, and standard deviation to describe the *distribution* of the sample durations.

You clearly have a design task to ensure that, most of the time, this computing process completes in around 100 milliseconds. Every system and software designer is aware of and understands this portion of the problem; this is what I called *getting the system to do what you want under approximately normal conditions.*

But you *must also implement* the other half of my design goal too: *preventing the system from doing horrible things under severely off-nominal conditions.* For example, what could happen to the physical devices that your system is controlling if a data instance takes 200 milliseconds to be processed, rather than 100 milliseconds? or 1000 milliseconds? or 10 000 milliseconds? In the amazingly large sample sizes that we deal with in today's engineered systems, these "really far off-nominal" data are not nearly as rare as you think. A statistician considers something really rare if it is what he/she calls a "6-sigma" event; "6-sigma" indicates that such an event occurs no more frequently than about three times out of 1 000 000 samples. There will, however, be *three such events every single second* in my 1 000 000 data-instances-per-second system.[7] In the world of complex engineered systems, 6-sigma does *not* signify a rare event.

And what if something actually physically explodes if the processing for some data instance takes 10 000 milliseconds, rather than 100 milliseconds? What if people die as a result? If you are building the automated controls for an oil refinery, a chemical plant, a power plant, or a facility making medicines or packaging food, such fatal outcomes are completely possible.

And, of course, there can be a variety of really bad outcomes that don't kill people, but will still damage your customer's mission and your company's reputation (and potentially your company's financial well-being too, not to mention your own reputation and career prospects).

Here is the lesson to absorb: because of the vast number of data instances that today's complex engineered systems are required to process (what I call the problem of *scale*), there will inevitably be a significant occurrence rate where things will be very far from nominal. Normal statistical methods will neither inform you nor protect you; they assume symmetric distributions and skinny "tails" on the distribution. Both of these assumptions are *wrong* for our engineered systems: they have *asymmetric distributions* (that is, more samples with worse-than-average outcomes than samples with better-than-average outcomes; just like in our previous example of the distribution of commercial airline flights: there are many more flights that arrive late than arrive early), and also have a lot of *outliers* (a statistician would say that the "tails" of our system's adverse event distribution are "fat"). You must *protect your system and your users from bad outcomes* caused by these *asymmetric distributions* and these *outliers*; this is a vital goal for your project's design activity.

---

7 If our input data have a Gaussian distribution. If our input data have a non-Gaussian distribution (which is very common for engineered systems), there could be many more off-nominal data instances.

8.2.4 Signal and Noise

When we measure something, the value returned by the measurement is affected by two factors:

1. The actual value of the item being measured.
2. Errors in the measurement process.

For example, if we measure something on 1 March and the answer is 301, and we measure it again on 1 April and the answer is 302, then:

- All of the change might be due to a real change in the item being measured.
- All of the change might be due to errors in the measurement process.
- Or (most likely) ... some combination of the two!

That part of the change in the measurement that is due to an *actual* change in the underlying item is called the *signal*. That part of the change in the measurement that is due to *errors and variation* in the measurement process is called *noise*. *Every* measurement contains *both* noise and signal; in order to get meaningful data from our measurement, and to allow us to draw valid conclusions from the measurements, we must learn to *separate* the *signal* from the *noise*.

There are many sources of noise; they combine to create a *random variation*.

Most people, unfortunately, do *not* separate signal from noise. This is an opportunity for you to be better than your competitors.

Example: Last month, your staff predicted that you were going to finish your project on budget. This month, your staff's prediction says that you are going to be 5% over budget. How will you react?

Most project managers will ask "What went wrong?" They *assume* that the data are *valid*.

Instead, you should be asking "How much of that change is random variation (e.g. noise), and how much is an actual change in the signal?" I will show you one way to accomplish this – one that works particularly well on engineering projects – in just a moment. But first, we have one more piece of groundwork to establish: the strength of a statistic.

Statistics is the branch of mathematics that deals with the collection, organization, analysis, and interpretation of numerical data. A *statistic* is a measure that can be used to make a prediction. A statistic is said to be *strong* if it has good predictive power; it is said to be *weak* if it has poor predictive power.

The most common example of a statistic is the *comparison of a single current measurement to a single previous measurement*. You weigh one pound more today than you did a week ago (the two measurements are subtracted). Today's stock price for Google is 2% higher than yesterday (the two measurements are divided, one by the other).

Unfortunately, it can be shown mathematically that this form of statistic – the *comparison of a single current measurement to a single previous measurement* (whether by subtraction or by division) is a *weak* statistic. That is, such a statistic has almost *no predictive power* – it cannot separate signal from noise. Despite this weakness, this statistic is used all the time (think of quarterly company reporting: "profits are up 2% compared to this quarter last year"). This is another opportunity for you to be better than your competitors!

If one is trying to derive signal from multiple measurements of the same item, two measurements are *not enough*. One usually cannot separate signal from noise in a measurement until one has more measurements; seven is a typical minimum number. You must also ensure consistency in the methods used to make all of the measurements, and you must wait long enough between the measurements to allow for the possibility that something real will actually have changed. If you weigh yourself once per minute, most of the change that you might see from measurement to measurement is simply noise.

Once you have a series, there are simple tests (described below) that can then separate signal from noise; if the data do not pass at least one of these tests, the changes in the data are likely just noise.

Let's look at Figure 8.4. This is called a *control chart*, or a *process behavior chart*.[8] The chart looks a little complicated, but really isn't. First of all, the chart represents a series of measurements of the same item; in the example, it was the monthly prediction of some technical capability for an engineered system. The prediction was made during the first week of each month for 36 consecutive months. The sequence of months is shown on the X-axis; the monthly prediction is the line with the small circles at each monthly measurement, labeled as "monthly results" in the legend. Remember, this measurement includes a mixture of the true *signal* and the random variation that I called *noise*. To use the measurement, we want to separate what changes in the measurement are due to an *actual change* in the signal, and what changes are due instead just to *noise*.

Every one of the other lines on the chart is *calculated*; only the monthly results is an actual measurement. The purpose of all the calculated lines is to help us *separate signal from noise*. There are calculated lines that represent what statisticians call 1-sigma up and 1-sigma down from the measurement. There are also calculated lines that represent 2-sigma up and 2-sigma down from the measurement. And there are calculated lines that represent 3-sigma up and 3-sigma down from the measurement, which receive the special names *upper natural process limit* and *lower natural process limit* (these names are abbreviated in the legend). There is a line that represents the average of the measurements to date, and a line that represents the desired or contractually mandated value (called in the legend the *minimum spec*). At the bottom of the chart are three separate lines, called *moving range, average moving range*, and *moving range limit*. To separate signal from noise, you look at the most recent set of measurements (e.g. the item labeled "monthly results") for five different effects, which I call the *Wheeler/Kazeef tests*. If any of these five tests result in a positive outcome, then that data point is likely an actual signal, and not just noise. Here are the five Wheeler/Kazeef tests:

1. There is one point outside the upper or lower natural process limits.
2. Two out of any three consecutive points are in either of the 2-sigma to 3-sigma zones.
3. Four out of any five consecutive points are outside the 1-sigma zones.
4. Any seven consecutive points are on one side of the average.
5. Any measurement of the item called *moving range* is above the item called *moving range limit*.

8 I learned about this type of chart from the writing of Donald Wheeler, SPC Press, and from Mike Kazeef, a professor at the University of Southern California.

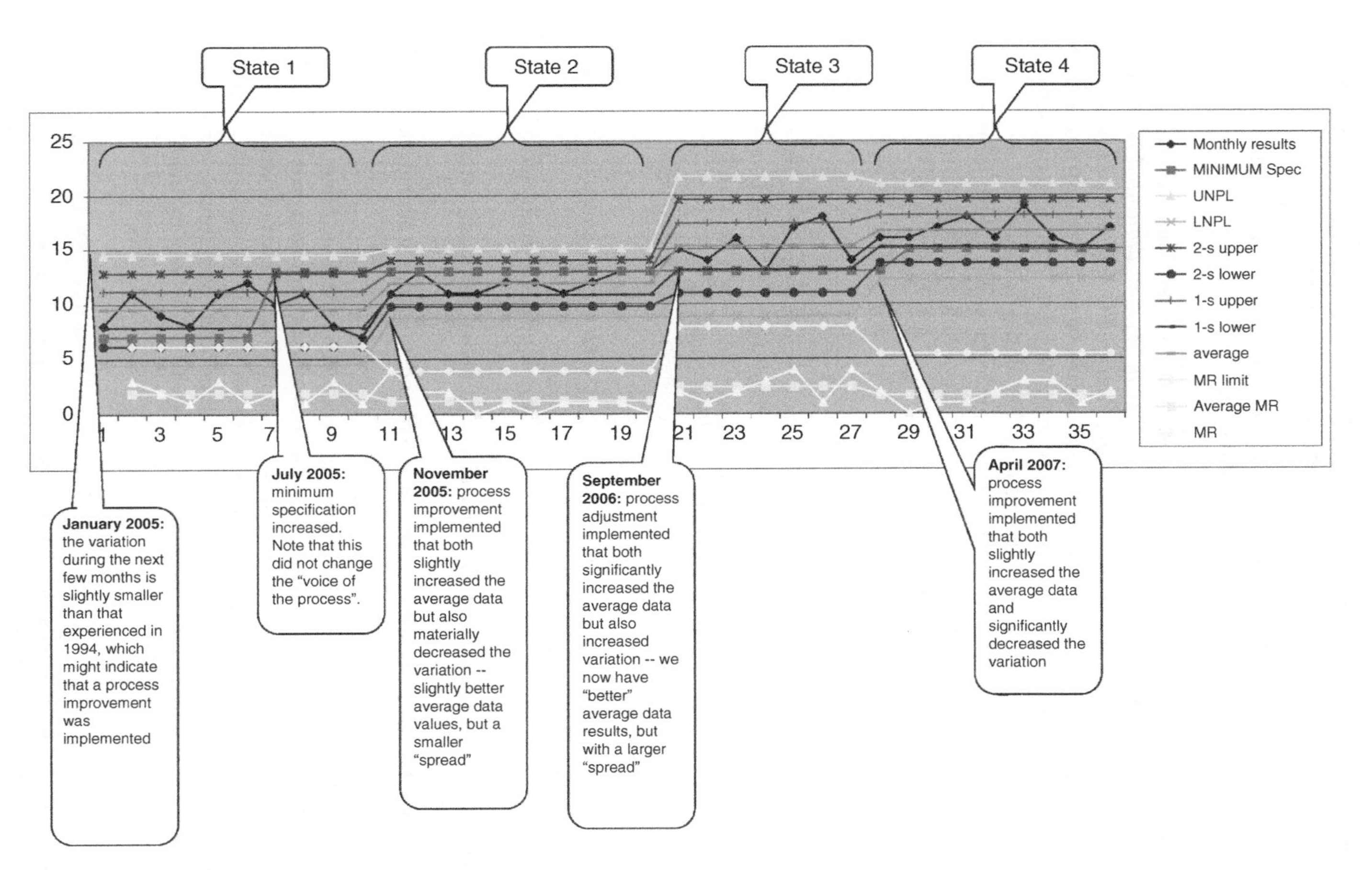

Figure 8.4 The process-behavior chart (also called the control chart).

That's it. If *none of these tests comes back positive*, the *changes in the measured data are likely due only to noise*; there has been no actual, meaningful change in the item being measured. That is, there is no actual change in the *signal.*

With a little practice, you can learn to apply these tests in seconds.

When you see the big changes in the calculated lines, that is another clue that a real change may have taken place in the state of the system, which is why I have annotated the chart with "state 1" and so forth.

As the project manager, of course, you almost never build and analyze a control chart yourself; what you actually do is *establish project policy that data used for decisions must be collected and analyzed using the control-chart methodology*, ensure that appropriate training is provided to your team, and that someone is checking up that everyone is actually following through. You also set an example by *rejecting* any analysis or any decision that is brought to you, unless you see that all the quantitative data have been processed via the control-chart methodology. This includes all of the information prepared by every cost-account manager as a part of the periodic management rhythm that we will discuss in Chapters 10 and 11.

8.2.5 A Special Type of Measurement: The Test

A *test* is a particular type of measurement that returns only a single, binary result: pass or fail. The key insight in this section is that the result of the test may be *wrong*; it may say "pass" when the true state is "fail," or it may say "fail" when the true state is "pass." And some of the time, the test result will actually be correct.

That is to say, each time we conduct a pass/fail test, there are *four* possible outcomes, not two:

- The test might say "pass," and the particular item under test might actually be good. This is a "*true positive.*"
- The test might say "pass," and the particular item under test might actually be bad. This is a "*false positive.*"
- The test might say "fail," and the particular item under test might actually be good. This is a "*false negative.*"
- The test might say "fail," and the particular item under test might actually be bad. This is a "*true negative.*"

That is, what the test *says*, and what is *true*, need *not* be the same thing!

This situation can be represented by a 2 × 2 matrix called a *Veitch diagram*[9] (Figure 8.5).

There are actually two Veitch diagrams in Figure 8.5, one that expresses the outcome in counts (e.g. the number of instances for each of the four possible outcomes) and another that expresses the outcome in percentages. Sometimes you might be given the counts and have to derive the percentages, or vice-versa; sometimes even a bit of each.

9 I learned about Veitch diagrams from George Friedman, formerly chief engineer of The Northrop Corporation, and later a professor at the University of Southern California.

	COUNTS			
	propulsion unit is ACTUALLY:			
	Good	Bad		
test says "good"	26	5	**31**	
test says "bad"	2	3	**5**	
	28	**8**	**36**	total events

	PERCENTAGES			
	propulsion unit is ACTUALLY:			
	Good	Bad		
test says "good"	72%	14%	**86%**	
test says "bad"	6%	8%	**14%**	
	78%	**22%**	**100%**	vertical check-sum
			100%	horizontal check-sum

P(G | test says good) =		**84%**		
P(B | test says good) =		**16%**		1 checksum
P(G | test says bad) =		**40%**		
P(B | test says bad) =		**60%**		1 checksum

Figure 8.5 The Veitch diagram.

At the bottom of the figure, underneath the two Veitch diagrams there are four additional lines. These are the *conditional probabilities*. The first line reads "the probability that the item is actually good, given that the test result says that it is good, is equal to 84%." You read the other three lines in a similar fashion.

The point I want you to understand is this: we may run a test and get an "answer," but that answer might be wrong! Just as we aspire to separate *signal* from *noise*, we aspire to separate true positive and true negative test results from false positives and false negatives. False positives and false negatives are, of course, themselves just a particular form of *noise*. When your staff analyzes outcomes, be sure that they are allowing for false test results. As we saw in the doctor example earlier in this chapter, if we do not account for these false test results properly, we are likely to draw an incorrect conclusion, and therefore make poor decisions.

8.2.6 The Decision Tree: A Method That Properly Accounts For Conditional Probabilities

A common management decision is to choose among various options for a course of action. One method of *informing* this decision (the decision itself should *always* be based on your judgment of the totality of circumstances, and not just based on the numbers – the proper role for data is to *inform* your decision, but not to *make* your decision for you) is to create a single numeric measure that can be used to compare the alternative courses of action. A typical such measure is *expected cost*.

There is a method called the *decision tree* that allows one properly to assess the expected cost of a set of alternative courses of action.[10] The decision-tree methodology deals correctly with dependencies (conditional probabilities) – something where (as we saw above in the doctor example) human intuition is *often* wrong!

A decision tree is a multi-step horizontal tree, branching out of an initial condition on the left. The tree has two types of *nodes*, one called a *decision node* and another called a *chance node*. Using these two types of nodes, a step-by-step depiction of multiple complex courses of action can be created from left to right, across the page.

At the decision nodes, multiple candidate next steps in the courses of action arise from the right-hand side of those nodes. Each line (e.g. candidate next step) that emerges from the decision node can have a cost assigned to it. For example, if one of the branches coming out of a particular decision node involves conducting a test, you can assign the estimated cost of conducting that test to that line.

At the chance nodes, you provide a probability that the step represented by the line going into the left-hand side of the chance node will work, or not work; or if there are *more* than two possible outcomes (e.g. the voltage produced will be more than 100 V, the voltage produced will be between 50 and 99.9 V, the voltage produced will be between 25 and 49.9 V, the voltage produced will be less than 25 V), you assign a probability to each of those possible outcomes. The probabilities, of course, must add up to 100%.

You can nest combinations of decision nodes and chance nodes to any extent that is required by the problem. At the very right-hand edge, you always end each branch with a chance node, and also assign a cost to each line emerging from these final chance nodes. For example, the entire decision tree might represent some alternatives for a key portion of your design, and your contract might call for a financial penalty if a certain requirement is not met (or a range of financial penalties, that vary with the magnitude of the shortcoming[11]). An example of a decision tree is provided in Figure 8.6.

The decision logic unfolds from left to right, but when you are done laying out the tree, and want to make the calculations – which will create an estimate of which of the potential courses of action has the lowest expected cost – you perform those calculations starting at the extreme right-hand portion of the tree, moving leftwards. At each stage, you multiply cost times probability to reach an *expected cost* and keep moving leftwards. At the chance nodes, you *sum the expected costs from all the branches* coming into that node from the right; at the decision nodes, you instead *select the smallest number* as you go left across the node.

I am not going to spend a lot of time on the details of how to do it, because like so many other things you – as the manager of an engineering project – do not actually build and evaluate the decision trees used to inform decisions; you establish a project

10 If you wanted to use a single evaluation metric other than expected cost, you could still use the decision-tree methodology. But expected cost is the most common metric used in the decision-tree method.

11 This type of varying financial penalty is very common, often embodied in a portion of the contract called a *service-level agreement*. For example, you might be penalized a certain amount if your system is available only 95% of the time during a certain month, penalized less if your system is available 97% of the time, not penalized at all if your system is available 99% of the time, and actually receive a financial reward if your system is available 99.5% of the time.

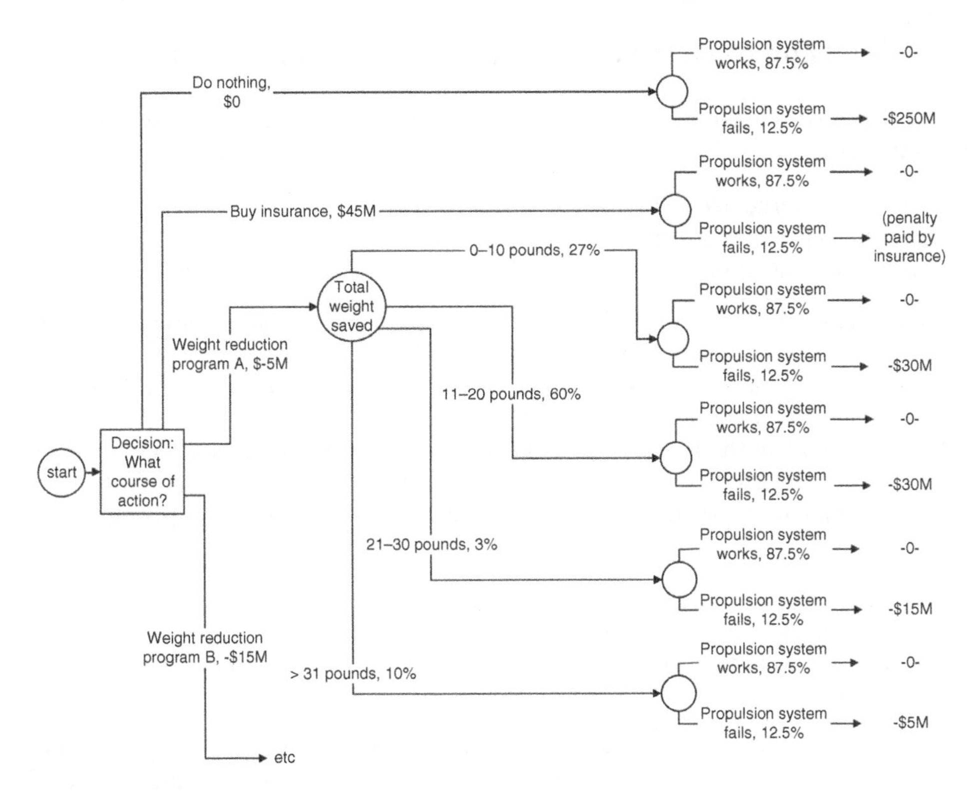

Figure 8.6 An example of a decision tree.

policy requiring the use of decision trees under the appropriate circumstances, ensure that people are properly trained, and so forth. Mostly, you look at the results; just as in the doctor problem, the results may be far different from your intuition.

The detailed instructions for how to do it properly belong in a different type of course, such as a course on systems engineering methods. Just in case you have not yet taken such a course, I do provide the complete problem statement, and the "solved" version of this particular example decision tree, in the accompanying teaching materials, which are available to your instructor.

There are some phenomena (e.g. failure rates of electronic parts) where an *expected-value* approach (such as the decision tree) is valid. There are other phenomena for which an expected-value approach is *not* valid; these may have unusual (e.g. strongly non-Gaussian) distributions, or the signal-to-noise ratio may be so small that the behavior is dominated by the noise, or have other characteristics that make an expected-value approach not valid.

One more hint: Don't fall into the trap of trying to predict something that is inherently unpredictable, because it is dominated by noise or random variation, or for which you simply do not have enough data to build a valid predictive model. The correct management strategy in those cases is *not* to try to make predictions, but instead to *take actions to (i) decrease your exposure to bad outcomes and (ii) increase your opportunity from good outcomes.* This is part of *risk management,* which will be the subject of Chapter 9.

8.3 What Engineering Project Managers Need to Measure

As managers of engineering projects, we need to have our team make three types of measurements:

- Technical performance measures
- Operational performance measures
- Management measures.

We discussed the technical and operational performance measures earlier. We will discuss management measures in Chapters 10 and 11.

As pointed out above, the making of these measurements, and the interpretation of the resulting data, are often done poorly.

Note that most of the technical and operational performance measures are *predictions*. We measure things *now*, but what is important is using the data from those current measurements to predict how the system will perform when it is finally completed. For example:

- What will the processing capacity of the system be, *once we are done*?
- What will the weight of the satellite be, *once we are done*?
- For how long will the battery be able to power the system without being recharged, *once we are done*?

We already talked about the necessity for you to understand how your customer determines value, and to understand what is *their* "coordinate system" of value. We also already talked about the necessity of then relating the *technical* performance measures (which represent our degrees of design freedom) to the *operational* performance measures (which represent the goodness of the resulting system, as interpreted by the customers). We can now see that doing this – both the making of the measurements and the method of relating the two coordinate systems of value to each other – involves processes that are *subject to all of the errors described in this chapter.*

Our third type of measurements are the management measures. Here are some examples of management measures:

- What will it cost, when we are finally done?
- When will we likely be done?
- How many requirements have we tested to date? When will we likely be done with testing?
- How many problem reports (at each level of severity) remain open? When will we likely reach a state of zero remaining level-1 open problems?
- How many unfilled positions exist now on the project? How many positions were we able to fill in each of the last three months? How is our staffing posture affecting our ability to complete the project on time?

We will have a lot more to say about management measures in Chapters 10 and 11.

What we just said about the technical and operational measures being subject to all of the errors described in this chapter is also, of course, true for all our management measures.

8.4 Implications for the Design and Management Processes

8.4.1 We Need *Measurements* in Order to Create Good Designs

I started my college education as a mathematician; I admired the formal rigor of mathematical proofs, and the theoretical insights that result. But after graduation, as I started building complex engineered systems, I learned that such systems cannot be built based exclusively on theory; there is too much uncertainty, the actual operating environment for the system is too complicated for any reasonable theory, there are too many competing interests among the stakeholders, and the conditions in the real world are never as clean as the assumptions in a mathematical proof.

As a result, we engineers must use empirical methods, and *tinker*; we can never create good designs based solely on theory, models, and algorithms. In order to create effective designs via tinkering, we need *measurements* from our models and prototypes, providing data that allows us to refine the design, and to guide the tinkering process. But those measurements – and their interpretations – must be reasonably correct! As we have seen, not only are there many ways in which errors can be introduced, but the magnitude of those errors can be stunning, and thereby drive us to completely erroneous decisions.

8.4.2 Projects Provide an Opportunity for Time Series

As we will see in Chapters 10 and 11, as part of routine project operations, we will be remeasuring all the operational performance measures, the technical performance measures, and the management measures every single month. Because of this, on an engineering project, we automatically obtain the data needed to create a *time series*.

I showed you that we can analyze the data in a time series using the process behavior chart (also called the control chart); this chart – and the associated five Wheeler/Kazeef tests – will help us separate *signal* from *noise*, so we can try to make decisions based only on the signal. As the project manager, you must *create a project policy* that requires our measurements to be gathered and analyzed as *time series*, using the *process-behavior chart* and the *five tests*.

Most projects do *not* do this. As a result, the managers of those projects make decisions based on data that contain *noise*. Those tend not to be good decisions!

8.4.3 Interpreting the Data

We must not assign too much confidence to the detailed accuracy of the specific predictions though. We are looking for *jumps*, not little incremental changes; if your satellite must weigh less than 1000 pounds in order to be launched into the correct orbit, you will likely design it to weigh somewhat less, perhaps 950 pounds. We use the time series and the control chart of the predicted weight for each subsystem (and hence the predicted weight of the entire satellite) to tell us if something has gone *drastically wrong*; the difference between a prediction of 914 pounds and 917 pounds is probably below the level of fidelity of the prediction. What we are concerned about is a signal that indicates something is out of control, that big changes are either possible or looming: something that might drive the weight to 1050 pounds. Don't try to get more accuracy from the predictions than is warranted.

We are looking for the things that will *break our design* and *cause the project to fail.* Time is a powerful design corrective, so finding these things early is *vital.* In addition, experience shows that it gets progressively far more expensive to fix things as you move from stage to stage of your project's development cycle. It costs a lot less – about 100 times less – to fix something during the requirements or design stage, as compared to fixing it during the integration or test stage. So, there is significant value in finding problems earlier, rather than later.

8.4.4 How Projects Fail

The following is a pretty typical failure scenario for an engineering project:

- During the *requirements, design,* and *implementation* stages, all of the measurements indicate that the project is completely on time and on budget. All is well.
- Then, the project enters the *integration* stage.
- As multiple pieces are put together into larger and larger subassemblies, things that appeared to work well in the individual parts start showing signs of very significant problems. The system crashes every 10 minutes. The system processes data 100 times slower than it is intended to do so. And so forth.
- We laboriously track down these issues one by one. Each problem takes far longer to find and fix than we thought it would; as a result, our predicted project end date starts "slipping to the right" as fast as the calendar progresses, or even faster. Each problem turns out to be a problem of *unplanned dynamic behavior*: not a problem with the individual parts, but with the way that the parts *interact.* Things occur out of their planned sequence. Things queue up unexpectedly. Off-nominal data (outliers) or user actions cause the system to behave badly.
- But there seems to be no end to the occurrence of such problems; fixing one does not prevent new examples of such bad behavior from occurring. No prediction about progress toward the completion of our project turns out to be justified; the predicted end date just keeps slipping. And people get discouraged: they work very hard and fix one such problem, but three days later a new, equally difficult and equally detrimental problem is found ... and none of the previous corrections fix that problem. We have to start the diagnosis process entirely afresh.
- The project is soon canceled, because the customer has lost confidence in your ability to manage and deliver. Or you are fired, and someone else is given a chance to finish the project. Neither of these outcomes is good![12]

How did this happen? They *did not measure the right things,* and therefore did not think about the right things in the course of their design. As a result, they did not actually know if their design was going to work or not.

12 There is a bit of engineering folklore called the "90/90 rule of project management," which says that "The first 90% of the project's work accounts for the first 90% of the project's schedule. The remaining 10% of the project's work accounts for the next 90% of the project's schedule." These paragraphs explain why this happens!

How do you avoid this? Through (i) paying attention to the dynamic behavior in the design, especially the part that I called "preventing the dynamic behavior that you do not want" (Chapter 2); (ii) the creation of good operational and technical performance measures (Chapter 4); (iii) creating a work plan that addresses the difficult portions of the design early in the project, rather than doing just the easy parts first (Chapters 2 and 7); (iv) employing accurate measurement methods and valid analysis processes (this chapter); (v) creating a strong risk-management process (Chapter 9); and (vi) employing good techniques for monitoring the progress of your project (Chapters 10 and 11). Most engineering projects that fail do so because they have *bad designs*; they had bad designs (in part) because they did not do these things well.

8.4.5 Avoid "Explaining Away" the Data

Having – through avoiding the mistakes described in this chapter – created all of these data, and having some understanding of the extent of validity, we must still be aware of our desire and ability to create retrospective and/or plausible explanations. If you observe (e.g. in a process-behavior chart) a big change that you did not predict, there is probably a real cause that is different from any of your conventional wisdom born out of prior experience. Look for a new cause. Beware the tendency to create a compelling story that "explains away" the significance of data that tells a scary story, or after-the-fact explanations that assert "yes, I did in fact expect that," when in fact the data are a surprise.

Try to obtain as much of the data that goes into your operational and technical performance measures as you can from experimentation, prototyping, and other methods of measuring actual operation. Decrease your dependence where you can on the predictions of theoretical models and algorithms. Favor tinkering over story-telling, actual data from past experiences over glib summaries of past history, and clinical lessons learned over theory.

8.4.6 Keep a Tally of Predictions

Here is one really effective way to "keep your feet on the ground," so to speak: take the time to keep a tally of the eventual outcomes of your predictions, and of the *predictions of others*, especially those considered "experts." Such a log ought to make you humble and wary about your predictions and the predictions of so-called experts (even the experts who work for you, on your project). I believe that such caution is both *useful* and *well justified*. Here's one of my experiences with such a tally of predictions:

- My wife and I do not have a television. So those few occasions when I find myself stuck in front of a television can be very interesting. Once, when I was on jury duty, I found myself sitting in the jury assembly room for days at a time. Of course there was a television in the room, tuned to continuous reporting from one of the Gulf wars. On an inspiration, I pulled out a piece of paper, wrote down the next 20 factual assertions by the TV reporters (no selection; no cherry-picking things that seemed the most important or the most interesting or the most likely to be wrong), and then posted that list to my calendar page for a date one year later. I figured that by one year later, I could assess the accuracy of those assertions.

- What did I find? One year later, exactly 2 of the 20 assertions turned out to be correct. Seventeen turned out to be completely wrong. One could be deemed either right or wrong.
- That's a "wrong" rate of 85% to 90%. Why so bad? I don't know. I suspect it is at least in part that the methods of television news reporting emphasize speed over accuracy (I also suspect that this tendency has only become more pronounced in the Internet era). What is important is that *television would appear to be nearly worthless as a source of information.* Despite the huge news-gathering budgets, despite the fame and reputation of those doing the reporting, these "experts" have a predictive ability *far less than if they just flipped a coin.* They probably make many of the mistakes that I have cited in this chapter, especially when it comes to quantitative information.

As should be apparent from the linguist and doctor stories with which I started this chapter, even highly trained and highly educated technical experts make bad predictions; the reason, as we saw, was that they (despite their training and expertise) make all of the mistakes listed in this chapter. Therefore, you as the project manager *cannot assume that your own experts will know how to avoid these mistakes,* or even if they know, that they will take the time and effort to do so. You must therefore create project policies that guide your people to do this work correctly, provide the necessary training and enforcement, and so forth.

Taleb[13] points out that when a set of winners of the economic prize in honor of Alfred Nobel tried to apply their financial predictions in a tangible way (by forming an investment company), they lost billions of dollars. Why should your systems engineers be better at making *their* predictions than those prize-winning economists were? My prescription: spend more effort protecting your project against fatal consequences (in Chapter 9, we will call this "reducing the potential impact if an adverse event comes to pass"), rather than depending on the accuracy of predictions.

8.4.7 Social Aspects of Measurement

There is an important social aspect to this too. Correct and valid methods of data collection and analysis are among the items to include in your efforts to achieve alignment among your people and customers. Good methods create credibility with your customers, and pride inside your team.

It is also the case that creating some of these measures – certainly most of the operational performance measures, and several of the management measures – requires us to understand the customer and (at times) people in general. So we, as engineers and managers of engineering projects, need to understand quite a bit about people and how they interact.

What is a good source of information about people and social structures? My conclusion from observation is that *people* and *social systems* change very slowly. We have a tendency, however, to want *recent* information. I believe that when dealing with the *technical* aspects of our project, that desire is appropriate (technical matters can change relatively rapidly), but I believe that depending solely on recent information may *not* be effective for understanding the people and social aspects of our project. Recent sources

13 In his book *The Black Swan.*

of information about people and social systems have not *stood the test of time*; they are simply of unproven worth (as we saw in my example about the validity of the information provided by television news). I believe that news is *actually* caused by factors that change slowly (culture, attitudes that were formed long ago because of long-ago events, and so forth) as much as it is caused by the things that change fast (like yesterday's events, or the advent of cars, electricity, plumbing, or technology in general). I therefore believe that I can understand the people, the sociology, and the world – even current events – *better by reading old history books and traditional cultural artifacts* (e.g. poetry that is considered culturally significant within a group) than by viewing television.

I always recommend that managers of engineering projects, and systems engineers, do vast amounts of eclectic reading; you never know what bit of remembered reading will trigger a thought or an insight during a crisis on an engineering project. I have certainly benefited many times by making such connections from my reading in many technical fields.

Furthermore, since (as noted above, and as we will discuss further in Chapter 13) the *people* and *social* aspects of project management are very important – taking up more than half of your time as project manager, if the project is of any size or complexity – this reading ought *not* to be confined just to technical readings. I myself read vast amounts of history, old novels, and the classics of Greece, Rome, and Persia. I firmly believe that Rumi, Austen, Bulgakov[14], and Aristotle have taught me more about how to deal effectively with people, and to identify and manage interpersonal conflict, than all of the many modern management books that I have read.

These books are *old*, I can hear you say; but my view is that if a book is still available hundreds of years after it was written, it is *more likely to have content of value* than a randomly chosen new book. It has *stood the test of time*. I urge you to read old books!

8.4.8 Non-linear Effects

I recommend that you use the data that you gather to look for *non-linear effects* in your design. In mathematics, we say that an effect is *linear* if a change in an input results in a change in the output that is, at most, a multiplicative effect (this is called *linear* because, if you plotted the relationship between the input and the output, the graph would be a straight *line*). Since multiplying a number by a very small number always results in a small number, a linear effect is also *stable*, in the sense that we defined that term in Chapter 2.

Such linear effects are intuitive and relatively easy to comprehend. But the behaviors of our engineered systems are *seldom* dominated by such linear effects. Instead, our engineered systems usual display multiple types of complex, unpredictable, *non-linear* behavior, where small changes in an input may result in huge changes to an output or a behavior. These are difficult for our intuition to grasp and predict, but, in my experience, such non-linear effects are *usually at the root of the difficulties* that cause big cost and schedule over-runs, and significant shortfalls in capability from that promised in the contract. Non-linear effects are also *unstable*, in the sense of Chapter 2 – small changes in conditions and inputs cause drastic changes in the behavior of our system.

14 His masterpiece is a novel called *The Master and Margarita*.

Non-linear effects can arise, however, from deceptively simple phenomena. Consider a rotating-media computer disk drive: the platter spins at a constant rate; the data are stored in a set of concentric circles of magnetic charge (called a *track*), each arranged in small angular sectors. If the goal is to read all of the information on one track in one revolution of the platter, the processing and information transfer processes must keep up with the angular rate of rotation. If either the processing or information transfer processes fall behind, the disk must rotate all the way around again before it can read the next sector. I once worked on a system with 32 such angular sectors per track; when we started, it turned out that a very tiny time-overage in the processing and information transfer processes caused the disk to have passed by the beginning of the next sector by the time the computer was ready to read that next sector, and as a result, the disk platter had to make an entire revolution before the next sector could be read. In this case, a tiny delay in timing – literally microseconds – caused the data transfer from the disk to take 32 times as long as planned, because the platter had to spin 32 times to read the entire track, rather than (as planned) just once. This is a classic example of a non-linear effect: a *tiny change in an input* (in this case, the speed of the processing and information transfer processes, taking 15 microseconds instead of the planned 10 microseconds, a mere 5 microsecond increase) caused an *enormous change in the behavior* (e.g. 32 times slower than planned).

Our engineered systems are full of such non-linear effects; in engineered systems, behavior can jump quickly from adequate to horrible, as some conditions change just a small amount.

These are the types of problems that kill projects. Don't be lulled by our intuition – which appears always to expect linear effects – into expecting only small deviations from planned behavior due to small variances from the plan. This warning applies to management measures, not just to operational and technical performance measures: a small increase in the duration of some task may result in a significant increase to both the predicted completion date of your project and the predicted cost at completion.

This is one of the reasons that good design practice has a large *empirical* component: our intuitions and our models tend to be linear, and we fail to imagine all of these non-linear effects. But if we use empirical methods (build prototypes, measure behavior under beyond-full-load conditions, and so forth), we have a chance to *observe the non-linear behavior early enough to try to fix it or account for it*, before it disrupts the progress of our project. It is therefore my experience that we cannot create good designs based solely on theory and algorithms; theory and algorithms can be effective for designing *data transformation*, but design is mostly about *structure* and *sequence control*, and not about data transformation. Structure and sequence control drive the quality of a design, and structure and sequence control are in my view exactly those aspects of a design that are most susceptible to non-linear effects. Therefore, creating good designs for structure and sequence control is largely an *empirical* activity, guided by past experience (via *design patterns*) and *measurements* (prototypes and actual full-load measurements). We must try alternatives and variants, and benchmark their differences. Such tinkering is important.

Be cautious too, in assigning *causality*. Causality may be far more subtle than you think, and even random chance (manifest through *outliers* in the data stimuli) may play a role in what you observe. There are probably many more explanations than you think for observed phenomena. Don't jump to conclusions, and never accept a single instance as "proof." Be curious, and not overly confident in your reasoning when drawing conclusions. Allow for the existence of multiple potential explanations, and use the design to protect your project from the implications if the explanation is not the one you at present favor. Through the design, you can protect against behavior, even without always understanding all of the potential causes!

Focus on using your measurements to find (as we discussed in Chapter 2) the potential sources of *unplanned adverse dynamic behavior*, and *prevent* them, rather than relying strongly on theoretical design constructs. Observe actual behavior, rather than depending on the predictions of theory.

8.4.9 Sensitivity Analysis

How can we find such non-linear effects? One way is by performing *sensitivity analyses* on critical predictions. Even though our models and predictions are based on assumptions that are not perfect, we can still investigate the *sensitivity* of that prediction: do small changes in the inputs and assumptions lead to small changes in the predictions? Or big changes in the predictions? These two conditions are very different.

We can create a graph of the predictions, as the inputs vary. Is the response to the varying inputs approximately *linear*? Or do we see the beginning of a sharp curving, a *non-linear* response? Is the response *symmetrical*, as we increase and decrease the inputs? Or is the response *asymmetrical*, that is, on one side of the variation the rate of change is much higher than on the other. We need to find such non-linear and asymmetric responses. We can find them by looking for variations of the inputs that cause the predictions to curve strongly up or down on one side as we vary the inputs.

When we find asymmetry and big variations in predictions due to fairly small variations in inputs, this should be investigated. Such *non-linear behavior* could become a catastrophic problem for our system; we may need to change that portion of the design to make it less sensitive to variation in the inputs and assumptions.

For example, *queuing* (such as waiting your turn to gain access to communication media, waiting your turn to gain access to a computing service, and so forth) is a common situation that we encounter in the design of our engineered systems. Queuing often displays highly *non-linear behavior*: below certain levels of activity, things proceed very smoothly, and the *wait times* to gain access to the communication media are quite predictable, but when the activity increases past some hard-to-predict threshold, the wait times suddenly jump by a considerable amount, and display very large variations, becoming very hard to predict.

Such discontinuities are not always predicted by our models (remember our example about cooling water). Hence my preference for actual *benchmarking* as a predictive method.

8.4.10 Keep it Simple

This need to prevent inadvertent adverse, unplanned dynamic behavior is why *simpler is usually better* in design. Everyone advocates the KISS[15] principle, but what is it that

15 "Keep it simple, stupid."

you actually measure in order to achieve a simple design? The sources seldom tell you what to measure, in order to see if your design is simple or not. Here is my favorite example of a tangible design parameter that you can aspire to keep simple: *the number of independently schedulable software entities within the mission software* for a large, complex system. I *always* measure this parameter when I design or evaluate a system. One of my best systems (still in use 30+ years later – and that is an eternity in the software business!) had only *seven* independently schedulable software entities in the mission software. At the same time, the same customer had another company building a system for a slightly different mission, but one that shared many of the same operational conditions and constraints. That contractor was having problems, and the customer asked me to take a look at their work. It turned out that they had no count or list of the independently schedulable software entities within their system! How could they expect to control adverse dynamic behavior? At my suggestion, they made such a list; it turned out that they had more than 700 independently schedulable software entities within their system's mission software. What human being could understand the potential interactions and implications of so many independently schedulable parts in a complex system? Their system was *never* fielded; it was about 100× less reliable than needed (and more than 1000× less reliable than my similar system). In the end, the next system that I built for this same customer (which had nine independently schedulable software entities within the system's mission software; I was unhappy that we went from 7 to 9!) was eventually adopted to take over the mission intended for the other company's project. The other company spent nearly $1 000 000 000.00 and yet produced nothing useful. They failed to implement the KISS principle.

8.4.11 Modeling

I wish to re-emphasize the importance of having the actual *model creators* be the ones who run their models, and the importance therefore of what I called in Chapter 2 the *segmented model* for system performance prediction. These model builders often spend years collecting data, and gathering real measurements and observations; they understand how their piece of the system really works. You therefore want those same people to run the models that form the core of your predictive process, because they will know how to keep the model from straying into predictive territory that extrapolates beyond credibility.

By the way, it is my experience that these individual models will *not* be full of *algorithms*, but instead will be full of *tables of actual measurements*. The model builder will know and remember the conditions under which each measurement was made too, and therefore has a chance to avoid unjustified extrapolation.

8.4.12 Ground Your Estimates and Predictions in the Past

As we discussed in Chapter 7, good companies have archives of *what actually happened on past projects*. Use them! Those archives provide real metrics about how much time and money it cost to do certain things, and what were the levels of technical and operational performance achieved by each of those previous designs. This is a precious resource, because it is factual.

The world of technology is full of fads about improvements. The improvements in technology can be measured, but my advice is not to be swept up in enthusiasm for tools that promise productivity breakthroughs in the efficiency of your *people*. I *never* incorporate such productivity improvements in my project estimates. By all means, use the breakthrough (if you have reason to believe that it will work, and will not introduce too many adverse side-effects and risks), but have a "plan B" to use an older, proven method, and create your schedule and cost estimation based on proven past levels of productivity from the company archive, not the promised improved levels of productivity. Do not try to capture the schedule and cost benefits until that particular breakthrough has been used a few times (not just once!); note that by the time that productivity improvement has been used and succeeded several times, it will have *automatically* become a part of your company's archive of past performance – the company's baseline of productivity will have incorporated the effect of that improvement.

Consider, for example, software development productivity (e.g. how many lines of software code per month the average software programmer can produce). It has not changed in the 60 years of the entire life of the software industry! Why should it? People are still people. We *do* get *more work* from each line of software, now that we write software in programming languages (like Java) that encapsulate more work in each line than we did when we used assembly language and FORTRAN. But the number of lines of code produced by a software programmer each month has stayed essentially constant for decades. Here's the potential trap: every decade, breakthroughs are announced that promise more lines of code produced by an engineer each month. They have all failed.

8.5 Your Role in All of This

You must yourself acquire some basic knowledge of the errors of logic and analysis described in this chapter, sufficient for you to spot these errors, and avoid making them yourself (or allowing your staff to make them).

You need to establish a formal, written project policy about how quantitative decision-making is performed on your project. I recommend that you mandate all quantitative data used for decisions be collected as a time series, and analyzed via the control-chart methodology. You also must build up the courage to *reject* any analysis brought to you that fails to comply with this guidance, no matter what the time pressure may be to make a near-term decision.

Your team will need training in how to do all of this. Your project or your company can create a training course that covers all of the materials described in this chapter (you can even use this book and this chapter as written guidance for such a course). You yourself should take the course, and do so in company with your staff. The shared experience of the course, and the implicit sign that you consider this important by spending your own personal time on the course, will be a very important team-building experience that will pay dividends for the duration of the project. Such a training course should include both practical team exercises and individual tests that must be passed at the conclusion; these are a sign of seriousness.

8.6 Summary: Drawing Valid Conclusions From Numbers

We are engineers ... so we must make measurements, and base our decisions (in part) upon these measurements. Measurements, however, inevitably entail *errors*. This is because measurements *always* contain *both signal* and *noise*.

One cannot make effective decisions from data unless one can separate the noise from the signal.

Most people and most projects do this very badly. But there are techniques – based on the use of *strong statistics* – that can actually accomplish this separation for us. Many of the statistics in actual use (such as comparing a current measurement to a single past measurement) are too weak to have any predictive power, and their use leads to poor decisions.

Conditional probabilities are everywhere; you must learn to recognize them, and not allow your team to fall into the trap of throwing out vital data, picking what they consider to be the "most important" item of data and ignoring the others. In fact, it is not very difficult to handle conditional probabilities properly (through the use of the Veitch diagram, the decision tree, and so forth).

Projects measure the same things (operational performance measures, technical performance measures, management measures) every month, so we have a natural opportunity to organize our measurements as time series and use the process-behavior chart (control chart) to help us separate signal from noise.

We must recognize the customer's "coordinate system of value" in our decision processes; the operational performance parameters help us to do that. But those operational performance parameters are quantitative data, and therefore subject to all of the errors described in this chapter. So are all of the management measures that we will discuss in Chapters 10 and 11. Just because they are not technical parameters does not prevent them from being handled in an invalid way.

8.7 Next

Now that we know how to handle measurements and analysis of quantitative information correctly, we next apply these methods to the problem of characterizing what could go wrong on our project (we call these *risks*), and how we can go about mitigating the potentially adverse effects of these risks.

8.8 This Week's Facilitated Lab Session

There is no facilitated lab session this week. The time slot for this week's session may be used for a mid-term examination.

9

Risk and Opportunity Management

Some things can (and probably will) go wrong on your project; how can you still get the job done? Every human endeavor entails uncertainty; things may not go as planned. In light of that uncertainty, how do you get the project done within the promised parameters, such as schedule, cost, technical capability, and so forth? In this chapter, I show you how to combine good engineering and good statistics in a manner that allows you to cope with these uncertainties.

9.1 Things Can Go Wrong With Our Project: How Do We Cope?

In any human endeavor, and most especially in a complex engineering project, things may go awry and not according to plan, and thereby cause problems. These problems might manifest themselves as problems in any of various ways: as problems in achieving the technical capabilities that we promised in the contract, as an increase in the predicted cost of completing the project, as a delay to the predicted completion date of the project, or as any of a variety of other problems, such as safety, reliability, our ability to staff the project with appropriate numbers and skills of personnel, and so forth; and of course, a problem might manifest itself as a combination of these symptoms.

Since our engineering projects are both *important* and *expensive*, we endeavor to think *in advance* of these problems arising about what might go wrong, and to take actions in advance that might lessen both the *likelihood that such a problem will actually occur* and the *adverse impact on our project and our company* if (despite our attempts to lessen the likelihood of occurrence) the problem comes to pass. At the point in time when we start thinking about the potential of these problems arising, they have *not yet come to pass*; they are only *potential* problems. We call a potential problem that has not yet occurred a *risk*.

We use the term *risk management* for this process of characterizing what could go wrong on our project, and for planning how we could go about both lessening the *likelihood* of problem occurrence and mitigating the potentially adverse effects of problems (the *impact*) should they in fact occur. Risk management is an important portion of your responsibility as the manager of an engineering project.

Engineering Project Management, First Edition. Neil G. Siegel.

Companion website: www.wiley.com/go/siegel/engineering_project_management

As noted above, every complex human endeavor is likely to encounter problems. You need not be afraid or embarrassed by the fact that there is a chance of such problems occurring on your project. What you must do, however, is lead your team to perform the appropriate *risk management* activities, so as to mitigate their likelihood and impact.

In this chapter, we will teach you something about risk management. We will show you how to identify your project's risks, how to evaluate those risks – both for likelihood and impact, and how to establish a set of reasonable mitigation measures. In Chapter 11, we will show you when and how you make decisions about if and when to actually implement those mitigation measures.

If *risks* are things that might go wrong with the project, and thereby cause an *adverse* impact on the established plans, there is, at least in principle, a set of symmetric potential events: things that might occur which, if they do occur, will cause a *positive impact* on the project. We call these *opportunities.* Like risks, we start thinking about them *in advance of their occurrence.* We look for actions that we could take that might *increase the likelihood* of their occurrence, and we also look for actions that we can take that will *maximize the positive impact* on the project should they occur.

We therefore call the combined process of managing both risks and opportunities *risk and opportunity management.* We will talk about opportunity management, as well as risk management, in this chapter.

I have, at times, managed non-engineering projects: construction projects, scientific research projects, artistic projects, and so forth. These can be challenging, but I have found that managing such projects is *materially different* from managing engineering projects. I characterize this difference as follows: engineering projects require significant *invention,* but not just invention (as is likely the case for an artistic project), but significant invention *simultaneously* with an objective standard of rigor (because the system must work in accordance with its requirements), and also an expectation that the project will succeed. There is seldom such an objective standard of rigor for an artistic project. There is seldom such a material level of invention for a construction project. A scientific research project may require invention and have some sort of objective standard of rigor (but usually, one that is far more flexible than that for an engineering project), but it is understood and accepted that a scientific project may not succeed! In my experience, it is *only* engineering projects that have all three characteristics: significant invention, an objective standard of rigor defining success, and an expectation of success.

This *combination* of characteristics creates a set of risks that are *different in kind and scope* from the risks generally encountered in non-engineering projects. This makes the management of risk (and opportunity) central to the management of engineering projects.

There is a potential trap entailed in the use of the word "management" in the phrase "risk and opportunity management": this phrasing sometimes creates the impression that we can address and solve these problems solely through the use of "management" actions: planning, organizing, structure, assigning tasks, monitoring task progress, and so forth. That is, in my experience, not correct! We must use *engineering* activities as the solution for many, if not most, of these problems; most of the time, we must "engineer our way out of the problems," and not just "manage our way out of the problems."[1]

1 This phrasing comes from my friend and former supervisor Barry Press, who said this to me in 1979.

As you will see in this chapter, engineering is deeply involved in every step of the process that we call "risk and opportunity management."

Technical and engineering risks can arise from all of the obvious sources:

- The invention required to create the desired system.
- The lack of maturity of some of the selected technologies and products; there is often a gap between what a product is advertised as capable of achieving, and what it actually can achieve.
- The scale and complexity of the project, including the problem of managing the dynamic behavior of the system (discussed in Chapter 2).
- All of the uncertainty and errors that can arise from quantitative measures and analyses (discussed in Chapter 8).
- ... and many others too.

Technology and engineering, however, are not the *exclusive* sources of risk on engineering projects. Risks can arise from other matters too, such as:

- Poorly defined or changing goals and requirements.
- Tensions between stakeholders: the buying customer, the eventual users, the paying customer, the regulators, and other stakeholders may want slightly different things, and trying to satisfy all of them may over-constrain your solution.
- Tensions within your development team: people may have different ideas about the appropriate design; if you have subcontractors, they may have different business aspirations from you and your company; there will always be some of the ordinary issues of people not getting along, and/or having conflicting personal goals and aspirations (see Chapter 13); and so forth.

The existence of all these potential sources of risk is, of course, one of the very aspects that makes managing engineering projects so interesting!

The data[2] indicate that huge numbers of engineering projects fail, and many more under-deliver on promised capability, take far longer than promised to complete, and/or cost far more than promised to complete. All of this indicates that most engineering project managers are not doing an adequate job of risk management. Here is yet another opportunity for you to be better than your competitors!

As stated above, we endeavor to minimize *both* the *likelihood of occurrence* of an adverse action and the *impact of occurrence* of an adverse action. We treat these as separate endeavors because experience indicates both that (i) *likelihood* and *impact* are fairly separate matters and (ii) they usually must be *mitigated through separate actions*.

Here's an example that illustrates this separation:

- If we want to decrease the *likelihood* that our house will burn down, we can take actions such as trimming the foliage away from the house, replacing the roof with completely non-flammable materials, replacing defective wiring, enclosing the eaves, and putting ember-resistant screening over the attic vents.
- But if we want to decrease the *impact* to our family in the case (despite our measures intended to lessen that likelihood) that the house does catch fire, we can install smoke

2 For example, Robert L. Glass in *ComputingFailure.com* (Prentice Hall, 2001) indicates that far more than half of all engineering projects fail.

detectors (to increase the chance that we are alerted early enough to get our family out of the house) and buy fire insurance (so that we will have the funds with which to rebuild the house, in case it does somehow burn down).

Buying insurance does *not* decrease the *likelihood* that the house will burn down, it only decreases the *impact* in case it does. Similarly, replacing the roof with completely non-flammable materials does not decrease the adverse impact to us if the house does happen to burn down anyway, but it does decrease the likelihood that it might. We can see that, in this case, managing *likelihood* and managing *impact* are fairly separate from each other.

So it is with the typical engineering project: managing the likelihood of various risks is fairly separate from managing the adverse impact that would occur if one or more of those risks came to pass.

The basic process of risk management works like this:

- Before we start the project, we try to figure out what might go wrong with our project, and we make a list of these items. We also list the *symptoms*, that is, what we will see if this risk starts to come to pass and threatens to become an *actual* problem, and not just a potential problem.
- We identify and write down ideas about actions which would lessen the likelihood and impact of those problems, but usually we *cannot afford either the time or the money to fix all of them.* Nor might we wish to spend that money, even if we had it: these are only *potential problems,* not confirmed problems. Furthermore, we recognize that as we progress with the project, we will uncover new potential risks; we cannot be content with doing this assessment just once at the beginning of the project. We need instead some sort of on-going activity to assess and manage risk.
- So, for most of the items that we identify as potential risks, instead of fixing them immediately, we try to figure out what we would do if, at some point in the future, it looked as if one of our identified risks was coming to pass, and we write that down on a list.
- We institute a *periodic process* (usually monthly) to look at the items on our list of risks. We add or eliminate things from that list, as seems appropriate in light of new data. We look for items on the list that seem to be coming to pass (by looking for the *symptoms* that we identified), and make a decision if this is the time at which we will elect to initiate the planned mitigation actions (which, of course, require people, time, and money to implement). We may also decide that we no longer require a mitigation action to continue that which previously we had started.

The key to this process is the *periodic assessment* of whether risks on our list seem to be coming to pass or not. The efficacy of this assessment will, of course, be driven by a combination of (i) how good your list of risks is and (ii) how good your list of symptoms is. On an engineering project, the list of problems and the list of symptoms are likely to be dominated by issues associated with *engineering* and *technology*. This is why, in my view, the manager of an engineering project should be an *engineer*, and why the project manager must take a personal role in the risk management process.

It is tempting for most people to say "Well, if anything goes wrong, it will show up eventually as an increase to the estimated cost to complete the project, or a delay to the estimated date of completion of the project, or both of these items. So, I can make most of my list of symptoms be statements about such increased cost or delayed completion date." Unfortunately, this does not work very well. Adverse schedule and cost impacts are the *result* of these underlying technical and engineering issues, and hence (in my view) are *lagging* indicators (that is, they show up later, giving you less time to respond). This makes schedule and cost predictions actually a *weak metric* for risk management – we are better off if we can create *leading* indicators. In my experience, you create such *leading indicators* by having your symptom list consist of statements about off-nominal *technical and engineering* properties, rather than statements about adverse schedule or cost impact. This is why we engineers – rather than, for example, the project financial manager – must be personally in charge of the project risk management process.

9.2 The Steps of Risk Management

Figure 9.1 presents the steps that I advocate as the basics of risk management.

I will now discuss each of these items in some detail.

9.2.1 Step a: Identify the Potential Risks and Opportunities

We start the risk management process by creating a list of potential risks, which we call a *risk register*. Where do we get ideas for what might constitute risks to our project?

- Through brainstorming sessions by the team members, and outside advisors.
- From lessons learned in previous projects (you ought to be able to learn these from your company's archive of historical project data).
- From consultations with experienced project managers, project chief engineers, and other experts.

You must hold brainstorming sessions about risk, starting long before you even begin to write a proposal; you will need the information about risk in order to get authorization from your company to submit a bid (as discussed in Chapter 5), and of course you must also include the risk register and the risk management plan in your proposal.

These brainstorming sessions are always fruitful, not only for the content they create, but also for the team-building role they play. You, as the project manager, should attend these sessions in person. I advocate, however, that you not chair these meetings yourself, but have someone else, probably your project's chief engineer, chair them. This demonstrates your confidence in your chief engineer. It also gives you the chance to listen; I always find that my ideas are changed by these sessions. If I speak too early, my status as the project manager may discourage some of the other participants from voicing their honest opinions. This notion of *the boss speaking last* for each topic is always a good meeting-facilitation strategy; you may ask questions at any time, but be very careful not to reveal your own opinions until you think that all of the ideas are already on the table.

As we discussed in Chapters 5 and 7, mature organizations ought to have an *archive of historical data from past projects*, and this archive ought to include the risk registers from those projects, together with a list of those risks that actually came to pass, and an assessment of the actual performance/cost/schedule impact that occurred as a result of

Step	Discussion
a. Identify the potential risks and opportunities	We call these lists the *risk register* and the *opportunity register*
b. Identify the symptoms	What will we see that will indicate that this risk or opportunity may be coming to pass? These become the triggers for potential future mitigation / exploitation actions (below)
c. Select the item to be measured, and the measurement methods	What do we measure, in order to look for the symptom? How do we make that measurement?
d. Score each risk and opportunity for both *likelihood* and *impact*	Evaluate every risk and opportunity in two ways: a numeric assessment of the likelihood of this risk coming to pass and a numeric assessment of the impact, if this risk or opportunity in fact does come to pass
e. Create mitigation and exploitation plans	For each risk, create two mitigation plans: one for reducing the *likelihood* that this risk will come to pass and another for reducing the *impact* of this risk, should it in fact come to pass. For each opportunity, create two exploitation plans: one for increasing the *likelihood* that this opportunity will come to pass and another for increasing the *impact* of this opportunity, should it in fact come to pass
f. Create triggers and timing requirements for those mitigation plans	For each mitigation plan, select a specific measurable value for the symptom that will become the trigger threshold for taking action. Also, estimate the time required to implement each mitigation and exploitation plan. We use the combination of the trigger and the timing requirement to make decisions about starting mitigation and exploitation plans
g. Create a method to aggregate all risk assessments into a periodic overall project impact prediction	A typical engineering project will end up with many items on the risk register. It is tempting to look at just a few risks that you consider the most material, but sometimes the aggregated effect of lots of small risks can result in a big adverse impact. So, we need a method to aggregate the totality of risks into some type of project-wide assessment
h. Create and use some sort of periodic "management rhythm" wherein you periodically make decisions about risk-mitigation and opportunity-exploitation actions, based on the periodic assessment	We institute a *periodic process* (usually monthly) to look at the items on our list of risks and opportunities. We add or eliminate things from that list, as seems appropriate in light of new data. We look for items on the list that seem to be coming to pass, and make a decision if this is the time at which we elect to initiate the planned mitigation actions (which, of course, require people, time, and money to implement). We may also decide that we no longer require a mitigation or exploitation action to continue that we previously started
i. When risks actually occur (and therefore transition from *risks* to *issues*), perform a *root-cause analysis*	Once a risk has become realized (e.g. transitions from a risk into an issue), it is often appropriate to do a *root-cause analysis*. That is, why did it occur? This might decrease the likelihood that you will have a similar problem occur in the future

Figure 9.1 The basic steps of the risk management process.

those risks transitioning into actual issues. In my view, these example artifacts are among the most important inputs to help us construct our new project's risk register. Some of the risks may be directly applicable to your project, but in many other instances I have found that I simply get ideas about the risks that might occur on my project by considering those risks that were either considered, or actually came to pass, on prior projects. Even if your project is creating something completely new, many of the parts and processes will still be similar to those prior projects, and you can therefore learn a lot from their experiences.

I often get asked to come and help a team that is getting ready to write a proposal. They always show me their risk register; it is almost always full of superficial statements of risk, far too general, and not sufficiently reflecting the specific details of their project. An example of a poor (but very common!) sort of risk statement that I have actually seen time after time is "The software could be late and/or over budget." This is a poor risk statement for many reasons:

- If a project has a lot of software (and these days, most do), the software is highly likely to be the source of a lot of risks. These need to be enumerated separately, not as one "catch-all" category. This is so that they can be assessed and mitigated individually.
- A statement of risk such as the above only lends itself to a symptom of late delivery or significantly increased cost estimate. Such symptoms tend to come *fairly late* in the project life-cycle (think of the *how projects fail* scenario presented in Chapter 8), and as such are *lagging indicators* that probably occur too late for easy correction. We want *earlier* indications of our problems (we call these *leading indicators*), and as noted above, these usually have to be technical or engineering in nature. Therefore, our risks must be stated in sufficient detail to allow for the identification of a symptom (which can become a trigger event) early enough that we can credibly solve the problem.

Here is an example of a good software-related risk statement: "The XYZ mechanism could fail to meet its requirement to complete port-to-port processing in less than 100 ms at least 99% of the time, and could also fail to meet the requirement that no instance of port-to-port processing ever exceed 250 ms." This statement makes it very clear what needs to be measured, and what the trigger thresholds for initiating mitigation actions will be.

These days, most organizations that perform engineering projects have guidelines regarding the creation of a risk register; where they fall short is whether the risk statements themselves are any good.

The entire process described above should be mirrored to look for opportunities (e.g. changes that can *improve* project outcomes – whether technical, operational, safety, reliability, cost, schedule, etc.). I call the resulting list an *opportunity register.*

9.2.2 Step b: Identify the Symptoms

Once we have the beginnings of a risk register, we turn to the problem of how we determine if a risk (or opportunity!) is coming to pass, that is, transitioning from a risk (a *potential* issue) into an *actual* issue (or from a potential *opportunity* into an actual improvement). We start this process by identifying the symptoms, that is, for each risk or opportunity, what is it that we will see (or could measure) that will indicate that this risk (or opportunity) may be coming to pass.

These need not be esoteric or difficult. If you have created suitably specific statements of risk, as described above, it may be fairly obvious what the symptoms are going to be. Consider the example from the previous section: "The XYZ mechanism could fail to meet its requirement to complete port-to-port processing in less than 100 ms at least 99% of the time, and could also fail to meet the requirement that no instance of port-to-port processing ever exceed 250 ms." There are two separate potential symptoms in that statement. The first is that port-to-port processing may not complete in less than 100 ms often enough, and the second is that port-to-port processing may sometimes take too long.

From these symptoms, we will later derive the triggers for potential future mitigation actions (see below). Note that we have not yet decided exactly what numeric values will constitute such a trigger; we have only identified the category of behavior that we will watch for. The exact numeric value that constitutes a trigger will come later.

9.2.3 Step c: Select the Item to be Measured, and the Measurement Methods

What do we measure, in order to look for the symptom(s)? How do we make that measurement?

For each item on the risk register, we identify the operational measures, technical measures, and management measures (which includes schedule and cost) that will help us look for the occurrence of the selected symptom(s).

Some of our selected symptoms will make it clear what needs to be measured (as was the case in our example above). But this is not always the case; sometimes, the selection of the item to be measured is subtle enough that determining what needs to be measured can constitute a separate step.

We must also determine how we will go about making the measurement. This starts with a determination of how accurate a measurement we need. If the symptom is the weight of a part or subassembly, the measurement method may be straightforward if the accuracy required is to the nearest pound; it may be more complicated if the accuracy required is to the nearest milligram.

Nor is it always the case that the measurement method itself is always straightforward; for example, in our port-to-port timing example above, it may be difficult to separate the timing of the section we are concerned about from other sections of the product. This is especially true when the item whose timing we want to measure consists of software. Making the timing measurement might require *special measurement equipment*; in the port-to-port timing example, this might include special software to log events and the time of their occurrence, and might include actual measurement equipment, separate from the mission equipment. We must also be aware of the potential that the process of simply making the measurement will change the behavior; for example, if capturing the data requires that you insert extra software code, the insertion of this extra code will change the timing and other factors of the execution.

9.2.4 Step d: Score Each Risk for Both *Likelihood* and *Impact*

Not all risks are equally likely to occur, nor will they all have the same impact if they do occur. We therefore evaluate each risk on the risk register in two ways: a numeric assessment of the likelihood of the risk coming to pass and a numeric assessment of the adverse impact, if in fact the risk does come to pass.

In general, we will find that we can make more credible estimates of the *impact* than of the *likelihood*. That is not a problem; the method does *not* depend on our being able to make highly accurate estimates of likelihood; all we really need to do is try to separate those more likely from those less likely, and even if we get the likelihood estimates dramatically wrong, the method still provides a fair amount of value. I explain why this is so below.

Notice also that our "risks" fall into very different categories:

- When we assess risks to the *design*, we are actually assessing the *characteristics of a thing that exists now* – the design exists *now*. We are trying to assess its current characteristics (e.g. can it process 1 000 000 data-input instances per second)? We call this a "prediction," but it really is not a prediction in the sense that this term is normally used; it is instead an assessment of a current state of being, and not a prediction about a future event. We probably ought to use a term other than *prediction* for this process, but it is the term that you will hear being used, and I have reluctantly decided not to try change the world in this regard. No matter what term is used, it is not really a prediction, but rather a statement about the *uncertainty* of our analysis about the current design.
- In contrast, when we predict the schedule and cost of our project, we are in fact making *predictions about future events*. The business of making predictions about the future is always fraught with error; the track record of even the most eminent specialists is absolutely awful. I will go so far as to say that predictions about the future are practically never correct. At least we have this comfort: the future events that we try to predict as managers of an engineering project are far more structured than, say, the chain of events leading to the next big earthquake in southern California. We likely understand at least *some* of the causality that drives the schedule and cost of our project, although we certainly do not understand it nearly as well as we would like. So, while our predictions about schedule and cost ought to be better than someone's prediction about when the next big earthquake will occur on the San Andreas fault, our predictions about schedule and cost are still *predictions*. Therefore, they will likely display far more uncertainty than our "predictions" about the performance of the design, because – as noted above – we are not actually "predicting" anything about the design, but instead are just continually improving our assessment of that design. Much of the variation of our "prediction" of design performance comes from the fact that each day and each week, we are continually adding details to that design, and presumably improving it. This is quite different from our predictions about the schedule and cost of our project: those are actual statements about future events. The reason that these predictions for items like schedule and cost have any utility at all – in light of the fact that the business of predicting the future is hard – is that, as you will see in a moment: (i) we use a process that *continuously uses our improved knowledge* to reassess the schedule; (ii) we focus on *very near-term events*, which we likely actually understand reasonably well; and (iii) we are looking as much for *indications of significant change* as we are trying to predict an absolute date.

So, bear with me here. I fully recognize that predictions about future events are almost always wrong. But that does not mean that you cannot learn from the process of making those predictions. That is what we are trying to do: learn from our process for making those predictions, and use that learning to improve the way we execute the project.

We are *not* in the business of being seers or necromancers, or even economists, all of whom pretend to predict the future. You saw earlier in this chapter that although we attempt predictions for schedule and cost, the items upon which we prefer to depend to be our *leading indicators* of potential problems are the *technical and operational performance measurements* (such as the characteristics of the design). Note that I just pointed out that in some real sense these are not actually predictions at all; they are measurements of a current state. Our concern is with the uncertainty of that measurement. You will also see (a little later in this chapter) that I tell you essentially to ignore the estimate of likelihood if the estimate for impact is severe. These mechanisms take us out of the realm of seers and necromancers, and into the realm of engineers making periodic measurements, and using the insights from those measurements to tinker incrementally with our design.

The numeric assessments are going to become entries in some sort of database, so that we can easily perform assessments with them (as will be described below). But projects often find it useful to make use of a graphic depiction of this scoring as well, which we call the *risk matrix* (see Figure 9.2). This is a graphic depiction of those two dimensions, likelihood and impact. Each risk is placed at its appropriate location in the matrix.

It has become common to color-code the squares within the risk matrix, as indicated in Figure 9.2 (I know that this book is printed in gray-scale, so you cannot see the colors, but imagine red in the upper-right-hand region; green in the lower-left-hand region; yellow in the remainder). Clearly, if something is both likely and of high impact, that is a more serious risk to manage than one that is both unlikely and of low impact. But the color-coding (which I include here, because you will likely see it on your projects) *contains a trap*: it tends to *mislead* us about how correctly to respond to those risks that we have rated unlikely to occur, but severe in their impact if they do occur (as is the case for the risk identified as "NGT24" in the figure). We will return to these types of risks later in the chapter.

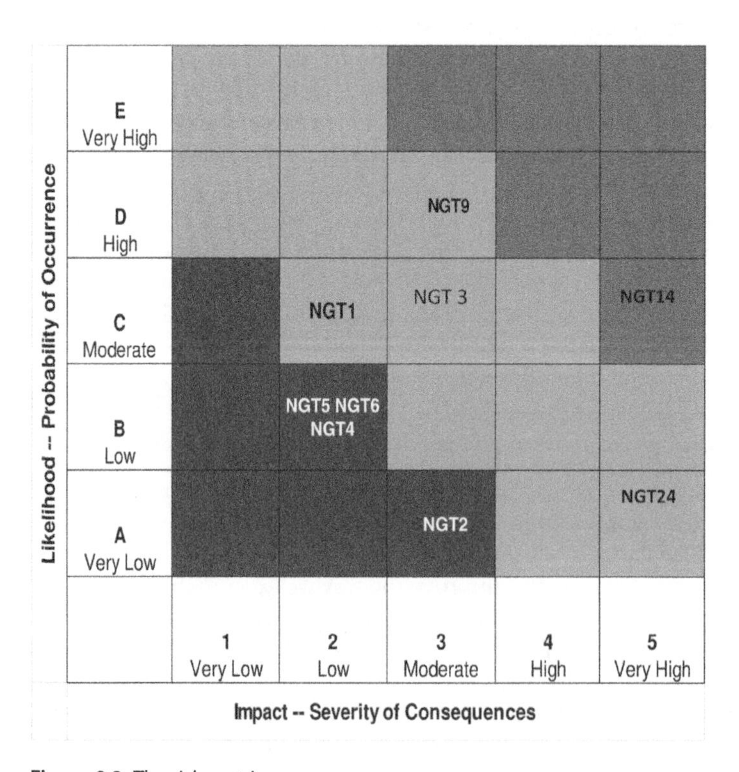

Figure 9.2 The risk matrix.

Also, always remember that these assessments are *estimates*, so all of the warnings from the previous chapter apply. Furthermore, they are often quite subjective; therefore, transparency is vital – each score should be accompanied by a description of the estimation method, assumptions, and so forth. These should all go into that same database.

It is useful if these assessments of likelihood and impact can be expressed as *distributions*, rather than just as *point estimates*. If nothing else, expressing these assessments as distributions (which could be simple *three-point estimates*, like the ones used in Chapter 7) expresses the uncertainty in those assessments. When there is uncertainty, it is always better to make that uncertainty *visible*, rather than to hide it.

9.2.5 Step e: Create Mitigation and Exploitation Plans

For each risk, we next develop approaches (for some risks, we might in fact develop more than one such approach) for how we could decrease the *likelihood* of that risk coming to pass. We also, for each risk, develop approaches (again, perhaps more than one) for decreasing the *impact* of that risk, in case it does come to pass.

Think of the example we discussed above about the risk of our house burning down. We had multiple approaches for *lessening the likelihood of occurrence* (e.g. trimming foliage, replacing the roof with non-flammable materials, and so forth) and a *separate* set of approaches for *lessening the impact if in fact something* (despite taking the steps intended to lessen the likelihood) *caused the house to burn down* (e.g. purchasing a policy of fire insurance, etc.). For each entry on the risk register, we now create such separate lists of mitigation, one list about how we will lessen the likelihood and a separate list about how we will lessen the impact. The lists of mitigation approaches will naturally be longer and more involved for risks considered serious (e.g. either high likelihood *or* high impact) than for risks considered minor (e.g. low likelihood *and* low impact).

As in our house example, these may well proceed in multiple steps of mitigation. Perhaps we decide that – if and when we make a decision that it is time to actually start implementing these mitigation steps – we will start to lessen the likelihood by trimming the trees and other foliage overhanging the house, as the first mitigation measure. If we decide that additional mitigation of likelihood is required, we select screening the attic vents as the next task, in order, and replacing the wood-shingle roof with non-flammable concrete tiles as the third.

For each such candidate mitigation step, we define: (i) what you will do; (ii) why it will help; (iii) how you measure its effect; (iv) what the resulting position on the risk matrix will be at the end of this step; (v) cost and duration estimates for this step. You also, as described above, place these candidate mitigation steps into a priority sequence: first, second, third, and so forth.

You can also depict the progress of mitigation for this risk through the risk matrix as shown in Figure 9.3.

As noted above, it may be too expensive to *mitigate* every risk from your project's budget. In fact, we have other options that can be considered.

We have four responses that we can make for any identified risk. These are sometimes called the potential *dispositions of risk*:

- *Accept*. We can *accept* the risk and its consequences. In which case, we need undertake no mitigating actions, at least for the present. We always reserve the option of changing our mind at a later date.

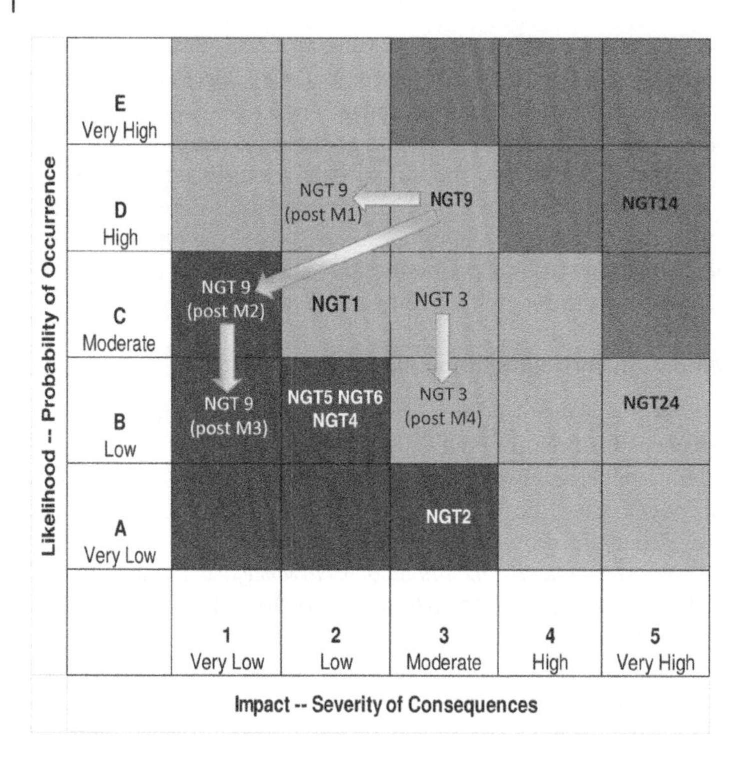

Figure 9.3 Depiction of the progression of mitigation.

- *Mitigate*. We can *mitigate* the risk, either in likelihood or impact, or both. This involves assigning people and facilities to the problem, and spending money.
- *Share*. We can *share* the risk with another party, such as our customer, or one of our subcontractors. This is usually done through a formal contract modification.
- *Transfer*. We can *transfer* the risk to another party, such as our customer, one of our subcontractors, or an insurance company. This is also done through a formal contract modification, or through the purchase of insurance.

We must be careful, however, about *share* and *transfer*; we may have shared or transferred the financial and/or contractual aspects of this risk, but we likely have *not* been able to transfer the *reputational* aspects of this risk. If this risk comes to pass, perhaps some other party will bear the financial cost, but the reputation of *our* company – and of *you*, as the project manager! – is still likely to be negatively affected. Reputation is *very important* (the trendy term among consumer-oriented companies is "brand"), so be aware that one can seldom actually share and transfer reputational risk. People almost always hold the *prime contractor and the project manager accountable for failures*, even if some responsibility has been formally shared or transferred. Your company may have purchased an insurance policy protecting you financially against the risk that the rocket explodes upon take-off, and the satellite you just delivered is destroyed, but your reputation may still suffer, causing you to lose future business, making it harder to attract employees, and so forth.

As noted above, the entire risk management process can (and should) be mirrored to create and exploit *opportunities* (e.g. changes that can *improve* project outcomes – whether

			Opportunity Benefit					Risk Consequence ($)				
		% Cost:	>10%	10%	5%	3%	1%					
		Unfactored Impact $:						< 100k	100k – 400k	400k – 800k	800k – 1000k	>1000k
			−5	−4	−3	−2	−1	1	2	3	4	5
Likelihood	>95%	5 Very Likely										7
	71% – 95%	4 Very Likely										
	26% – 70%	3 Possible			2				3,4		1	
	5% – 25%	2 Unlikely				3	1	10	5	6	2	8
	< 5%	1 Very Unlikely						9				

Figure 9.4 Matrix depicting both risks and opportunities.

technical, operational, safety, reliability, cost, schedule, etc.). We can mirror the risk matrix with an opportunity matrix. See Figure 9.4.

Whereas we assess and attempt to mitigate *risks*, we assess and attempt to exploit *opportunities*.

9.2.6 Step f: Create Triggers and Timing Requirements for Those Mitigation Plans

We have now created a set of candidate mitigation plans for each entry on the risk register. But, as noted above, we usually *cannot afford either the time or the money to fix all of them*. Nor might we wish to spend that money, even if we had it: these are only *potential issues*, not confirmed issues. Furthermore, we recognize that as we progress with the project, we will uncover new potential risks; we cannot be content with doing this just once at the beginning of the project. We need instead some sort of on-going activity to assess and manage risk.

We will discuss this on-going activity in a moment, but we need to set up just two more preparatory steps. The first one is to establish specific triggers – quantitative thresholds – that tell us that it may be time to consider initiating mitigation actions against a specific risk, and estimates of lead time.

To illustrate, let us return to the example of the satellite. We know that its launch weight cannot exceed 1000 lb; that is the limit for the weight that the rocket can lift to the correct orbit. We will not design our satellite to be 1000 lb, of course; if we did, even a very small overage would be catastrophic for the project. So, we pick some smaller weight – for example, 950 lb – as our target design to weight. Each month, we ask each of the managers – the person leading the team that is building the structure, building the solar panels, building the batteries, building the processors, building the sensors, building the cable assemblies, building the attitude-control system, and so forth – to update their own estimate for the predicted weight of their component when they are complete. Each month, we then add all of these weights up to create this month's estimate for the predicted weight of the satellite when it is complete. As discussed in Chapter 8, we can create a *time series* out of these numbers, and use that time series to separate signal from noise, and tell us whether an apparent trend up or down in the predicted weight at completion is a real signal, or just random variation.

We may be aiming for 950 lb, but perhaps we have decided that if the de-noised prediction ever rises to 970 lb, we will start some mitigation actions. Perhaps there is a portion of the structure that we are making out of aluminum; we have a mitigation plan that says we can make it out of carbon-fiber composite instead, and thereby save 30 lb. But making it out of carbon-fiber composite might cost $300 000 more than making it out of aluminum. Since we have a plan to bring the satellite to completion at 950 lb even with the bracket consisting of aluminum, using aluminum is our *baseline design*, and thereby we hope to save that $300 000. But if our de-noised prediction about the weight at completion ever reaches 970 lb, we have decided, as a risk mitigation step, that we are going to start the design and fabrication process for the carbon-fiber composite version of that part. We will not spend all $300 000 at once; we are just starting that process. It might be that three months later, we have saved enough weight somewhere else in the design that we can elect to stop the design and fabrication process for the carbon-fiber composite part, and save whatever money out of the $300 000 we have not yet spent.

We say that having our prediction exceed 970 lb is our *trigger*; if the trigger occurs, we will consider starting one or more of our mitigation actions.

There is a *time aspect* of the trigger process as well: perhaps we estimate that it will take eight months to design and fabricate that carbon-fiber version of the part. We therefore would create a dual trigger: exceeding 970 lb would cause us to consider starting a mitigation action, but so would passing the time where only eight months remain before we need that part! Even if we are at 960 lb when we pass the only-eight-months-remaining milestone, we would use that time milestone as a trigger to consider whether we want to start this particular mitigation action. We don't want to lose the chance to perform a mitigation action just because we run out of time!

Of course, we perform a symmetric process to look for the triggers and timing associated with our identified opportunities.

9.2.7 Step g: Create a Method to Aggregate All Risk Assessments Into a Periodic Overall Project Impact Prediction

A typical engineering project will end up with many items on the risk register. It is tempting to look at just a few risks that you consider the most material, but sometimes the aggregated effect of lots of small risks can result in a big adverse impact. So, we need a method to aggregate the totality of risks into some type of project-wide assessment.

Such an aggregated assessment should incorporate all of the following:

- The likelihood of occurrence of every risk on the risk register, with their distributions.
- The impact of every risk on the risk register, with their distributions.
- The cost and schedule to implement the mitigating actions (which might be separate actions for decreasing likelihood and decreasing impact!) and their effect on the overall schedule. This can be accomplished by updates to the *activity network*, as previously described in Chapter 7.
- The tiering/cascading/combining impact on the system and its technical and operational performance. This is usually accomplished through *tiered systems performance modeling*, as described in Chapter 2.

We have already described a method to implement the first two items on this list: the S-curve of Chapter 7. We can use this depiction to allow us to analyze the aggregate impact on various parameters (schedule, cost, weight, and so forth), and thereby make sure that our project is not endangered by the combination of the totality of our risks. Many of the risks may appear small, but when combined, the effect may be serious. The phrase "death by a thousand cuts"[3] is sometimes used to describe such a situation; each individual cut may be survivable, but the combination can be fatal.

9.2.8 Step h: Create and Use Some Sort of Periodic "Management Rhythm," Wherein You Periodically Make Decisions About Risk Mitigation and Opportunity Exploitation Actions, Based on the Periodic Assessment

We already noted that we often *cannot afford either the time or the money immediately to fix all of the items on our risk register.* Nor might we wish to spend that money, even if we had it: these are only *potential problems,* not confirmed problems. Furthermore, we recognize that as we progress with the project, we will uncover new potential risks. We therefore cannot be content with going through the process of identifying and scoring risks just once at the beginning of the project. We need instead some sort of on-going activity to assess and manage risk.

As the final step in our risk management process, therefore, we are now ready to talk about this on-going activity. Someone who once worked for me[4] called this the *monthly management rhythm*; on most of the large projects that I managed over the years, we performed once per month the process that I will describe below. I used that title for this process for many years, but it is the case that on some occasions, you might do it more often than once per month. So, I have renamed it the "periodic management rhythm," which is to my ears less catchy, but covers all cases.

We therefore institute a *periodic process* (usually monthly) to look at the items on our list of risks. We add or eliminate things from that list, as seems appropriate in light of new data. We look for items on the list that seem to be coming to pass (or any deadlines for a mitigation action decision that are coming up soon), and make a decision if this is the time at which we elect to initiate the planned mitigation actions (which, of course, require people, time, and money to implement). We assess mitigation actions that are already in progress – should we continue them, beef them up, or do we no longer require some particular mitigation action to continue?

We can then take actions:

- Adjust the contents of the risk register
- Adjust the contents of the mitigation plans, exploitation plans, and their associated triggers
- *Stop, start, or alter the mitigation and exploitation actions.*

In Chapter 11, we will see that this risk-related process is just one of five key steps in the larger periodic (monthly) process. I therefore introduce the term *periodic management rhythm* here, but will actually use this term for the complete five-step process described in Chapter 11.

---

3 I long thought that this expression was merely poetical. I have discovered that the term actually comes from a form of torture used for more than a thousand years in China.

4 Thank you, Eldon Christensen.

9.2.9 Step i: When Risks Actually Occur (Transition from *Risks* to *Issues*), Perform a *Root-Cause Analysis*

Once a risk has become realized (e.g. transitions from a risk into an issue), it is often appropriate to do a *root-cause analysis*. That is, why did it occur? This may help you prevent recurrence of similar problems in the future. Such prevention could save you headaches, and save the project time and money. Furthermore, if your project displays a pattern of having similar problems recur again and again, this will cause your customer and your management to lose confidence in your ability to manage the project.

Note that you may discover that the cause of the problem represented by the risk was different from the mechanism that caused you to identify it as a risk in the first place!

There are many methods for doing a root-cause analysis; two common ones are the *5-Why's* and fishbone diagrams.

The *5-Why's* refers to the practice of asking, five times, why the failure occurred; each time, you move one layer closer to the root cause. Of course, it need not be exactly five iterations; the point is to *reach the root cause.*

Here's an example.[5] Problem: The Washington Monument was disintegrating.

- Why? The use of harsh chemicals on its surface.
- Why? To clean pigeon poop off the surface.
- Why so many pigeons? They eat spiders, and there are a lot of spiders at the monument.
- Why so many spiders? They eat gnats, and there are a lot of gnats at the monument.
- Why so many gnats? They are attracted to the lights on the monument that are lit up at dusk.

The concept of a *fishbone diagram* is depicted in Figure 9.5.

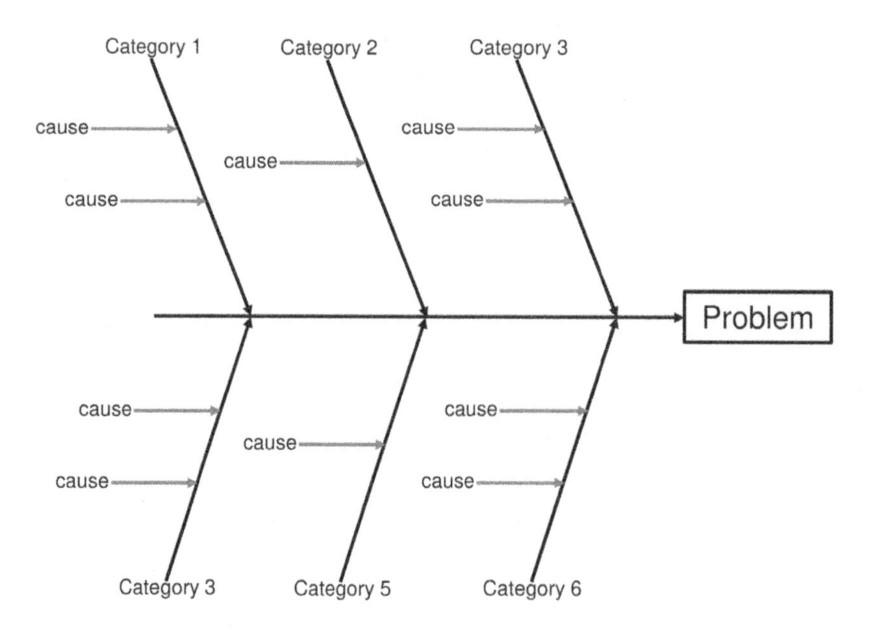

Figure 9.5 The fishbone diagram.

5 I first heard this particular example from Ursula Knopp-McKendree, in 2008.

- You name the problem or issue at the "head" of the fish.
- Label each of the main "bones" of the "fish" with a category. Typical such categories might include:
  - Methods, machines, materials, man-power
  - Place, procedure, people, policies
  - Surroundings, suppliers, systems, skills.
- Have a team brainstorm to identify the factors within each category that may be causing the problem or issue.
  - Causes can be nested.
- Continue until you no longer get useful info as you ask "Why is that happening?"
- Analyze the results and determine which items are the "most likely causes."

This concludes our walk-through of the risk/opportunity management process introduced in Figure 9.1. We have, however, *one more important aspect of the risk management process to consider*: there are some special types of risks that I believe must be handled somewhat differently than the approach described above. We now turn our attention to these special types of risks.

9.3 Two Special Types of Risks

9.3.1 The Low-Likelihood, High-Impact Event

After you build your risk register and your risk matrix, if there are items that appear in the high-likelihood/high-impact portion of the matrix, clearly you must spend time developing and implementing mitigation strategies for those risks. This is a management imperative, but spending time and money to lessen both the likelihood and impact of these events will be supported by both your customers and your management; they will likely insist on such actions, because of their position within the risk matrix.

You will also have a number of items on your risk matrix that have low-to-moderate impact, at any level of likelihood. You can use the risk management processes described above, including the periodic risk assessment and the aggregated assessment – both of which are a form of *expected-value* analysis (we calculate an expected value by multiplying the likelihood by the impact) – to help you determine when it is appropriate to spend time and money to mitigate these risks.

There remain, however, *two other, special types of items* in the risk matrix:

- Those items that you have scored as low likelihood but high impact, and
- Those items that you did not yet identify at all!

See Figure 9.6.

These two other, special risks are the ones that you really must focus on, because these are the ones that *kill projects* and *kill reputations*.

Let's start with the items that you identified as *low-likelihood, high-impact*. These items can be a trap, luring you and your team into a false sense of complacency; after all, you scored them as unlikely to occur, and their expected value (likelihood times impact) is low, so why devote a lot of time and money to assessing and mitigating them?

The problem is that you may well have *underestimated* the likelihood of this event. Psychologists say that if an event is rare, we humans tend even further to underestimate its likelihood; if an event is rare, we essentially are wired to believe that it will *never occur*.

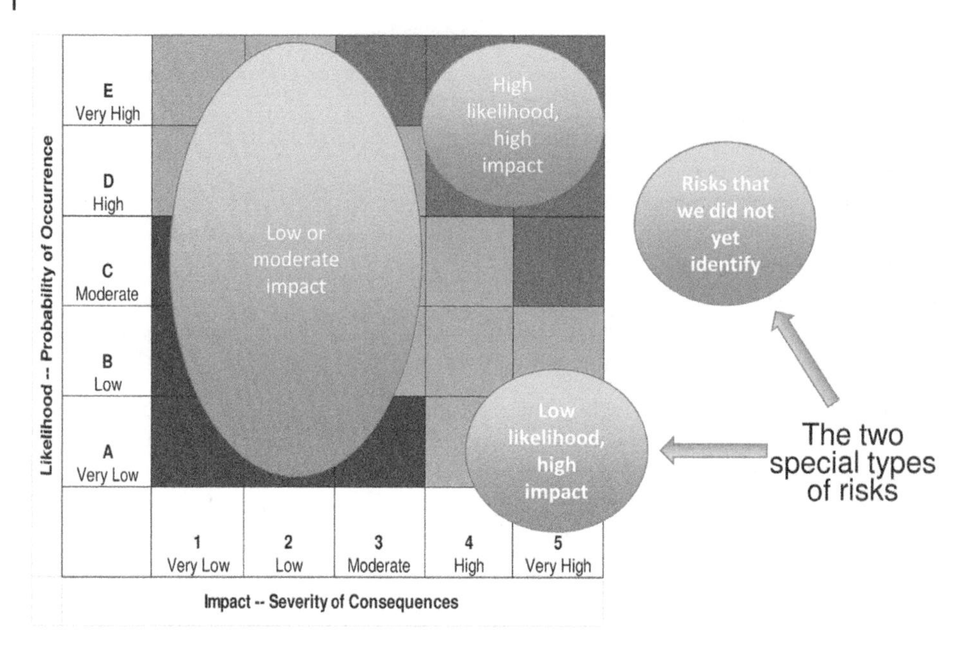

Figure 9.6 The two special types of risks.

My favorite example of this is something called the *100-year storm.* As the name implies, this is a rain storm of such severity that the weather service estimates that it ought to occur only once every 100 years. In fact, storms of the magnitude identified as 100-year storms occur far more frequently: about every 10 years. The event *is* truly rare – occurring only once every 10 years. But even the professional weather forecasters *underestimate the likelihood* of this rare event, assessing that it will occur only once every 100 years, rather than the correct likelihood of once every 10 years. Note that "once every 100 years" is, for most people, a statement very similar to "never"; few of us live to be 100.

Furthermore, there is the problem caused by the fallacy of the silent evidence (Chapter 8): we are more likely to see the good-news examples than to see the failures. Explorers who die during their exploits don't write books about their adventures afterwards. We therefore underestimate the likelihood of these types of events, because we tend not to hear about the relevant occurrences. They, after all, were *failures*, and people do not usually broadcast their failures. Rare events do occur!

Having now recognized that we might have underestimated the likelihood of the event, we are left with an event that may well occur, but if it occurred, would have a high adverse impact on our project.

Here is my philosophy about such events. There are some risks that, if the event they represent were to come to pass, would have a catastrophic impact on your project, your company, or your employees. In my view, it is ***not*** adequate to handle these risks via expected value (e.g. assigning them a small likelihood of occurring, and therefore ignoring them, or indefinitely deferring mitigating actions). You ***must*** *take actions now* that will, in the event of one of these events coming to pass, *prevent the impact from being catastrophic.* That is, you *must spend time and money on mitigating the impact,* even though you scored this event as unlikely to occur. This might take the form of extra safety equipment, training, and procedures. Or insurance against currency fluctuations. Or changing some aspect of the design to be completely non-flammable. Or (my personal favorite) spending extra time and money early in the project building prototypes and full-scale performance threads so that you can really work out the details of the design that pertain to the dynamic behavior of the system. Can we really control and prevent adverse unplanned dynamic behavior in our system?

If the adverse impact is severe enough, my advice is that you *must mitigate the impact, no matter how unlikely you deem its occurrence.* Do not handle these types of risks through expected-value approaches; you will be trapped by our tendency to underestimate the likelihood of rare events. If a rare event can kill one of your employees, you need to spend time and money to mitigate that potential impact.

Another way to think of this is that if the impact is severe enough, you do *not* need to work very hard to estimate the likelihood. Any estimate that is created of the likelihood of a rare event is likely to be wrong anyway (as noted above, we tend to underestimate the likelihood of rare events), so my approach is not to worry about trying to predict the likelihood very carefully. What you must do instead for these types of risks is to take actions that prevent the impact – if the event happens to occur – from being catastrophic.

Interestingly, this also works for rare events, that if they happened to occur, would have staggeringly good impacts. In engineering projects, my experience is that there are more rare events with really bad impacts than rare events with really good impacts, but rare events with good impacts do occasionally happen; I have had some in my career. The strategy here is if you have identified rare events with potentially game-changing *good* impacts, take actions that will, if the event happens to occur, maximize the benefit. This is the *opportunity* side of the risk/opportunity matrix. You do not need to spend a lot of time trying to predict the likelihood; just take actions that will allow you to be ready to benefit.

9.3.2 The Risks That We Have Not Yet Identified

The second special type of risk is *the item that you failed to identify*, and therefore *does not appear on the risk register at all.*[6] This is also depicted in Figure 9.6. There will *always* be such items; our systems are too complicated for you and your team to be able to create a complete and perfect risk register during the proposal. So, you must *find them as the project progresses*, and hopefully before they actually occur. How can you do that?

I have found that the answer is *measurements*, taken from actual, empirical experiments. Build prototypes and exercise them. Build segments of your system and let real users experiment with them. Increase the load on your system until it breaks. Increase the operating temperature beyond the specified limits, both up and down. Subject your components to more shock and vibration than the specified limits. Do these types of things *early in the project*, and *continuously*, as a regular practice; that way, you might find the issues early enough that you will still have the time and money to fix them.

6 My senior military customers call these "unknown unknowns." You don't even know that this is a problem, in contrast to "known unknowns," where you know there is a problem, even if you have yet to identify the solution.

Don't wait until the end of the project, during the formal test program, to try to find these problems. Also, don't succumb to the temptation to interpret away measurements that show failures or unexpected behavior; the entire point is that these measurements might be signals of *heretofore-undiscovered risks* to your design or implementation.

Use real measurements on real operations when doing this, not computer models. Your computer models did not predict these risks in the first place – that's why these items are not yet on the risk register. Running the same models again a few months later is not going to give you a materially different result. Exercising segments of your actual system as they first become available, especially if you push them past their specific limits and find what they do when they fail, will teach you real and important lessons, and can create new and useful insights about your system.

In essence, you must contrive assessments that will *cause failures under safe and controlled conditions*, so that you can *find new things that belong on the risk register* that you failed to place there originally.

In other words, we ought to pursue a *bifurcated risk management strategy*: the ordinary risk management strategy (based on expected value: probability of occurance x impact if it occurs) for regular risks (those that may hurt us, but not kill us), but a completely different strategy for the events that would cause our project or our company to fail, or would harm a person.

We can and ought to use data from past projects to deal with ordinary risks, but we ought not to depend entirely on such past data to deal with those low-likelihood risks that, if they came to pass, would cause your project to fail, or people to be harmed. This is because, since they are by definition rare events, our sample of past data may not include them at all, just because they are rare.

The real world works in a bifurcated manner too. For example, the physics of very small objects (the size of atomic nuclei or smaller) operates according to completely different laws of physics than everything else that is larger. Really and truly different. Scale matters. If the scale of the adverse impact of an event is sufficiently bad, we must deal with that event differently.

In my view, interpersonal relations are bifurcated too. A policy of *forgiveness* works pretty well at times among a small circle of people truly bound together in mutual love (e.g. a husband and wife, a cohabitating family), but on a larger scale, such a policy has proven time and time again to be ineffective. The strategy that appears to work for larger and looser-bound groups is a policy of *justice*. Again, we see that scale matters.

Scale matters in many ways for the complex engineered systems we are considering in this book. The large scale of our engineered systems is itself the source of many of our risks: large numbers of users, large numbers of transactions, large numbers of opinions among the stakeholders (not everything that needs to get onto the risk register is related to technology!), large numbers of interfaces, large numbers of interacting parts (e.g. the problems of dynamic behavior discussed in Chapter 2), and so forth. Risks associated with large scale are often those that can cause a disaster, so they are among the red-flags that should cause you to examine those risks to see which risk management treatment (regular risk management or mitigation against a catastrophic impact, no matter how unlikely the event) is appropriate.

There is an important social aspect to this recommendation. Finding new things to add to the risk register is, in my view, a sign of *effective project management*, and *not* a sign of weakness. Your team can do some large portion of this investigation each month as part of the data gathering for the periodic management rhythm (Chapter 11). But the precursor need is for you, as the project manager, to create an environment where people *feel free to conduct such assessments*, and *feel that it is safe to find and bring forward such new risks and/or issues*. Creating such an environment is a leadership challenge for you, and we will talk about this aspect more in Chapter 13.

9.4 Lessons Learned From Risk Management

- The first version of the project's risk register is often too high-level and too simplistic:
  - For example, "the software might be late."
  - Examples from real previous projects (successful and failed) can really help you build a better risk register. Use your organization's historical archives.
- It is very common to plan to make use of new technologies on an engineering project. My advice is that there ought always to be a risk-register entry noting that each specific new technology might not work out as planned. Have in the risk register (as a mitigation plan) a back-up approach that uses a proven technology. Utilize what the financial people call a "stop-loss" approach for deciding when to drop the new technology and fall back to the proven approach; that is, define your quantifiable "pain" threshold in advance, and stick to it. Trust me: in the midst of your project, your team will be reluctant to make the switch. I have therefore found that it is better to select the criteria for switching to the older but proven technology *in advance*, before you become psychologically wedded to the new technology.
- Risks often arise from interactions across work-package boundaries. Therefore, gathering and assessing risk based only on assessments by the work-package managers (e.g. the work-breakdown structure) *is insufficient*. The program-level aggregated assessment described above – driven from the task inter-relationships embodied in the *activity network* (Chapter 7) – is essential.
- Schedule and cost are usually dependent factors, and therefore form a *lagging* management indicator. It is therefore usually *inadequate* to depend on schedule or cost predictions as the triggering event for deciding when to initiate mitigating actions. Instead, we want *leading* indicators, which are usually based on the *technical and operational performance parameters*.
- We need to use valid/strong statistics in our assessments. However, note that the statistic of comparison of the current prediction with the prediction from one prior period is just about the weakest possible statistic … and yet, this is the statistic that is in most common use. *Direct your people to use control charts and the five Wheeler/Kazeef tests* from Chapter 8!
- There are two special types of risks which must be mitigated: those that you deem of low likelihood but of high adverse impact; and those that you failed to predict (and therefore, never got onto the risk register). I strongly urge the use of what I call the *bifurcated risk management approach*: use the expected-value approach for ordinary risks, but the different approaches for the two special categories of risks – an approach of always mitigating the impact of low-likelihood, high-adverse-impact events and continually looking for new items to go onto the risk register, through empirical experimentation, and pushing your system and components until they fail.

9.5 Your Role in All of This

You must be a personal participant in the risk management process. You probably delegate day-to-day operational authority to your project's chief engineer, but you read the materials, attend the meetings, listen to the discussions, and make the final decisions about what risks to accept and what risks to mitigate, when to spend money on implementing risk mitigation approaches, and so forth.

9.6 Summary: Risk and Opportunity Management

- Every project entails *risk* – things that can go wrong.
  - Risks may derive from technical and engineering matters, but they may also derive from other matters: contracts, personnel, facilities, and so forth. It is the project manager's responsibility to identify and mitigate risks. We call this process "risk management."
  - In parallel, one ought also to be looking for *opportunities*; that is, deviations from the nominal plan that could make the outcomes *better* (in performance, in cost, in schedule, etc.).
- Risks (and opportunities) must be identified and scored, mitigation plans created (including triggers and timing requirements) and aggregated periodically to an overall risk assessment.
- Decisions to execute (or terminate) specific elements of the risk management plan can then be made.
  - This must be done periodically, as part of a "management rhythm," not just during the proposal phase.
- We must use strong and valid statistics and statistical methods to guide these decisions.
  - These must be based on distributions, not just point estimates.
- There are two *special categories of risk* that must be treated differently.

9.7 Next

We will spend weeks 10 and 11 learning how to monitor the on-going progress of our project. During week 10, we will learn a technique called *taking earned value*. During week 11, we will learn a technique that I call the *monthly* (or *periodic*) *management rhythm*. We will discover that the activity network (Chapter 7), drawing valid conclusions from numbers (Chapter 8), and risk management (Chapter 9) are all essential elements of this monitoring process.

9.8 This Week's Facilitated Lab Session

We will do a team exercise on risk management. This will become a section in your team project report, and team presentation at the end of the course.

- Assignment:
    - Refer back to the project that your team selected during the previous facilitated lab sessions.
    - As a team:
        - Create a risk register of at least six items for that project.
        - Create at least one mitigation plan for three of them, and two mitigation plans for three of them.
        - Create triggers for each of these mitigation plans.

10

Monitoring the Progress of Your Project (Part I)

Once you have created a great project plan, you can start work on your engineering project. You now need a set of tools and mechanisms to allow you to monitor what is going on, and to determine if progress is as expected (or not!). In this chapter, I show you how to assess progress on schedule and cost. I also introduce you to the principal financial measures that your company will use to measure the business performance of your project.

10.1 Monitoring Progress Via Updated Predictions to Schedule and Cost

In Chapter 7, we learned to build an initial version of an activity network for our project, and then, using the data in that activity network, to develop an initial version of the estimate of the cost of our project. These activities are done *before* the project starts, as part of your proposal to the customer, and those same data are also used in order to get permission from your company to submit your proposal to the customer.

After you win and start work, as the project progresses, you will of course learn much more over time about your project. This new information needs to be used to *improve the estimates for schedule and cost.* And of course, there will be actual changes in status of the project – progress or lack of it, and also changes to the requirements and to the design – that will affect these predictions too.

Therefore, on most large engineering projects, we do such an update to the schedule and to the cost estimate *once per month,* although on some projects it might be accomplished at a different time interval. This periodic update to the predictions for schedule and cost is the subject of this chapter.

The title of this chapter is *monitoring the progress of your project.* The goal of monitoring your project is to *prevent adverse "surprises."* If we are falling behind schedule (or our cost is increasing, or we are going to fall short on technical capabilities), we want to know *now,* and not wait until very late in the project to notice it. We have already stated the reasons for wanting to know *now,* rather than *later*: it is very useful to have as much time as possible to correct any emerging problems, and it costs much less to correct a given problem early in the project life-cycle than to correct it later in the project life-cycle.

Engineering Project Management, First Edition. Neil G. Siegel.

Companion website: www.wiley.com/go/siegel/engineering_project_management

By preparing such an update to your predicted schedule and to your predicted cost at project completion, you are in fact performing one portion of the task of monitoring the progress of your project. We assess the status of our project, and then use that revised assessment of status in order to update our schedule and cost estimates. Those updated estimates comprise a statement about our progress. For example, it might be that we now estimate that we have fallen a bit behind and will need three more weeks, and $100 000 additional, in order to complete the project.

These monthly revisions to the predicted schedule (e.g. when we will be done) and the predicted cost (e.g. how much it will cost by the time we are done) are two of the tangible representations that are produced as a result of monitoring the progress of our project.

These periodic updates to the predictions for schedule and cost are among your most important and most watched tasks! But *you* do *not* perform all of it; your subordinate managers actually do most of it; your role is largely to coordinate. But there are also some important aspects that you do yourself; we will describe these later in the chapter.

Since your subordinate managers will be preparing the updates for their particular portions of the project, just like for the original estimates (discussed in Chapter 7), they must believe in those estimates! Therefore, just as for the original estimates:

- You must allow them to create their own updated estimates, and not dictate the answer.
- You can mandate their use of particular estimating methodologies (including calibration against previous actuals).
- You can provide goals (e.g. urge them to bring their portion of the work to completion within the previous estimates for schedule and cost that they have created).
- You can negotiate with them about the values contained in their updated estimates.
- But, in the end, they must be allowed to work to estimates which they psychologically "own."

As you did for the original estimate, you have your financial manager roll up all of the individual updated estimates (using the appropriate statistical methods, as discussed in Chapter 7), so as to create an updated estimate for schedule and cost for the entire project. You then report progress on a periodic basis (again, usually monthly) to your customer and to your company's management.

Whereas the original estimates for schedule and cost consisted *entirely of predictions*, these updated progress reports will now consist of a *combination of actual results to date and predictions for the future.* As the project progresses, the time period covered by the predictions will get shorter. Expect that most of the attention by your management and by your customer will be on the *predictions* for both the schedule (e.g. the date by which I expect to complete the work on the project) and the cost (e.g. the amount that I expect the project to cost when I have completed). What will you be reporting? Items such as these:

- When will we be done?
- How much will it cost when we are done?
- Will it do everything it is supposed to do?
- And ... what else could go wrong?

This chapter is about the *first two items* on this list; in Chapter 9 we discussed the third and fourth items on this list. In the next chapter, we will integrate all of these activities into a single, cohesive management process that I will call the *periodic management rhythm.*

But remember the warnings from Chapter 8. These predictions are based on *extrapolating forward* from measurements about the progress made by the project to date. But:

- All of these measurements about progress to date have errors in them; they contain both *noise* and *signal.*
- All of the statistics and extrapolations that we use in order to predict our status at the completion of the project introduce additional sources of error.

10.2 Making the Updated Predictions

How do we make these updated predictions, and do so in a manner that will allow them to be deemed credible? Here is my approach:

- We start by creating an estimate for the schedule ("when will we be done?") (Chapter 7)
- We then use the estimated schedule to estimate the cost at completion (Chapter 7)
- We do this many times:
  - Create a baseline at the beginning of the project (Chapters 7 and 12)
  - Create updated estimates on a periodic basis (usually monthly) (this chapter)
- Remember that in Chapter 7, we talked about the necessity to *calibrate our estimates to previous data,* in order to improve the realism of our predictions. We showed you how to do this by drawing upon data from your company's archive of historical project results, and also by using parametric models. You might think that on our project, we would undertake a systematic program of comparing our predictions to the unfolding reality, and try to adjust. You can do that sometimes. For example, if you discover that some task takes twice as long to perform as you planned, and the project has several more similar tasks, you would certainly consider adjusting the predicted duration to those similar tasks. In general, however, a single project is too short a duration, and too small a sample, from which to draw valid inferences by comparing predictions to actual results. It is best for the data from your project to go into that archive of historic data that is kept by your company, where it can be assessed in light of all those other data too, and used to calibrate future estimates.
- Each time we create a new estimate, we then try to draw inferences from these data by creating statistics that compare current estimates to a *set* of previous estimates. We don't want to compare current estimates to a *single* previous estimate; remember in Chapter 8 we said that was a weak statistic, with almost no predictive power. By doing the updates every month, we automatically create a time series of measurements, and by using the *entire set* of measurements within each time series, we can obtain more predictive power than we could by just comparing the current estimate to the previous estimate, because the time series allows us to separate signal from noise.
- Because we need a time series to make credible predictions, we cannot really make such predictions until we are a few months into the project; we need those seven consecutive monthly data points that we discussed in Chapter 8 before we can use the

process-behavior chart methodology to separate signal from noise. So, for the first few months, we make the estimates and start plotting the time series, but cannot really make credible predictions. We can still look for *outliers* in the data though; remember (Chapter 8) that some of the Wheeler/Kazeef tests only require one or two data points.

Let's go through these steps in more detail.

10.2.1 Creating the Updated Prediction for the Schedule

We start by creating an updated estimate for the schedule. Remember what we discussed in Chapter 7 (and will cover in more detail in Chapter 12): prior to the commencement of the project, we establish a *baseline schedule.* This consists of a set of tasks, each with a predicted duration (optimistic, most likely, pessimistic) and a set of linkages that define how the tasks link together to form the *activity network.* From this, we can predict statistically when we are likely to reach various milestones, including *project completion.*

The baseline schedule gives us the *first point* in what will become a time series of predictions of the date on which we estimate that the project will be complete. Each month, we *update* the estimated schedule:

- The cost-account managers update the *predicted duration* of their tasks (each duration ought still to be expressed as a three-point estimate: optimistic, nominal, and pessimistic). Each cost-account manager may also update the way their tasks interact with each other, and the way that their tasks interact with those tasks managed by other cost-account managers.
- At times, the cost-account managers may also choose to *break one of their tasks into multiple tasks,* which allows us to "replumb" the task interconnections. The reason that we might break tasks apart into such smaller tasks is to look for the actual task-to-task dependencies, and see if we can find pieces of work that, if separated, could proceed in parallel with certain tasks, rather than having to proceed in sequence. This might lead to shorter schedules; that is, such "replumbing" might provide us a feasible way to achieve *shorter schedules,* which in turn might lead to *lower costs.* We will discuss this aspect of the process later in this chapter.

Each monthly update creates a single additional point on our time series of predictions about the project's likely completion date. As soon as we have enough such points, we can start using the process-behavior chart and the Wheeler/Kazeef tests to separate signal from noise, and thereby make better predictions about our completion date. Furthermore, once you have an actual signal that the schedule is showing a (statistically valid) delay, you can look inside the schedule to try to determine the *root cause* (Chapter 9) of that schedule delay, and take corrective action.

The description above says that a cost-account manager may choose to *break one of their tasks into multiple tasks,* which allows him/her to "replumb" the task interconnections. Let's examine this process in detail, through an example.

Take a look at the top half of Figure 10.1. It shows a simple example of a baseline schedule for one cost-account manager. As you can see, this schedule consists of three

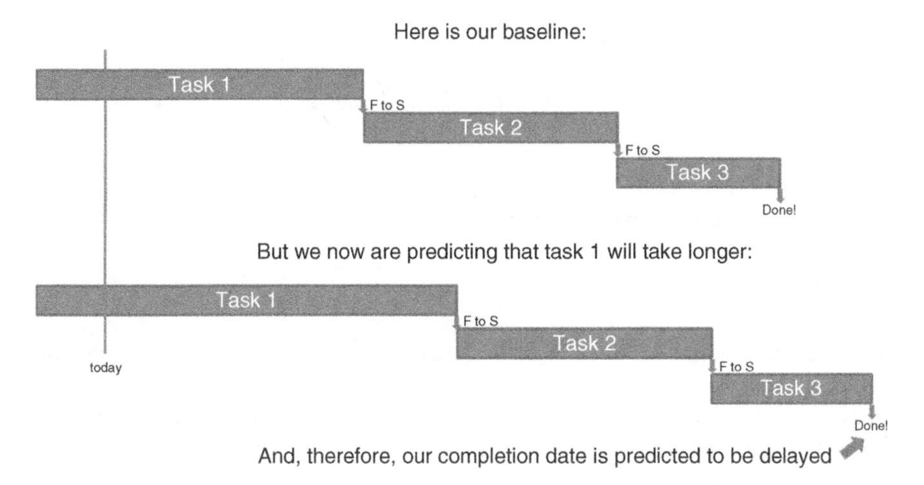

Figure 10.1 A chain of three tasks, and a delay to the completion date of the first of those tasks.

tasks (labeled 1, 2, and 3) in a simple set of finish-to-start relationships (labeled F to S in the figure), forming what I called a *chain*. You can see where we are in terms of the calendar, by the vertical line marked "today" on the figure. We have started to work on task 1, but not yet started either task 2 or task 3.

Now take a look at the bottom half of the figure. In our most recent monthly update, we have estimated that task 1 will in fact take longer to complete that we originally estimated. Therefore, the commencement of task 2 is delayed, and so is the completion of task 2. The duration of task 2 remains the same. Of course, the delay in the completion of task 2 also delays the commencement (and hence also the completion) of task 3. The date by which we predict that this cost-account manager will complete the work represented by this chain of tasks is therefore delayed.

Imagine the reaction of a typical project manager to the news about such a delay; this manager will say "Please try to get this work back on schedule!". Some managers might even not say "please."

The situation might be much worse too. What if the factor that led to task 1 taking longer than planned also applies to task 2 and task 3, and actually all three of these tasks are now predicted each to take a lot longer than originally planned? In such a case, the completion date would become delayed even further.

I want to point out that monitoring the progress of your project is *not just reporting that we hit a snag*. Monitoring progress *also* includes *figuring out how to recover from the problem(s)*, as much as is realistically possible. In this example, the problem is the delay in our project's predicted completion date, and figuring out how to recover from the problem would consist of finding a credible method to get the work completed sooner, if possible, by the original (baseline) schedule estimate, despite the new information that task 1 is likely to take longer than planned.

Note that this delay is also likely to cause an increase to the estimated *cost* for this chain of tasks; people need to be paid every week, and we are now estimating that the work will take longer. So, we likely also have a second problem: how to get the cost estimate back as close as possible to the original (baseline) cost estimate.

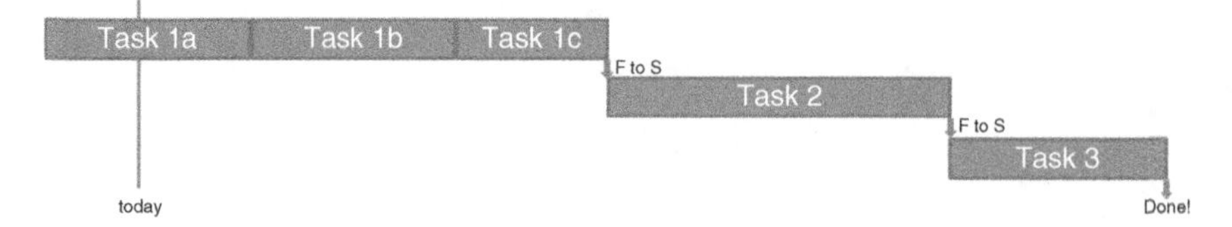

Figure 10.2 Task 1 actually consists of three subtasks.

What can we do?

- We can urge our people to work harder. But they are probably already working pretty hard.
- We can ask them to work overtime, but that is only effective for short durations, so if our delay is big, we cannot catch up via overtime.

Here's a different type of approach for catching up that can actually work!

We start by asking what is inside those tasks that we now predict to be late? Look at Figure 10.2, where we discover that task 1 actually contains *multiple parts.*

We further discover that it is only subtasks 1a and 1b that are needed before task 2 can actually start; subtask 1c is needed, but not in order to start either task 2 or task 3. So, we can replumb the schedule by removing subtask 1c from the dependency relationship with task 2 (Figure 10.3). We will of course have to figure out what other tasks actually require the outputs of task 1c, and make that connection in our activity network.

Such replumbing will likely, however, affect the *staffing profile* (Chapter 7), so we need to do the "smoothing" of the staffing curve again after we have done the replumbing. This smoothing of the staffing curve might constrain the way you implement the replumbing; you might not get as much benefit as you originally thought you might.

But you still have a good chance of gaining back some schedule. And since, as we have noted before, shorter schedules generally cost less to implement, you may well gain back some of the predicted cost increase too.

In order to make the example simple, we restricted the scope of the proposed replumbing to a set of tasks managed by a single cost-account manager. But naturally, in real situations, there are potential benefits from such replumbing to be obtained from looking at relationships over a set of tasks managed by multiple cost-account managers. We will take up this aspect of the problem in Chapter 11.

The above actions create updated versions of the *activity network* for our project. We will do this every single month for the duration of our project, or whatever the selected reporting period is for our project (I have never seen a real project of any significant size

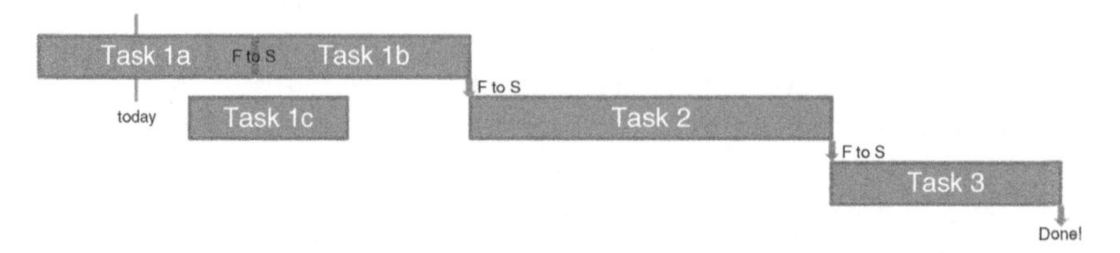

Figure 10.3 We can replumb the schedule, and thereby gain back some time.

where the reporting period was longer than a month, and few where it was shorter than a month, except during times of special risk or stress).

10.2.2 Preview: Variance Analysis

We can now analyze the information provided by that new activity network, especially the predictions it is making about the completion dates for every major milestone along the path to completion of the project. We do this analysis in a format called *variance analysis*. What is variance analysis? It attempts to answer the following types of questions:

- What is different in the current estimated schedule than in your previous predictions, all the way back to the baseline schedule ... and why?
- Are the positive differences sustainable?
- Are the negative differences recoverable?

Since we are also going to do a variance analysis on both schedule and cost, we will defer the discussion of how to do such a variance analysis for a few pages. We will first discuss the periodic update to the cost predictions, then introduce the concept of *earned value*, and then we will be ready to talk about how we perform variance analyses on both the schedule and the cost predictions.

10.2.3 Creating the Updated Prediction for the Cost

Now, having updated the schedule (remember from Chapter 7: schedule estimation must always precede cost estimation), we are ready to do a similar update to the project's cost estimate.

Each month, you can use the new assessment of the duration of each task (this comes from the activity network above) in order to update the estimate of the cost of completing each task. Remember, from Chapter 7, we determined that there are multiple elements of cost. Some of these elements are affected by the schedule duration, and hence these elements change when the estimated duration of the task changes. Some of these elements (such as travel and materials) may not change when the estimated duration of the task changes. The cost-account managers (and their business specialists) use the procedure already described in Chapter 7 to turn the updated prediction for task duration into an updated prediction for task cost at completion.

Then your business specialist can roll all of those updated task-cost-at-completion predictions together, and you have an *updated prediction for cost at completion* for the entire project.

10.2.4 Taking Earned Value

It turns out, however, that there is *additional, very useful information that can be obtained from these updated predictions*, over and above the predictions about the *end states* (e.g. the predicted completion date and the predicted cost at completion). This additional information is a statement that *compares how much progress we have made so far, to the amount of progress that we expected* that we would have made by this date. We can make this comparison literally for every task on the project, if we elect to do so.

Here's why this additional information is so useful: remember how we said that the goal of monitoring your project is to prevent adverse "surprises," and that if we are falling behind, we want to know *now*, and not wait until very late in the project to notice it. It might be the case that we have made a lot less progress so far than we expected we would, but for some reason our prediction at completion is not yet showing this lack of progress. By looking at this comparison of progress to date against the amount of progress that we planned to make by this date, we have another view of progress in our project, in addition to the view of progress provided by the predictions about schedule and cost at completion. These in-progress comparisons often turn out to provide an earlier indication of problems than the at-completion predictions. So, good projects *always* do *both* types of analyses.

We have already stated the reasons for wanting to know about problems *now*, rather than *later*: it is very useful to have as much time as possible to correct any emerging problems, and it costs less to correct a given problem early in the project life-cycle than to correct it later in the project life-cycle.

Here's what we are looking to discover. We estimate the percentage completion rate of a task. Our business specialist knows how much money the team has already spent on the task, and also knows how much the total budget is for that task. *If your estimated percentage of work completed is markedly different from the percentage of the budget that the team has spent, this is an indication that the estimated schedule and cost for that task are likely wrong!*

So, now we will describe the procedure for determining how much progress you have made on a single task.

We start by noting that at any given point in time, every single task on our project is either (i) *completed*, (ii) *in progress*, or (iii) *not yet started*. There is no other possibility.

Clearly, it is easy to make a high-quality estimate of your progress on completed tasks: they are 100% complete.

It is also clearly easy to make a high-quality estimate of your progress on tasks that are not yet started: they are 0% complete.

Therefore, the only difficult part in estimating your progress lies in the tasks that are *actually in progress* at the time of the (monthly) estimate/progress assessment.

Notice that this implies it is desirable to have as *few tasks* as possible *actually in progress at any given time*, because these in-progress tasks are the only tasks for which you have to create periodic assessments of progress. It is also the case that it is *desirable that these tasks be as short and as simple as possible,*[1] because this leads to the least amount of estimation uncertainty, and the highest-credibility estimates. See Figure 10.4.

As the figure depicts, you can create more credible estimates by making your tasks short in duration. When you use short tasks, more of the work will lie in tasks that are complete; there is no argument about progress on those tasks. Since the tasks that are in progress are short, you can use the simple *0–100 method* to estimate progress on those tasks (described below). Using short tasks for the near-term portion of the project will slightly increase the portion of the project that is embodied in tasks that have not yet started; again, there is no argument about progress on those tasks either. You will

1 Note that these two goals might be in tension with each other; we will discuss resolving this tension in a moment.

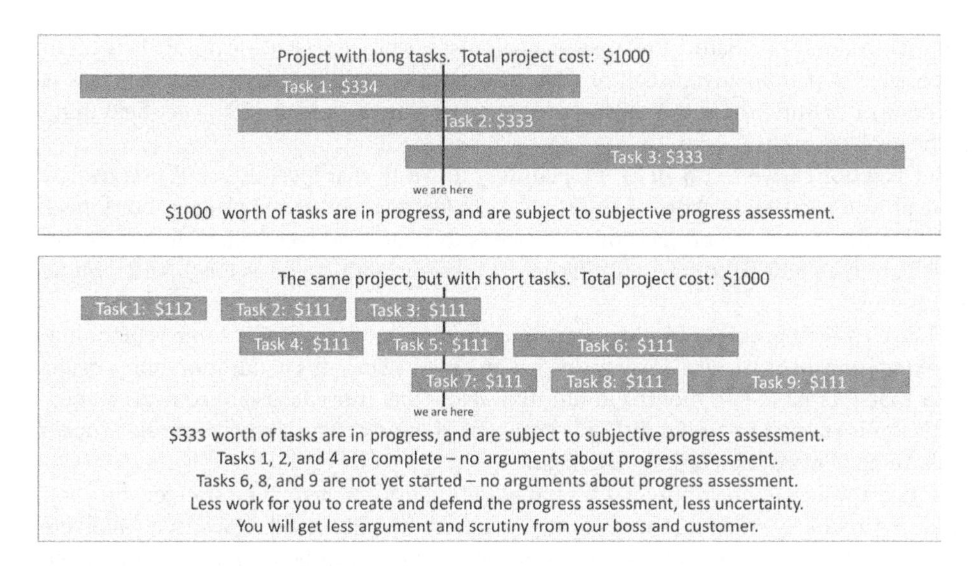

Figure 10.4 You create more credible estimates by making your tasks short in duration.

end up having more total tasks, but I will show you how to cope with that below, when I introduce a technique for planning called the *rolling wave.*

This process – estimating how far along we are toward completion of those tasks that are actually in progress this month – is called *taking earned value.* There are many methodologies offered about assessing estimated completion of an in-progress task; entire books have been written on this subject. Some of them are quite complicated, and it would take you quite a while each month to perform the necessary evaluations using these methods.

But ... notice that if (as I suggested above is desirable anyway) the work is decomposed so that every task is *short* – just one or two months in duration – you can use the simplest and most credible method of estimating progress on an in-progress task. One that is called *0–100*:

You rate a task as 0% complete, until it is 100% complete.

This estimation method is quick, effortless, and transparent.

Many books discourage the use of the 0–100 method, because in essence it gives you *no* credit toward completion for tasks that are in progress. If your work consists of a small number of long tasks, this will likely result in a significant underestimation of your actual progress.

But if, instead, all of your tasks are just one to two months in duration, you don't "lose" much task credit at in-between estimation points. Under such circumstances, the advantages of the 0–100 method (e.g. it is quick, effortless, and transparent) are available to you with little corresponding disadvantage.

10.2.5 The Rolling Wave

So, the next obvious question is: Can we really plan a large, complex engineering project in advance to a level of detail such that every single task is no more than one or two months in duration? Sadly, the answer is almost certainly that we cannot. It is often impossible to plan in advance all of a big project in enough detail so that every task is only one or two months in duration – we just do not know enough at the very beginning of the project.

But you don't have to do all of the planning down to that level of detail *in advance*. Instead, you can define short tasks for the first few months of the project, but longer tasks for the rest of the project; my practice is to issue a project policy allowing tasks which are not planned to start within the next three months to be as long as 12 months in duration.

Then ... as you progress and learn more, each month, you perform some replanning, breaking some of those near-term tasks that are up to 12 months in duration into smaller tasks that are one to two months in duration. Now that those tasks are so near, we certainly know enough to accomplish this type of replanning. We call this periodic process replanning of near-term tasks a *rolling wave*.

Notice another important benefit: by replanning into such smaller, shorter-duration tasks, we are likely enabling the replumbing activity we discussed above. We will likely find work that was buried inside those longer tasks that can be performed in parallel, rather than in a rigid sequential fashion. We saw above that this replumbing can lead to shorter schedules!

10.3 Using the Updated Predictions

10.3.1 Calculating the Schedule and Cost Variances

Now we are at last ready to discuss calculating the *schedule and cost variances*.

Customers and companies often require that two specific statistics – called *variances* – be calculated, reported, and analyzed. These two specific statistics are the *schedule variance* and the *cost variance*.

By the way, the term *variance* is *not* used here in the same sense as it is in statistics – don't get confused by this! Here, the term merely signifies that the schedule (or cost) varies from the value which we expected. In statistics, the term has a more specific meaning.[2]

Before we get to the definitions of the schedule variance and the cost variance, it is important to understand why there are two separate variance calculations. The reason is that *what you spend* and *what you accomplish* are separate matters. Spending is *not* an indication of progress!

Consider this example:

- If a task is supposed to cost $100 (baseline budget), when you finish it, you get $100 worth of credit for *work accomplished*.
- But it might have cost you $120 to do that work! Or only $80.
- Therefore, *work accomplished* and *amount spent to accomplish that work* are two separate measurable items.

2 When a word or phrase is used to mean different things in different contexts, computer scientists say that this word or phrase is *overloaded*. In this context, we can therefore say that the word *variance* is overloaded.

Schedule variance: The deviation (if any) from the baseline budget plan of ***work accomplished:***

The value of work accomplished to date, which is called *earned value* (EV)

less

The value of the work we *planned* to get accomplished by this date, which is called *planned value* (PV)

= ***schedule variance*** (SV)

Figure 10.5 Definition of the schedule variance (SV).

Cost variance: The deviation (if any) from the baseline budget plan of the ***cost of work performed:***

The value of work accomplished to date, which is called *earned value* (EV)

less

The cost of the work performed to date, which is called the *actual cost of work performed*, or usually, just *actual cost* (AC)

= ***cost variance*** (CV)

Figure 10.6 Definition of the cost variance (CV).

- Because of this, we calculate *two separate progress statistics,* which we call the *schedule variance* and the *cost variance.*

In Figure 10.5, I present the formal definition of schedule variance. In Figure 10.6, the formal definition of the cost variance.

In making these two definitions, we introduced several terms. Let's discuss them. Refer to Figure 10.7.

There are three lines in the graph that forms Figure 10.7.

- **Middle line: budget.** Let's start with the middle line, which is labeled *budget.* When we planned our task, we laid out the work month-by-month, and also determined how many people of what skill level would work on the task each month. Perhaps we even identified all of these people by name. You will remember from Chapter 7 that our business specialist converted all of that information into dollars. In Chapter 7, we focused our consideration on the *total dollar value* we thereby estimated our task would cost. But, of course, that analysis also produced a *month-by-month expenditure plan.* It is this month-by-month expenditure plan that is depicted in the middle line of the graph. You can see that the depicted task is going to spend $12 the first month, an additional $12 the next month (resulting in a total expenditure to date at that point of $24, which is the amount depicted at month 2 on the graph), and so forth. We can see that the expenditure rate for this task appears to remain about the same until month 7, at which point it appears that we plan to spend a bit less each month, until the task is complete at the end of month 10. Here is a subtle point that sometimes confuses people: this line, since it is a *plan,* actually represents *two separate things* – the

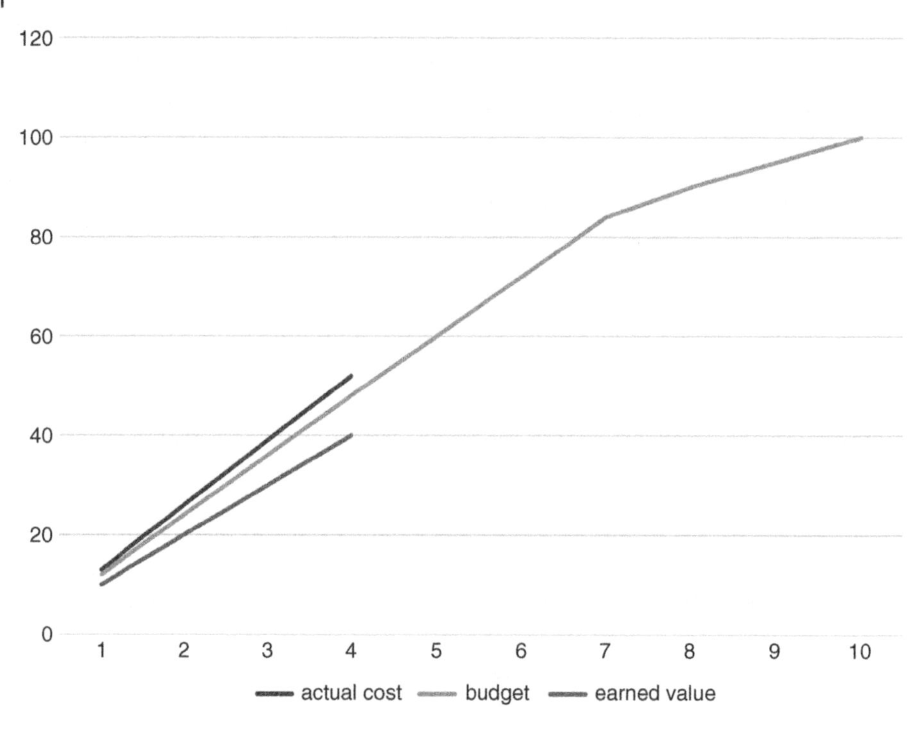

Figure 10.7 The budget, earned value, and actual cost.

rate at which we plan to expend money, but also the *rate at which we plan to accomplish work* (below, we will introduce the term *planned value* for this item).

As you can see by looking at this line, by the end of month 4, the line indicates that we expected to perform $48 worth of work on our task(s). We also expected to expend $48 in performing that work. Of course, once we leave the world of *plans,* and enter the world of *actual results,* these two items may diverge; as we discussed above, in the real world of actual results, *what we accomplish* and *what it costs to achieve those accomplishments* may not be the same thing. When we finish the tasks that we planned to accomplish during the first four months, we earn $48 worth of *accomplishment* (in a moment, we will introduce the term *earned value* for this item). But the amount of money that we actually spent to perform that work could be more or less than $48. We may also have taken more time or less time than four months to finish that work; no matter, we still get exactly $48 worth of accomplishment (*earned value*) whenever we finish that work, because *that is what we planned for that work to be worth.*

This line represents a *plan,* so it depicts information for the entire planned duration of the task(s), all the way through month 10.

- **Bottom line: earned value.** There is a second line of the graph in Figure 10.7, labeled *earned value* (which appears as the bottom of the three lines on the graph). But we already know what *earned value* is: we defined it above as our progress toward completion, or as it says in Figure 10.5, the *value of the work accomplished to date* on our task(s). We only calculate earned value on the work as we perform it, so this line on the graph does not extend all the way to the end of the entire planned duration of the task; it only extends as far as the last month of completed work. We can tell from the graph that we have just completed month 4. We can also see, by the fact that the last point depicted on the earned-value line (which corresponds to the end of month 4) is below the end-of-month-4 point on the budget line, for some reason the total amount of work on our task(s) that we have accomplished to date is *less* than the amount we originally planned to accomplish.

- **Upper line: actual cost.** The upper of the three lines on the graph in Figure 10.7 is labeled *actual cost*. We can understand what actual cost means: it is simply the amount of money that we actually spent doing work during each of the first four months of the project. As we discussed above, *what you spend* and *what you accomplish* are two separate items. This line depicts *what we spent*; the earned value line depicts *what we accomplished*.

Of course, I drew this example graph with the lines not crossing so as to make it easy to see. On an actual engineering project, the lines could be in any order, and may actually cross.

Having discussed these three lines, we are ready to depict in graphic form what the schedule variance and the cost variance represent. In Figure 10.8, we show the same graph as in Figure 10.7, but I have added some annotations. You can see the actual cost (AC), planned value (PV), and earned value (EV) at the end of month 4 indicated by arrows on the graph.

- If you refer back to Figure 10.5, you will see that we defined schedule variance as earned value minus planned value. We can see on the graph in Figure 10.8 that this amount (earned value minus planned value) is depicted by a curly bracket with a legend that says *schedule variance*.

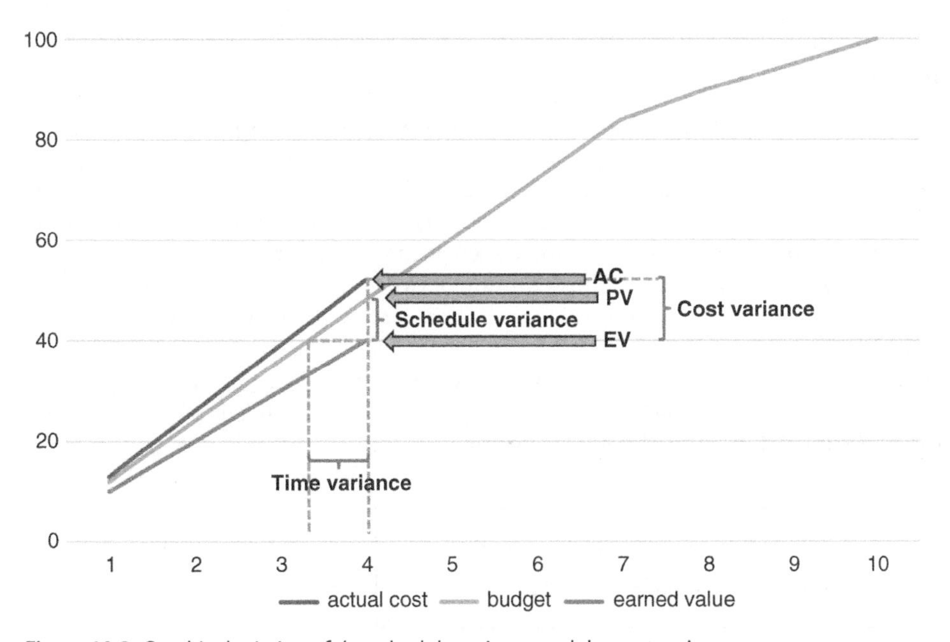

Figure 10.8 Graphic depiction of the schedule variance and the cost variance.

- If you refer back to Figure 10.6, you will see that we defined cost variance as earned value minus actual cost. We can see on the graph in Figure 10.8 that this amount (earned value minus actual cost) is depicted by its own curly bracket, and a legend that says *cost variance*. Note that the cost variance is *not* PV – AC; when looking at the graph, this is a mistake that many people make.

Usually, projects will each month calculate schedule variances and cost variances in two ways: (i) for the previous month and (ii) for the period from inception of the project to date.

10.3.2 Time Variance

One matter that is often confusing to people at first is that earned value (EV) and planned value (PV) are both expressed in *dollars*, and therefore schedule variance (SV = EV – PV) is *also expressed in dollars*. Most of us do *not* normally think of quantifying our schedule status in terms of money. Instead, we think of our schedule status in terms of some measure of *time*, such as hours, days, or weeks; we might be three days ahead of schedule, or two weeks behind schedule. For completeness, Figure 10.8 therefore also depicts one more variance, called the *time variance*. As you can see from the depiction, this is a measure (in units of *time*) of how far you are behind (or ahead) of schedule. The *schedule variance* – despite the use of the word *schedule* in its title – is expressed in *dollars*, not in *time*. If we were spending on average $XX dollars per day, you see that you can make the conversion from *time variance* to *schedule variance* through simple multiplication:

$$\text{time variance (in days)} \times \text{average expenditure per day (in \$ / day)} = \text{schedule variance (in \$)}$$

You can also convert in the other direction, that is, from *schedule variance* to *time variance*:

$$[\text{schedule variance (in \$)}] / [\text{average expenditure per day (\$ / day)}] = \text{time variance (in days)}$$

As we discussed, both the schedule variance and the cost variance are expressed in dollars. Sometimes, our company and/or our customer wants also to see these variances expressed as *percentages* (e.g. we are x% ahead of schedule, or y% over cost), in addition to dollars. This leads to the definition of two additional statistics, the *schedule performance index* (SPI) and the *cost performance index* (CPI), which are defined in Figure 10.9.

There is nothing complicated about these two additional statistics; they just express the variances as *ratios*, rather than as *dollars*. Notice, however, that as pure ratios, they express no units; they are no longer in dollars.

Schedule performance index (SPI) = EV / PV

Cost performance index (CPI) = EV / AC

Figure 10.9 Definitions of the schedule performance index and the cost performance index.

Like the schedule variance and the cost variance, SPI and CPI can be calculated both for inception to date, and also just for the previous period.

10.3.3 Variance Analysis

Having created these statistics, your customer and your management will now want you to explain your status *in words*. You will refer to the statistics (the schedule variance, the cost variance, and so forth), but the essence of the explanation is *not* the numbers, it is your discussion of the *causes behind the numbers*. This explanation is called the *variance analysis*, and it attempts to answer questions such as these:

- What is different in the current estimated schedule than in your previous predictions, all the way back to the baseline schedule ... and *why*?
- Are the positive differences sustainable? Why or why not?
- Are the negative differences recoverable? Why or why not?

On most projects, there will be numeric thresholds that act as triggers; variance analyses will not be required unless the variance for a cost account exceeds one of these thresholds. Typically, there are thresholds expressed both as:

- *Absolute dollars* (e.g. you must write a variance analysis for this month if either the schedule variance or the cost variance for the previous month exceeds $XXX, or if the inception-to-date schedule or cost variance exceeds $YYY), and as
- *Percentages* (e.g. you must write a variance analysis for this month if either the schedule variance or the cost variance for the previous month exceeds A%, or if the inception-to-date schedule or cost variance exceeds B%).

Whereas the *variance statistics* – schedule variance, cost variance, SPI, and CPI – are created for the cost-account manager by the business specialist, the variance analysis (which is the textual explanation of the causes behind the numbers) is *authored by the cost-account manager himself or herself.*

You might also be asked to create statistics about, and do variance analysis on, *staffing levels* (e.g. Are you finally fully staffed? Are you catching up on a previous staffing shortfall?), or on other parameters that seem particularly important to your project, at this particular time.

10.4 Financial Measures About Which Your Company Will Care

Your company and its senior management – starting with your direct supervisor – will also be interested in a series of *financial* measures. After all, the company is a business, and your engineering project is a vital portion of that business. So, in addition to metrics that measure the progress of your project toward completion (that is, a set of metrics that measure things that are first and foremost of interest to the *customer*), you will be expected to create a set of metrics that measure the quality of your project as a part of the business of your company. I will describe a few of these here.

10.4.1 Sales

The first of these financial metrics is called *sales*. This refers to the total dollar value of work that is performed under your contract: every dollar that is properly billable to the customer. Your company will want to know:

- Your *inception-to-date sales*; that is, the total dollar value of work performed under your contract since the award of the contract to your company. They will probably want to see this expressed not just as a total single number, but also on a month-by-month or quarter-by-quarter basis. These are measures of *actual performance*, rather than *predictions*. They are also likely to ask for these actual sales to be compared to an original plan for the month-by-month or quarter-by-quarter sales; that is, they will want you to compare the *actual sales you achieved* to the *sales that you promised to deliver* to the company when you signed the contract.
- Your *projected sales under the current contract*. These are your predictions for sales in the future. They will be most interested in your *estimated sales at the completion of the contract*, but will likely also want to see these predictions for sales on a month-by-month or quarter-by-quarter basis, as well.
- Your *potential sales*. It may be that, if you perform well on the contract, there is additional work that your company may receive from this same customer, perhaps even in the form of sole-source add-ons to your existing contract. For example, if you deliver an engineered system, you may hope to receive a modification to your contract in order to maintain and enhance that system. Your company will expect you to predict these sales too. They will usually ask you to multiply each such source of additional sales by a probability factor. That is, since these items are not yet on the contract, no matter how sure you are that the customer intends to add them to the contract, something might go wrong. So, we create an *expected value* for these potential sales by multiplying the dollar value times an estimated probability of occurrence. For example, if there is a potential contract modification worth $10 000 000, and you estimate that there is a 60% chance that it will actually be added to the contract, you would show these future sales as having an expected value of $6 000 000. Your company will likely want to see these predictions for sales on a month-by-month or quarter-by-quarter basis, as well.

10.4.2 Profit

The second of these financial metrics is called *profit*. A company is morally and legally entitled to charge a customer more than the cost of doing a piece of work, both to compensate them for the cost of capital, and also in compensation for the company putting their reputation on the line in order to create something of value for the customer. Your company will want you to make reports about profit, as well as sales.

- *Inception-to-date profit*. This is usually just a bookkeeping measure; profit is not certain until the contract is completed. But your contract will likely have terms that will allow you to predict how much profit you will make if your performance (schedule, cost, capability) continue in the future at the same level as thus far on the project.
- Your *projected profit upon completion of the project*.
- Your *projected profit on* those *potential sales* we discussed above.

10.4.3 Cash Flow

The next of these financial metrics is called *cash flow*. We alluded to this way back in Chapter 1: projects expend funds before they deliver benefits. Similarly, your company must usually incur expenses before it gets paid for its work on a project. Even if your contract entitles your company to some payments along the way (rather than requiring your company to wait for all payment until project completion and delivery), those *progress payments* seldom cover all of the expenses incurred so far. Therefore, the company must either use its own funds, or borrow money in order to meet current expenses. Your people must be paid every week. The rent, electric bill, property tax bill, and so forth must also be paid continuously. Therefore, your company will expect you to estimate your cash flow – dollars out and dollars in, by month or quarter – so that they can understand how much cash they need to plan to (in essence) loan to your project's checking account. They will also expect you to monitor your actual cash flow, and periodically to update your predictions about future cash flow.

10.4.4 Day-Sales Receivables

Your management will also expect you to collaborate with the customer to make sure that payments from the customer to your company are made both on time and for the correct amount. In order to measure these factors, they will expect you to use a metric called *day-sales receivables*, which is a measure of how long you have been waiting for payments from your customer. Your contract may call for the customer to pay you, for example, within 30 days of each invoice. Day-sales receivables is the metric that the company will expect you to use in order to determine if those payments are on time. Day-sales receivables and cash flow interact; clearly, if the payments from your customer are late, your cash flow will become more negative, because of course you must still pay your employees, pay the rent and the electric bill, and so forth.

Your management will expect you to pay attention to *all* of these metrics, to expend effort by you and your team to collect and track them, and (most importantly) to take corrective actions (just like you would for a technical metric) if one of them starts to show a signal of a true problem. Most likely, you will actually create entries on your risk register that pertain to some or all of these metrics too.

Most of the actual leg-work to gather the data and calculate these metrics is done by your project's finance team. But you monitor the results, ensure that the right items are placed on the risk register, and if and when problems arise, you address them just as you would any other project problem.

10.5 Your Role in All of This

Much of the detailed work described in this chapter is performed by the cost-account managers, and the business specialists assigned to support them. However, you, in your role as the project manager, still have a vital role to play. You review all of this information, usually in face-to-face meetings, such as these:

- With each cost-account manager, to discuss their schedule variance and their cost variances, and (especially) their analysis of those variances.

- With groups of cost-account managers, to find tasks that can be broken apart into smaller tasks, so that the activity network can be *replumbed*, and thereby aid in attempting to recover schedule delays. Because the interactions in such a replumbing activity usually cross over many parts of the project, the personal participation of the project manager is important in order to motivate participation, reach timely resolution of different points of view, and encourage acceptance of the results (nobody likes to change their plan, especially if the main benefit of that change is another portion of the project, rather than their own).

In the next chapter, we will combine these activities with other activities, such as risk management (Chapter 9), into an integrated management process that I call the *periodic management rhythm*.

10.6 Summary: Monitoring the Progress of your Project (Part I)

- Schedule drives cost, so we create the schedule estimate before we create the cost estimate. We do the same for the periodic updates to the schedule estimate and the cost estimate; we create the update to the schedule estimate first.
- We have a well-defined method for creating both initial ("baseline") estimates for schedule and cost, and for updating them on a periodic basis (usually monthly).
- We have well-defined methods for *assessing progress* (which we call *taking earned value*) and for *analyzing progress* (which we call *variance analysis*).
  - Separating noise from signal (as we learned how to do in Chapter 8) is very important to this method.
- We even have a method for credibly figuring out how to recover some lost schedule (*replumbing the tasks*).
- Each periodic reporting interval (usually once per month), you will therefore do *all* of the following:
  - Update the schedule, as described above. This results in a new *predicted completion date.*
  - Perform an additional increment of the *rolling-wave planning process*, looking at tasks that will be starting soon, and breaking them into smaller tasks.
  - Look at tasks underway and tasks that will be started soon, and look for opportunities to break those into smaller tasks, in order to identify work that can be removed from sequential constraints and therefore performed in parallel with other tasks, rather than in strict sequence. This *replumbing of the schedule* may result in a shorter schedule for your project.
  - Update the cost estimate, as described above. This results in a new *predicted cost at the completion of the project.*
  - For every in-progress task, compute the two variances defined above (*schedule variance* and *cost variance*). This is another reason to have short tasks – you will have fewer (and simpler) variances to explain.
  - Compare the current variances to the sequence of previous variances, so as to try to determine if there is a meaningful trend (*separate signal from noise* – remember Chapter 8). Most projects use the weak statistic of comparing the current variances

to last month's variances. This will result in a lot of "changes" that are actually just *noise*. You will be writing a lot of useless variance analyses. So, it is better to use a *stronger statistic* to determine if a change to either schedule or cost is a true signal, or just random variation (*noise*). The process-behavior chart and the five Wheeler/Kazeef tests provide you with such a stronger statistic. Apply the five tests, and you will eliminate the need to write variance analyses about changes that are simply "noise."

- Write a text explanation for all variances that are actual meaningful signals (the *variance analysis*):
  - If you mistake noise for signals, you will be writing a lot more variance analyses than you need to!
- Actually, almost all of the above is done by the *responsible cost-account manager* – the person whose name is in that box on the work-breakdown structure. This is done at every level of the project
- As the project manager, you *personally* will:
  - Review (usually in person) *all* of the variance reports with each responsible cost-account manager
  - Participate in the *replumbing* sessions
  - Write the project-level variance analyses.

- The periodic update of our schedule and cost estimates is actually part of a larger process that I term the *periodic management rhythm*, which we will discuss in the next chapter. This larger process combines earned-value assessment and variance analysis (this chapter) with assessment of technical and operational performance measures and risk/opportunity management (Chapter 9).

10.7 Next

Next, we will continue learning how to monitor the on-going progress of our project by focusing on a technique that I call the *monthly* (or *periodic*) *management rhythm*. We will discover that the activity network (Chapter 7), drawing valid conclusions from numbers (Chapter 8), risk management (Chapter 9), and earned-value/variance analysis (Chapter 10) are all essential elements of this monitoring process.

10.8 This Week's Facilitated Lab Session

In this week's facilitated lab session, you will work *by yourself*, not in your teams. Our topic is problems and exercises about earned value and variances. The problems are below.

You will prepare the answers to these problems, using either my lecture slides or the textbook as a reference. The answers should include an explanation of how you arrived at the answer, and cite the appropriate pages in the lectures or the textbook. The answers will be turned in as an individual assignment.

1. You are the manager of an engineering project that is five days behind schedule at day 25. It has a planned value of \$525 000 for this point in time, but the actual cost in

fact so far is only $425 000. Determine the schedule variance, the cost variance, the SPI, and the CPI.

2. Your engineering project has, at the end of month 24, an actual cost of $325 000, a planned value of $425 000, and an earned value of $275 000. Determine the schedule variance, the cost variance, the SPI, and the CPI. Is the customer happy or unhappy? Why?
3. Find the schedule and cost variances for a project that has an actual cost at month 25 of $500 000, a planned value of $480 000, and an earned value of $490 000.
4. An engineering project at day 70 has an actual cost of $68 000 and a planned value of $72 000. The relevant cost-account manager's estimates for earned value total $70 500. Determine the schedule variance, the cost variance, the SPI, and the CPI. Create and explain a method to calculate the time variance, and then use that method to estimate the time variance in work days.
5. Consider an engineering project consisting of the five tasks listed in the table below. Determine the schedule variance, the cost variance, the SPI, and the CPI. Create and explain a method for estimating the total predicted cost of the project at completion, and then use that method to calculate the total predicted cost at completion.

Task name	Predecessors of this task	Duration of this task (weeks)	Budget for this task	Actual cost, at the end of week 6	Percentage complete, at the end of week 6
Task 1	(none)	2	$300 000	$350 000	100%
Task 2	(none)	3	$225 000	$170 000	100%
Task 3	Task 1	2	$150 000	$300 000	100%
Task 4	Task 1	5	$800 000	$400 000	20%
Task 5	Task 2, task 3	4	$500 000	$200 000	20%

6. A project at month 7 had an actual cost of $44 000, a planned value of $50 000, and an earned value of $47 000. Determine the schedule variance, the cost variance, the schedule performance index, and the cost performance index.

11

Monitoring the Progress of Your Project (Part II)

There is much more to monitor on your engineering project than just schedule and cost (although sometimes it will seem like those are the only things your boss cares about!). In this chapter, I teach you a comprehensive method for monitoring schedule, cost, technical capability, and risk … and making sure that all of these parts fit together correctly. Since on most projects your contract will require you to assess progress at least once a month, I used to call this method the monthly management rhythm, but have (reluctantly) switched to the more general phrase "periodic management rhythm."

11.1 How the Manager of an Engineering Project Ought to Allocate His/Her Time

How do we allocate our personal time as the manager of an engineering project? There are, after all, only 40 to 50 working hours per week.

There is a strong pull to spend it on activities *inside* the company; you are there, and easily accessible. Your boss will ask you to attend meetings. People who are managing other projects will want to talk to you about what you have learned. Your team needs you. The president of the company will ask you to work on some long-term planning committee. You will be invited to an endless number of meetings inside and outside your project. This pull toward inside-the-company activities can easily use up all of your working hours.

I believe, however, that it is very important that you spend significant time on other constituencies too. The first of these is *your customers*: you must spend time meeting with them, talking to them, learning more about them, thinking about them. Therefore, *customers* are another one of the things on which we, as the manager of our engineering project, spend our own time.

You, of course, need to spend time with *your project team*; they do need a portion of your time.

So does *your supervisor or boss*.

Your company most likely has a number of specialist staff organizations: law, human relations, marketing, strategic development, and others. You need to spend time with these people too.

Engineering Project Management, First Edition. Neil G. Siegel.

Companion website: www.wiley.com/go/siegel/engineering_project_management

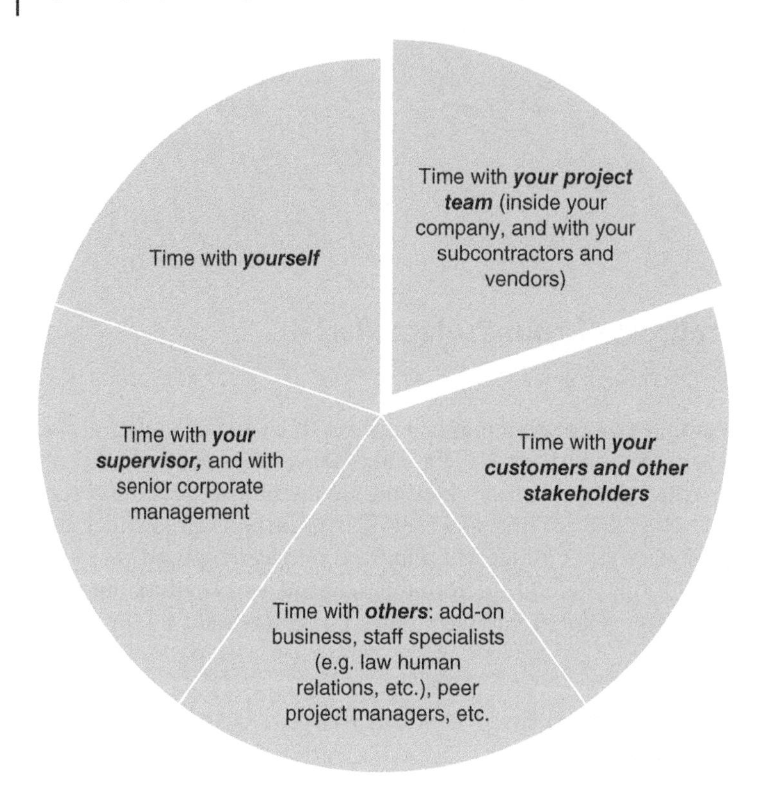

Figure 11.1 The time pie: constituencies to which you need to allocate significant blocks of your personal time.

In all of this, it is easy to forget one additional, but very important, constituency: *yourself.* You need time to yourself, so that you can think. Think about improvements to methods. Think about personnel assignments. Think about what could go wrong. Think about what the customer said last week, and whether you truly understood it. And many other things.

Figure 11.1 summarizes all of this in pictorial form, which, following a suggestion from my wife, we will call the *time pie.* You may well have things to add to this list, as you create your own version of this pie. But, however you spend your time, you need to be sure that you have prepared such a list, have thought about how much time you need to be spending on each of these constituencies, and make sure that the demands of one does not shut out the need and the opportunity to spend time with *all* of them.

11.2 A Big Claim on Our Time: The Periodic Management Rhythm

In this chapter, we will talk about something on which I believe you need to spend considerable personal time. This is a specific aspect of monitoring the progress of your project, which I call the *periodic management rhythm.*

This activity will *all by itself* require *10 or so hours per week* of your personal time; on a large project, perhaps even a bit more. In some sense, *this is the core task* that is personally performed on a regular, repeating basis by the manager of an engineering project.

Fortunately, most of the time needed by this task is amenable to *scheduling in advance*, so it can form a skeleton around which you can manage the allocation of the remainder of your personal time, so that you can still be sure that you spend adequate time on *all* of the items listed above in Figure 11.1.

We have already learned about all of the individual pieces of the periodic management rhythm; all we need to do now is put them together.

1. In Chapters 2 and 4, we talked about the need to establish *operational performance measures* and *technical performance measures* to guide decision-making, the mapping between these two types of measures, and the two coordinate systems of value that underlie these measures.
2. In Chapters 7 and 10, we talked about *estimating schedule*; in Chapter 7, creating our *initial* estimates for these parameters and in Chapter 10, how to *update* those estimates on a monthly basis (in Chapter 12, we will add one more step: the formal *baselining* of the schedule and cost estimates after you win the competition and start work on your new engineering project). We then monitor our progress through the mechanism of calculating and analyzing *variances* between our predicted schedule performance and our progress to date.
3. In Chapters 7 and 10, we talked about *estimating cost*. Just as we did for estimating the schedule, in Chapter 7, we talked about creating our *initial* estimates for these parameters, and in Chapter 10, we talked about how to *update* those estimates on a monthly basis (and in Chapter 12, we will add the step of *baselining*). We then talked about how we monitor our progress through the mechanism of calculating and analyzing *variances* between our predicted cost performance and our progress to date.
4. In Chapter 9, we talked about the establishment of a risk register, and how to perform *risk management* as a continuous process.

This week, we put all of these pieces together into a single integrated, periodic process that we use to monitor the progress of our engineering project.

Let's review each of these items, starting with item 1: the need to establish *operational performance measures* and *technical performance measures* to guide decision-making, the mapping between these two types of measures, and the two coordinate systems of value that underlie these measures (Figure 11.2).

As summarized in Figure 11.2, we make regular periodic measurements of both operational and technical performance parameters, as a way of monitoring the technical progress on our engineering project. Although our *degrees of design freedom* mostly reside in the *engineer's coordinate system of value* (and are therefore measured by our technical performance measures), we must also confirm the utility of our design by considering measures of predicted *operational performance*, which form the *customer's coordinate system of value*. We therefore need some mechanism to convert the predicted technical performance measures into predictions for the operational performance measures (such as the nested models discussed in Chapter 2), and even our non-technical stakeholders must be able to understand enough about this mapping to allow them to believe the predictions about operational performance.

The *customer's* coordinate system of value	The *engineer's* coordinate system of value
Usually non-technical in nature	Usually technical in nature
Characterizes something about the *mission*	Characterizes something about the *technology that implements the system*
Captured in what I call *operational performance measures*	Captured in what I call *technical performance measures*
Measures the *goodness* of the design for the users	Measures whether or not the design is *feasible*, and helps us estimate the *schedule* and *cost* needed to build the system
The two types of metrics should be *linked*; that is, we need to credibly and transparently show how changes in design (which change the *technical performance measures*) cause changes to the *operational performance measures* Our *degrees of design freedom* are located in the engineer's coordinate system of value, but we *measure the goodness of the resulting design* in the customer's coordinate system of value	

The mapping between the two coordinate systems must be transparent and credible–***even to our non-technical stakeholders***

Figure 11.2 The two coordinate systems of value, the two types of measures, and the mapping between them (repeat of Figure 2.8).

We have a second reason, of course, for making the measures of technical performance: our design must be technically credible, and we assess and establish that credibility through the technical performance measures.

Item 2 on our list was to create predictions for schedule, and item 3 was to create predictions for cost. We prepare these predictions many times: we create initial estimates before the project starts, as part of our proposal effort; as we will discuss in Chapter 12, we update those into a formal *baseline* version of the schedule and cost right after we start work on the project; and we then *update* those estimates on a periodic basis (most likely monthly). We then *monitor our progress* through the mechanism of periodically calculating and analyzing variances between our predicted schedule and cost performance (Figure 11.3), and our progress to date. Specifically, each periodic reporting interval (usually once per month) we do *all* of the following:

- Update the schedule. This results in a new *predicted completion date.*
- Perform an additional increment of the *rolling-wave planning process,* looking at tasks that will be starting soon, and breaking them into smaller tasks.
- Look at tasks underway and tasks that will be starting soon, and look for opportunities to break those into smaller tasks, in order to identify work that can be removed from sequential constraints and therefore performed in parallel with other tasks, rather than in strict sequence. This *replumbing of the schedule* may result in a shorter schedule for your project.

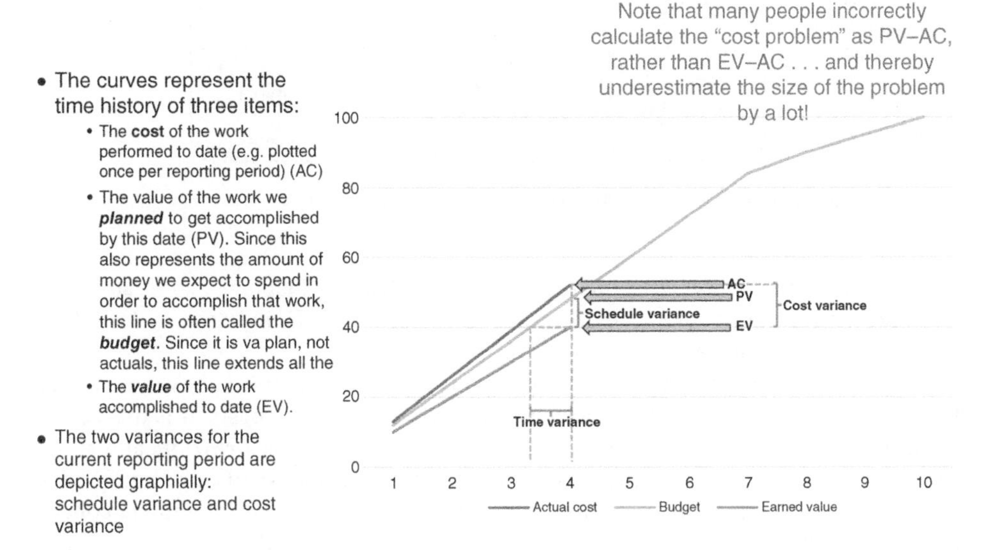

Figure 11.3 Schedule and cost variances.

- Update the cost estimate, based on the updated schedule. This results in a new *predicted cost at the completion of the project.*
- For every in-progress task, compute the *schedule variance* and *cost variance.*
- Compare the current variances to the sequence of previous variances, so as to try to determine if there is a meaningful trend (*separate signal from noise* – remember Chapter 8). Use the process-behavior chart and the five Wheeler/Kazeef tests, and you will eliminate the need to write variance analyses about those changes that are simply "noise." For those variances which are statistically significant (e.g. represent a real signal), write variance analyses that explain the root cause for the problem. This can lead to replumbing sessions to try to regain lost schedule.

Item 4 on the list was to create a *risk register,* and to conduct a periodic process to assess and mitigate risks, and to exploit opportunities. In Chapter 9, we identified these as the principal steps involved in accomplishing this:

- Identify the potential risks, capturing them on a risk register.
- Identify the *symptoms* that will allow you to notice if some particular risk is coming to pass.
- Select the *parameters to be measured,* and the associated measurement methods.
- *Score* those risks, separately for *likelihood* and *impact.*
- Create *mitigation plans,* which often will also be separate for *likelihood* and *impact.*
- Create *triggers* and *timing requirements* for those mitigation plans.
- Create a method to aggregate all risk assessments into a periodic overall project impact prediction.
- Create and use some sort of *periodic management rhythm,* wherein you make decisions about risk mitigation actions.

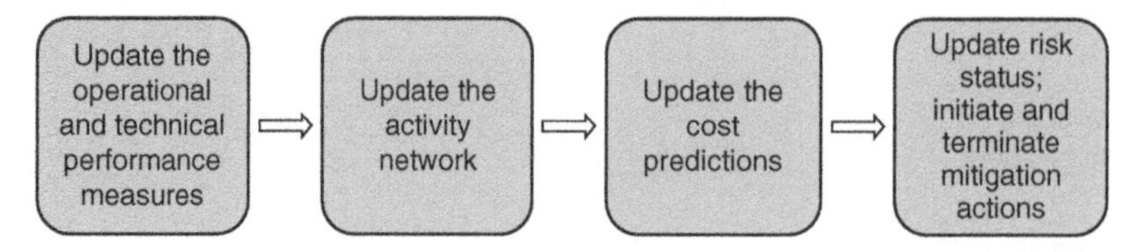

Figure 11.4 The four principal steps of the periodic management rhythm.

As you can see from the last bulleted item on this list, in Chapter 9, I actually gave away my plan for this chapter: the creation and use of a *periodic management rhythm*. What do I mean by this term?

I mean a process that integrates each of the four steps we discussed above into a single, coherent process, that we will perform on a regular, periodic basis. It is this act of repeating this process on a regular, periodic basis that makes this a *rhythm*, and since the purpose of the integrated process is to monitor the progress of our engineering project, and such monitoring is a *management* task, we can call this integrated, coherent process a *periodic management rhythm*. When I first adopted the term, I actually called it the *monthly management rhythm*, since all the projects I was involved in at that time did this process once per month; that was the *period*. It is possible, however, that some projects (especially very small, very short projects) might do this every two weeks. I have therefore changed the name for this book to the more general *periodic management rhythm*. I will not mind if you use the term *monthly management rhythm* instead; a monthly period is likely to be the best for almost all of the engineering projects on which you will work.

11.2.1 Sequence and Interaction of Steps

Each period, we do the four steps in order, as depicted in Figure 11.4.

Later in this chapter, we will actually show you one way to lay these four tasks out as a monthly calendar. We will also add a fifth task.

Figure 11.5 depicts the four steps of the periodic management rhythm in the form of an N^2 chart (remember, we introduced and described this format at the end of Chapter 2). This format shows what information flows from one step to another as we execute the periodic management rhythm. Note that there are a couple of items of information that flow backwards, from later steps to earlier ones. Those require either some reworking of the earlier step, or the incorporation of those inputs into the next round of the periodic process.

11.3 The Steps of the Periodic Management Rhythm

Now we are ready to discuss the steps of the process in detail.

11.3.1 Updating the Predictions of Operational and Technical Performance

The process starts with each cost-account manager updating their *technical performance predictions* and their *operational performance predictions*. These updated predictions are

Update operational and technical performance measures	Assessment of technical progress		Updates to risk register. Are risk mitigations still adequate?
	Update the activity network (predictions of schedule)	Update task durations. Cost impacts of mitigation actions	Variance analysis: does this change our risk assessment?
		Update the predictions of cost	Variance analysis: does this change our risk assessment?
Design alternatives from risk mitigations: request to update TPMs and OPMs	Add / subtract mitigation actions to activity network		**Review risk status; initiate and terminate mitigation actions**

Figure 11.5 The periodic management rhythm in N^2 chart format.

derived from actual measurements on laboratory and field equipment, from predictive models, from analytics, and so forth. The concept is that over the course of the last month, the team has probably learned more about their portion of the requirements, their portion of the design, and their portion of the implementation; this update allows them to incorporate that new (and hopefully, improved) knowledge into their predictions of operational and technical performance. In the middle portions of the project, they may well also have implemented some additional portions of the design, and can make actual benchmark measurements of how well those little portions of the system work; real benchmark data always trumps predictive models!

As we discussed in Chapter 9, each cost-account manager will have target values and/or thresholds of goodness for metrics that have been approved by the project. In other words, for each measurement, there will be *actual numeric values* that constitute the targeted level of performance.

For each measurement (whether operational or technical), this cost-account manager is using the sequence of measurements that emerges from the periodic process to build a time series of measurements. Each cost-account manager is analyzing each of those time series of measurements through the use of a process-behavior chart and the five Wheeler/Kazeef tests; these allow them to separate signal from noise in that time series of measurements.

Once we have determined which measurements are actual signals, rather than just noise, we can compare those measurements to our triggers, and thereby identify if any of these signals indicate actual problems. That is, do we have predictions of significant shortfalls in operational or technical performance? Are new problems arising that might prevent us from achieving our operational and technical goals?

If we planned our risk register perfectly, any problems identified will already be in the risk register, and therefore we will already have a set of candidate mitigation plans and triggers. Since we are not always perfect, we must allow for the possibility that a new risk has been identified, and this can be proposed as a potential new entry in the risk register.

In most projects, the actual process is a little more *socially complicated* than that described above. This is because it is likely not possible to calculate all of the operational and technical performance measures solely within the confines of a single cost-account manager's range of responsibility. More than likely – especially for the operational performance measurements – it will take a *combined* set of inputs *from multiple cost-account managers* in order to create the complete set of data that will allow for the calculation of an update to a predicted operational or technical performance measure. Not only must a set of cost-account managers collaborate to produce the *inputs* to the updated measurement process, but they must also collaborate in order to analyze the *outputs*, to assess technical progress, and to propose any updates to the risk register.

A common approach to handling this social complexity is to have a centralized systems engineering team perform the modeling, using inputs from each of the cost-account managers. The convenience of using such a centralized model, however, cannot be used as an excuse to relieve the cost-account managers of their responsibility for the outputs of this process: they must "take ownership" of the updated predictions, either agreeing with them or challenging them. They must also take ownership of the responsibility to assess technical progress, in light of the updated predictions, and take ownership of the responsibility to update the risk register, if that is warranted.

11.3.2 Updating the Predictions for the Schedule

Now we are ready for the next step. Each cost-account manager updates their *schedule predictions*, in the form of an update to the anticipated duration of each task that is not yet 100% complete. This anticipated duration ought to be expressed statistically, most likely in the form of the three-point estimate that we discussed in Chapter 7.

Each cost-account manager may also identify changes to the network topology, that is, to the arrows that interconnect the tasks within their portion of the work. They may also collaborate with other cost-account managers, and update the topology of the interconnections *between* their portions of the work.

Absolute dates are used only at the start of a new chain of tasks, or for external inputs over which the team has no control. An example of an external input over which the team has no control might be a piece of *customer-furnished equipment or data*; it is very common for the contract to identify some items that will be provided in this fashion to the contractor by the customer, and for each such item the contract will specify a specific date by which the customer is to provide that data or equipment. This date would then show up as an *absolute date* in the activity network; it will look like the start of a chain, or the start of a branch into a chain. If the customer *changes* the date by which they expect to provide this data or equipment, then that absolute date (once it has been formally changed in the contract) should also be changed in the activity network.

Of course, in order to do this effectively, your schedule must be constructed as an *activity network* – that is, every task has a *duration, preceding tasks* (e.g. those tasks that must complete before your task can start), and *successor tasks* (those tasks that cannot start until your task is complete).

Each task duration will reflect a *distribution* of predicted durations; at least the three-point estimate that we discussed in Chapter 7 (optimistic, nominal, and pessimistic). Remember that in most instances this distribution should be asymmetric in both range and probability! That is, the time difference between optimistic and nominal *will be less* than the time difference between nominal and pessimistic, and the *probability of the pessimistic outcome will be higher* than the probability of the optimistic outcome.

Then, use your schedule management tool that can use these distributions to create statistically valid predictions of milestones and end dates.

The milestones – predictions of absolute dates (with their distributions) – are therefore an *output*, not an input.

You may be shocked at the adverse effect on the schedule milestones that results from the incorporation of the underlying distributions (e.g. the three-point estimates for task duration). Failing to incorporate the underlying distributions of task durations (or creating unrealistic distributions, e.g. having optimistic and pessimistic be symmetric in range and probability) is a major factor in unpredicted project schedule delays. And remember that schedule delays almost always impose cost increases, so failing to incorporate the underlying distributions of task durations is also a major factor in unpredicted project cost increases.

The periodic predictions for each milestone form a time series; so do the periodic predictions for the project's end date. We can therefore create and use process-behavior charts in order to determine which of these measurements are likely to indicate an actual signal, rather than just random noise.

If we have a true signal, and if the magnitude of that signal (in either absolute dollars or as a percentage) exceeds the established variance thresholds for this cost account, the cost-account manager now writes *schedule variance analyses* and presents them in person to the project manager. This should always be done "bottom-up": each cost-account manager writes their own variance analyses, each presents them in person to the project manager in the presence of their entire management chain (e.g. every cost-account manager in the work-breakdown structure between them and the project manager), and the order of these presentations starts with the lowest-level cost accounts and works up systematically through the levels of the work-breakdown structure, so that each estimate is psychologically "owned" by the manager responsible for the work.

Most likely, at some point you will predict delays to the schedule. This is the time to investigate whether there is legitimate *replumbing* of the schedule (e.g. breaking tasks into smaller pieces, and thereby eliminating unnecessary sequential relationships; we discussed this in Chapter 10) that can help recover the schedule. Most of the time, such replumbing involves multiple cost accounts, and so this *replumbing* should be done in a face-to-face meeting with all of the cost-account managers present and participating. I always do this at the project level for the social reasons that we will discuss in a moment, but on a large project, you have the option to have some of the lower-level managers do it, and then present the results to the project manager. As I mentioned in Chapter 7, when I was the project manager, I always attended all of these replumbing sessions, but I usually let my chief engineer actually run the meeting.

As mentioned earlier, if you, as the project manager, disclose your opinion on a matter early in a meeting, it may intimidate others from speaking their mind. I therefore often find it advantageous to allow my chief engineer to run many sorts of meetings. During those meetings: (i) I listen; (ii) I encourage people to speak their mind forthrightly but respectfully; (iii) I (gently) drive the meeting back to the right subject if it appears to be moving off topic; (iv) I summarize all of the points of view expressed; and (v) I thank everyone for their participation and their opinions. If – and only if – I cannot tolerate the consensus decision toward which the group is moving will I speak out and offer my own opinion. I will first try to suggest options, premises, and analysis methods (that is, try to guide the team to create a larger trade space, as discussed in Chapter 2) that might result in a different decision. Only if that completely fails will I make a decision to override the group; that does happen at times, but not very often.[1]

11.3.3 Updating the Predictions for the Cost

Now we are ready for the next step. Each cost-account manager updates their *cost predictions*, using the methodology described in Chapter 10.

The periodic predictions for the cost at completion for each portion of the work, and for the complete project, form a time series. So, we create and use process-behavior charts in order to determine which measurements are likely to indicate actual signal, rather than just random noise.

If we have a true signal, and if the magnitude of that signal (in either absolute dollars or as a percentage) exceeds the established variance thresholds for this cost account, the cost-account manager now writes *cost variance analyses* and presents them in person to the project manager. Just like for the schedule variance analyses, this should always be done "bottom-up": each cost-account manager writes their own variance analyses, each presents them in person to the project manager in the presence of their entire management chain (e.g. every cost-account manager in the work-breakdown structure between them and the project manager), and the order of these presentations starts with the lowest-level cost accounts, and works up systematically through the levels of the work-breakdown structure, so that each estimate is "owned" by the manager responsible for the work.

There will often be an impact on total cost *over and above* that identified by the individual cost-account managers. For example, when person A is late, it will make my task cost more. To catch this, you need to use the activity network for the *entire project*, and not just for your piece of the project. *You need to have a mechanism by which each cost-account manager is presented with a list of impacts to the schedule of **their** tasks due to the delays in the schedule of the tasks of **other** cost-account managers!* Projects often overlook this aspect.

11.3.4 Updating the Risk Assessment and Initiating Risk Mitigation

Finally, we are ready for the fourth step: updating the risk assessment. Most projects use a top-N methodology: they create and update a list of the top N risks in *likelihood*, and a separate list of the top N risks in *impact*. In addition, every project must also perform and update the aggregated risk assessment described in Chapter 9.

---

1 I must admit that I actually found this hard – I often succumbed to the temptation to offer my opinion too early in a meeting. But what I describe is the ideal, even if I on occasion fell short!

The heart of the process is to use all of the information created by all four steps to decide if any mitigation actions need to be started or terminated at this time. Do this by using the previously defined triggers. Examples of potential mitigation actions include:

- *Adjust the activity network*, typically by breaking apart tasks, so that there are more opportunities for parallel execution, rather than serial. This is the *replumbing* that we discussed already.
- Release *management reserve funding* to do an assessment of a prediction/measurement.
- Release *management reserve funding* to start a mitigation action from the risk register, probably in response to a trigger having been set off.
- *Retire* an in-process mitigation action, and recover the unspent portion of the allocated management reserve funding.
- *Create a new item for the risk register*, when an issue has been identified that does not correspond to an item that is already on the risk register.
- *Do all of the same things with regard to opportunities.*

Don't forget to update the schedule and cost predictions *again*, based on the above actions. If, in particular, you have started or retired mitigation actions, those will have impacts on personnel assignments, and therefore might have an impact on other tasks (that is, the tasks that these people are nominally assigned to work on).

We will perform this entire process on a regular basis (monthly is common) for the entire duration of our engineering project. We can modulate the frequency of the process, if warranted. For example, during a particularly problematic or sensitive portion of the project, we might do it twice per month. I have never seen a large engineering project do this type of process less frequently than once per month.

11.3.5 The Monthly Calendar

I like to organize this through a monthly calendar:

- **During the 1st week of the month**, the systems engineering lead works with cost-account managers to update the predictions for each operational performance measure and each technical performance measure, look for signals of problems in the resulting time series of measurements, and when there are signals of actual problems, identify the associated entries in the risk register (or create appropriate new entries).
- **During the 2nd week of the month**, the cost-account managers update their schedule predictions (by adjusting predicted task durations and updating inter-task connection logic), prepare schedule variance analyses, and review these with the project manager.
- **During the 3rd week of the month**, the cost-account managers update their cost predictions, prepare cost variance analyses, and review these with the project manager.
- **During the 4th week of the month**, the systems engineering lead works with the cost-account managers to update the aggregated risk assessment, and makes recommendations to the project manager about starting and stopping risk mitigation activities. Adjust cost, schedule, and staffing plans, based on the selected risk mitigation activities.

- **During the next week**, the project manager (that's you!) prepares a *project-level* status report, incorporating the key variance reports, mitigation actions, updated schedule-at-complete and cost-at-complete predictions, and the project manager's overall assessment. This is the *fifth task* in the periodic management rhythm, to which I made reference above.

11.3.6 The Accounting Calendar

Most projects and companies operate on an *accounting calendar* that contains 13 weeks per quarter (a quarter is a three-month period, e.g. one-quarter of a year).

By convention, each 13-week quarter of an accounting calendar consists of three months, as follows:

- The first month of the quarter is four weeks
- The second month of the quarter is four weeks
- The third month of the quarter is five weeks.

Therefore, shortly before the end of each quarter, I have the project finance manager publish a daily calendar for the next quarter, showing each activity and each due date for the periodic management rhythm, especially the date on which each cost-account manager will be presenting their variance analyses (schedule first, and then – a week later – cost) to the project manager. This sets expectations, and allows for everyone to plan ahead. It also allows the project manager (and everyone else too) to build their calendar around these meetings, and thereby allows you, as the project manager, to allocate time on your calendar to all of those other constituencies that we identified in Figure 11.1, the time pie.

11.3.7 Management Reserve Funding

The above discussion introduced a new term: *management reserve funding*. We alluded to this back in Chapter 7, when we discussed estimating the cost of an engineering project. At that time, we said that the *price* that your company would bid for an engineering project would consist of the sum of three items.

The *estimated cost of actually doing the work* (labor + materials + facilities + overhead + all other items that can properly be attributed to the contract)

plus

Management reserve (a fund of money set aside to deal with problems as they arise on the project)

plus

Profit (an additional amount to be paid to your company, in compensation for its expertise, and for taking the reputational risk of undertaking the work entailed in this engineering project)

The *management reserve fund* on an engineering project typically consists of an amount equal to *10% of the estimated cost*. If, however, the work entailed by the project is particularly difficult or risky, or if the contract terms place an inordinate portion of the risk onto the contractor (which we discussed as part of risk-sharing near the end of Chapter 7), it might be appropriate for the management reserve to be a larger percentage of the estimated cost. But remember, if you designate a larger amount for management reserve, you will thereby be increasing the price that you bid, and this will probably decrease the chance that you will win the contract. As Robert Heinlein said, "There ain't no such thing as a free lunch."[2] As in so many other aspects of engineering and engineering project management, we strive for a *feasible balance*, rather than for some theoretical optimum. There are seldom opportunities to achieve an actual theoretical optimum in the real world of engineering, or for that matter in the real world of actual people.

Some contracting authorities do not allow you to have a separate line item in your bid that provides for management reserve. You still need management reserve, of course, in order to manage the contract, so in those cases you must incorporate your management reserve into your line items for estimated cost. Your finance and contract managers will figure out the proper and legal way to do this, given the laws, rules, and regulations that apply to your specific contract and this contracting authority. This might take the form of a percentage "tax," similar to the expenses for office space, vacation, medical care, and so forth, or it might take some other form that your contract manager has determined is the form allowed by this contract and this contracting authority. But you – as the project manager – must have a reasonable amount of management reserve. No real engineering project can be undertaken responsibly without it.

We have already described one of the principal ways in which management reserve is used: to fund the studies and risk mitigation activities that arise during the risk/opportunity management process. For example, if the data indicate that a particular risk appears to be coming to pass, you might release management reserve funding to start work on one or more activities intended to mitigate the likelihood and/or the adverse impact of that risk.

Similarly – but often neglected – are the corresponding actions regarding *opportunities* to improve the outcomes of your engineering project. If the data indicate that it might become feasible to realize a particular opportunity, you might release management reserve funding in order to start work on one or more activities intended either to increase the likelihood that the opportunity will come to pass, and/or to increase the positive impact to your project if and when it does come to pass. As is the case for risk, there are likely to be *separate* actions for increasing likelihood and increasing impact.

Sometimes, after you authorize the team to start spending money on some task that uses management reserve, conditions change, and that work is no longer justified. At that point, you may terminate that task, and reclaim for your management reserve fund any monies not yet spent by that task.

---

2 This wonderful phrase – and its acronym TANSTAAFL – appears in his novel *The Moon is a Harsh Mistress*, published in 1966. Highly recommended!

Also, if you successfully realize an opportunity, and thereby save money elsewhere on the project, you may be permitted (your contract manager and legal staff will advise you on the particulars) to allocate some of those savings to your management reserve account, thereby increasing the amount remaining for the rest of the project.

Of course – TANSTAAFL! – if you get hit by a problem that results in a cost increase to the contract, you may have to cover some or all of that increase by a corresponding decrease in your management reserve account.

You have to manage your management reserve funding like you would any fund of money: you must pick and choose among many items that appear to demand funding; there will always be more things available to fund with your management reserve than there are available funds. And those funds must last you all the way to the end of the contract!

I find it empowering to my subordinate managers to allocate some portion of the project's management reserve to them, rather than making them work through me (as the project manager) for every single request for management reserve funding. Not every project manager likes to do this.

11.4 The Social Benefits of the Periodic Management Rhythm

The *periodic management rhythm* process described in this chapter takes up a lot of your time, and a lot of the time of many of your subordinate managers. I have tried to show why this allocation of time (and money – every hour of every person's time on the project costs money!) is worthwhile, because it provides a structure that allows us to monitor the progress of our engineering project, and to make decisions that are both timely and appropriate to keep the project moving forward at the promised rate and cost.

There are, however, an entirely additional set of reasons why this *periodic management rhythm* process is worthwhile: the *social* benefits provided as a result of the face-to-face interactions that form the core of the process. Let's discuss these for a moment. We will discuss a larger set of social considerations for engineering project management in Chapter 13.

Recall that I described the periodic management rhythm entailing several face-to-face meetings between an individual cost-account manager – at any level of the work-breakdown structure – and the project manager. These take place during weeks 2, 3, and 4 of our monthly cycle (when I described the calendar, I used a monthly period, so I will continue to assume that monthly period in this discussion).

- During week 2, you (as the project manager) are meeting with the cost-account managers to listen to them individually discuss their schedule variances, and also you are participating in larger meetings that likely have many of the cost-account managers present, when you discuss candidate *replumbing* ideas.
- During week 3, you are meeting with the cost-account managers to listen to them individually discuss their cost variances, and also you are participating in larger meetings that likely have many of the cost-account managers present, when you discuss the interactions among many cost accounts that are contributing additional increases in cost.

- During week 4, you are meeting with groups, or perhaps even all of the cost-account managers, as the project chief engineer moderates discussions of risk, and the desirability of starting or stopping various risk mitigation and opportunity seizing activities (which might be funded out of management reserve).

These likely form the *largest single body of time* that you spend interacting with each individual cost-account manager, especially for those at the lower levels of the work-breakdown structure. This is your time to "see them in action," listen to them reason, see how they react under stress, watch them interact with their peers and their supervisors, and so forth. As we will discuss in Chapter 13, one of your cardinal needs as the project manager is to have a *list of people who are ready for more responsible assignments*, people that you can promote, or people to whom you can give important problem-solving assignments. You, of course, can depend on your direct-report managers to convey to you their views about which people within the project team are ready for such more responsible assignments, but there is no substitute for you having a basis for forming your own judgments too. This face-to-face time interacting with all of the people on your project who are already managers, at whatever level, is a vital management resource for *you*. And the *periodic management rhythm* is the activity that guarantees that you, no matter how busy you are, get some of this "quality time" with your cost-account managers on a periodic basis.

There is a reciprocal social benefit too: all of these managers get to see you in action. This is your chance to inspire them with your leadership and problem-solving skills, with the deftness of your people-management skills, and so forth. You have probably talked about how you want personal interactions on the project to proceed: everyone to be treated with dignity, always be calm and polite, and so forth (we will say a lot more about these aspects in Chapter 13). This is a chance for you to "walk the talk," and gain credibility with your team that you really do these things, and that you act in a way that motivates the other members of the project leadership team to do so too. This is very valuable and important!

These social motivations also apply to members of the project team who are not (yet) cost-account managers. Because of that, I like to extend the invitation to these meetings to some of the key technical performers who are not (yet) in cost-account management roles. I want to see them in action too, and I want them to see me and my leadership team.

11.5 Your Role in All of This

As you can see, you – as the project manager – are right in the middle of the activities that we discuss in this chapter. You cause the meetings to be placed on the monthly calendar, you and your staff provide written guidance for how you want the analyses conducted and the written materials prepared, you arrange for your staff to receive appropriate training to perform these processes, you attend many of these meetings in person, you interact with many of your people, you guide them to consensus and reasonable decisions, and on (hopefully rare) occasions, you must jump in and redirect the analysis or even make a decision different from the one that the team is tending toward.

In addition to the benefits to the monitoring of the progress of your project, there are important social benefits of this process. You get the chance to interact with all of your cost-account managers, and other employees too. You get the chance to watch them, and learn if and when they are ready for more responsible assignments. They too get a social benefit: they get to see you and your senior leaders "in action."

11.6 Summary: Monitoring the Progress of Your Project (Part II)

- We bring the techniques previously discussed for creating and using the operational and technical performance measurements, schedule prediction, cost prediction, variance analysis, and risk management together into a single integrated process that I call the *periodic management rhythm.*
- Most projects ought to do this activity *once per month*, so I will not mind if you prefer to call it the *monthly management rhythm.*
- Operating this process is a *key, central activity for you as the manager of an engineering project.*
  - Not only do you stay informed in some detail about what is happening, and get to influence corrective actions (e.g. schedule replumbing, initiation of risk mitigation actions, etc.), but you get to see all of your cost-account managers "in action" each month. This is some of the most "quality time" you will regularly spend with each of them.
- At times, you can modulate the duration of the cycle to be shorter.
  - I have never seen a project where the cycle was less frequent than monthly.

11.7 Next

We have now completed the portion of the course where we discuss how to monitor the progress of our engineering project. Congratulations!

Next week, we will cover four special topics:

- Launching your project: how do we get it started?
- Projects with lots of software
- The agile project management methodology
- Ending your project.

11.8 This Week's Facilitated Lab Session

We return to the activity network, and our example schedule-making software. As part B of this assignment (we did part A during week 7), we add asymmetric three-point estimates to our sample activity network, and note the impact to the end date.

12

Four Special Topics

(a) Congratulations! You have won a competition for a new engineering project. How do you go about getting this new project started? Starting a new project turns out to be a special problem that requires a special set of skills, which I will teach you in this chapter.
(b) Most systems today have large amounts of software, which provides particular benefits, but creates many particular liabilities and risks too. I show you how to deal with projects that involve large amounts of software development.
(c) People will come and ask if you are going to use *agile software development methods* on your project. I show you some of the key differences between agile and conventional development methodologies, how to decide if your project is suitable for agile methods, and how to cope with some of the most common risks and weaknesses of the agile methods.
(d) Projects are defined as temporary activities, so every project ends. I tell you what you need to do to end a project.

12.1 Launching Your Project

12.1.1 The Project Start-Up Process

Any new work assignment is interesting and difficult. For example, to take over – and run effectively – a project that is already underway and doing well is a difficult task. You have people to meet, many things to learn, yet the project's schedule advances every day, whether you are up-to-speed yet or not.

It is, however, *far* more "interesting" and difficult to start a big engineering project from scratch. Even though every project is different, I believe that there are a set of common themes that need to be addressed when starting a new engineering project. So, in this section, we will discuss how to go about getting a new engineering project started.

Back in Chapter 5, we discussed how to go about writing a winning proposal for an engineering project. Because you were paying attention, you have now actually won a competition for a new engineering project. Congratulations!

How do you go about getting your project started? The project may call for 300 people, but you had only 50 people working on the proposal with you. Furthermore, since you turned the proposal in to the customer for their competitive evaluation nine months ago,

Engineering Project Management, First Edition. Neil G. Siegel.

Companion website: www.wiley.com/go/siegel/engineering_project_management

40 of those people have gone off to work on other assignments; only 10 of the people who worked on the proposal with you are still working with you by the day of the announcement that your company has won the competition. How ought you to proceed?

First of all, there are a set of things that have to be selected or created, and somehow brought into being:

- A set of people, and some sort of organization
- Facilities and infrastructure
- A set of processes that define how certain things get done (e.g. who has the authority to purchase something, and how does that get accomplished?)
- Tools and methods, together with instruction and training
- ... and many more.

Most of these have to be created essentially *simultaneously*. But they are *not* independent; there are likely linkages between them. Here's an example:

- You need to recruit a large number of people to come and work on your project; remember, we said that you have 10 people today, but in just a few weeks, you need to have 300.
- Generally, you recruit people by offering them a *particular role* on the project. This implies that those roles are already defined. Those people, before they accept the position, probably also want to know something about the organizational structure into which that role fits. This implies that the organizational structure too must already be defined.
- But you need help from other people in order to define all of those roles – you cannot write all those job descriptions yourself – and to define the organizational structure into which all of those roles will fit.
- Furthermore, at times, you will want to create the organizational structure around the skills/weaknesses of a *particular person* (including yourself!).

This would appear to say that you cannot recruit the people without having already defined the roles and the organizational structure, but also that you cannot define the roles and the organizational structure without having already obtained the services of some of those people. This sounds like a circular dependency, a Gordian knot. How do we break out of this apparent dilemma?

You start by creating a list of topics for which you need plans and initial preparation. Figure 12.1 shows a candidate set of project start-up topics, which I provide in order to give you an idea of what you might need for *your* project.

That's a long list. No one can memorize even the top-level concepts for all of these topics. So, you need to *write down* your approach for each of these topics. We call such a written document a *plan*. It doesn't really matter whether you create a separate document for each topic, or combine all of the topics into a single document. On bigger projects, each topic is likely to have a separate author (or a small team), and most project managers find it easier to coordinate the work of preparing and reviewing these plans if the item produced by each author is a stand-alone document.

Of course, you need people to help you write all of these plans; you can't do it all yourself. But they some need guidance from you: how do you want to organize and operate the project?

Therefore, you and a small cadre of associates need to create – and document in writing – a set of top-level strategies and approaches for *everything*. Then you can bring

Project strategy	Risk management
Organization	Quality assurance
Staffing	Bill of materials
Team-building	Budget
Subcontract management	Systems engineering
Communications	Software development
Facilities and equipment	Integration
Configuration management	Testing
Data management	Acceptance
Scheduling	Delivery
Security	Additional ***technical*** topics

Figure 12.1 A candidate set of project start-up topics.

on additional people to start flushing those strategies out into actionable plans. You will need to interact with them almost daily to help them create what you need. I might schedule 15-minute meetings with plan authors every two days during the early portions of the plan preparation, concentrating on strategy, key decisions, and a few key tables and graphics that capture our emerging decisions.

Most likely, your company actually has a library that can provide you and your authors with outlines for most of these plans, and that library probably also has examples of each type of plan that was produced by a previous project. The outlines are useful; in my view, however, using the previous complete example artifacts can sometimes bog your authors down in too much detail. I prefer that my authors just work from the outlines.

You will need to prepare some of these plans for your proposal, in order to convince your customers that you have a credible approach. Also, the specificity of these plans allows you to develop cost estimates for implementing each of these start-up activities, which you need for the proposal. But it is likely that these proposal versions of the plans do not cover every topic, nor are they sufficiently detailed. Because of this, your team is likely to need to do some additional planning as the date of the anticipated contract award approaches.

Of course, writing a *plan* does not finish the job; it is just telling us what we need to set up for our project. Each plan needs to be *implemented*; the implementation can often wait until after you win and sign the contract.

For example, engineering projects need *infrastructure*:

- Electronic mail
- Directories of project personnel
- Desktop and/or laptop computers, with the appropriate software and network connections
- Security tools
- Financial reporting tools
- Schedule reporting tools
- Tools for other routine aspects of business: timecards, status reports
- Technical tools, for systems engineering, requirements management, software development, etc.
- ... and so forth.

You might be able to "inherit" some of this from your company, but likely much of it will need tailoring and adaptation (e.g. your company's email system does not include directories of customers and subcontractors, and in order to protect the company's network against intrusions and hackers, they likely place various limits on the use of these systems by non-employees), and some of it will be new. Your customer might have requirements in the contract about particular tools that they require you use: they might want you to use LaTex or Open Office for the creation of documents; they might require that you use a particular programming language and a particular compiler for that language; and so forth. Even if your company is going to provide this infrastructure, someone from your team needs to determine exactly in what configuration, how many, which rooms each item is to be installed in, and so forth. This all needs to be planned, items purchased, the financial aspects considered, the appropriate purchasing procedures followed, and so forth. Without some written guidance – the plan we talked about above – it can't be done.

On a big project (or even on a small project inside a big company), *every task* needs a *written process*. How do orders for computers get placed with a vendor? How do people record their weekly work time, so that they can be paid properly? Who is authorized to approve which type of expenditures? How do we move and store materials in a way that does not damage them, and is also safe? As you can see, a large portion of the items in Figure 12.1 address this aspect of the start-up challenge.

You will actually need more than a written process telling you how to do something too. For each *work process*, you will also need:

- Some sort of training program, so that people can learn about the existence of the process, and how to use it.
- A set of tools for implementing the process. These might be as simple as on-line forms, or complex software that performs some task relating to the process, or the tools could take many other forms. They might even be hardware tools: wrenches and so forth.
- Some sort of a database: who needs to be trained, who has been trained, and so forth.
- Some sort of an enforcement program; for example, if a process requires that people receive annual training in a subject, you must have some mechanism that verifies this is being done.

Here's an example. If you (or the customer) have a requirement that a particular construct in the chosen programming language *not* be used (this is actually fairly common), you must:

- Have a policy that documents this
- Have a mechanism to disseminate that policy
- Have a way to train your people about the policy
- Have tools and processes that audit software code turned into the project library for compliance
- Have a process to notify those whose code did not comply, and to track their corrective actions to completion
- Have a mechanism to track how often each person fails to comply, so that you can have a plan for retraining, or even discipline, if a given person is a repeat offender.

Since you are *starting* the project, you must create/borrow/steal/adapt this process, and all of those supporting steps, into being.

Figure 12.2 The Tactical High-Energy Laser, the world's first complete laser weapon. *Source:* The photo is by the US Army, and was released by them into the public domain.

And then, of course, you must do that for *every* item on your list (e.g. your own version of Figure 12.1).

The list provided in Figure 12.1 is fairly generic, but *your* project might need some processes or tools that are very specific to the goals of that project. For example, when we built the world's first complete laser weapon, it involved *crane operations*; many of the pieces were simply too big and too heavy to be moved and placed by hand (Figure 12.2). But some of these heavy pieces had optics, mirrors, and other fragile components, so the crane operations required a balance of capacity, safety, and not over-stressing the fragile components. This combination, however, was different from what the company's crane operators normally did; they had not worked in the past with heavy parts that also had large, fragile, glass components. We discovered that we had to create *new* processes and procedures for crane operations on these components. Your project too will likely have some unique aspects that will require such novel processes and procedures.

Good companies should have *project start-up procedures, check lists,* and *teams of people* who will come in on a temporary basis and help you get the project start-up defined and accomplished (*project start-up teams*). To give you an idea of the magnitude of project start-up, on a big project, start-up can be a three- to four-month task.

Figure 12.3 depicts an excerpt from an actual project start-up check list, as used by an actual company.[1] The complete list is 10 pages long; I just want you to get an idea of the level of detail that a good project start-up check list will include.

---

1 Kendra Risdon was the person at my former company who championed the creation of the project start-up process, check list, and start-up team that would actually go to where a new project was being started and help them through the process.

- Develop project concept of operations (based on the proposal). doc resp
- If project is being transitioned from one organization to another the assuming organization needs to understand project status and specifics (see checklist). chkl
- Ensure that all members of the project team are appropriately badged. doc
- Review this Business and Finance 101 material to understand organizational and accounting structures as helpful information for the project management team. trng

Call and Hold Joint Planning Session(s) with the Customer brfg brfg brfg

- Items that would be covered in these sessions would include review, update and approval of the contractual requirements, acceptance criteria, WBS, and the network schedule (IMS) at the appropriate level. (Also see Guidance). risk guid
  - In addition, the customer should understand and concur with the contractor organization (end-item), corresponding points of contact, communication methodologies and confirm billing methodologies and/or format.
  - Identify customer expectations for each major review/control gate.
  - Identify risk items in the critical path and develop mitigation for each risk. doc
  - Define business relationship with customer (i.e. "rules of conduct").

Plan and Manage Project Staffing resp resp guid guid

- Update Staffing Plan. resp guid guid guid
- Update staffing requirements and position descriptions.
- Using resource tracking tools for project staffing, identify filled and open positions, grades, salaries, target start dates, and organization sources. resp guid guid
- Update assessment tool for project staffing. resp guid
- Refer List of Experts to assist with project staffing. form resp help
- Determine personnel required from external sources such as VOLT, etc. See Guidance button.
  - As applicable, use Confidential Disclosure Agreements, i.e. "Non-disclosure" agreements. (See Forms next to each agreement).

Figure 12.3 An excerpt from an actual project start-up check list.

12.1.2 The Earned-Value Baseline: A Special Project Start-Up Task

We talked about earned value, and the creation of the schedule and cost estimates in Chapter 7. One special aspect of starting a project is the creation of a special version of those schedule and cost predictions, which is called the *baseline*.

You create, as we described in Chapter 7, an *initial* version of these predictions during the proposal activity. You use this version to obtain approval from your company to submit the bid to the customer, and of course you also provide this version to the customer as part of your proposal.

That initial version of the schedule and cost predictions is good enough for those purposes, but is *seldom detailed enough* actually to use in order to manage the project. Since at the time you prepare them, you don't know yet if you are going to win the competition or not, it may not be worth the effort and money to create them at that level of detail during the proposal. You probably don't have enough people available during the proposal to create schedule and cost estimates at that level of detail, anyway.

Therefore, right after you win the competition and sign the contract, you have to *create a second version of the schedule and cost predictions*, one that this time will be sufficiently detailed to use in order to manage the project. We call this second version – the one that we will actually use to manage the project, following the procedures of Chapters 10 and 11 – the *baseline*. We use that term because it is the *base set of values and measurements* against which we write the schedule and cost variances described in Chapters 10 and 11.

There is another reason to create this second version too: as discussed in Chapters 7 and 10, the schedule and cost predictions are worthless if each little piece is not created (and psychologically "owned") by the *actual cost-account manager who is responsible* for that piece of the work. But many of these people do not show up until *after the contract is signed*; remember, we discussed how the team that writes the proposal is likely much smaller than the team that will actually do the work, once we win the competition. So, it is important to allow the *actual cost-account managers* to prepare the schedule and cost predictions that will be used to manage the project (which will be accomplished using the methods described in Chapter 10).

But what if these new predictions are different from the previous predictions? After all, your company actually signed a contract based on the *proposal version* of those predictions. Well you, as the project manager, have the challenge of trying to negotiate with the actual cost-account managers to try to get the baseline version of these estimates to fit within the constraints of the proposal version. You may not succeed, in which case you will be starting the project "in the hole" in schedule and/or cost.

Since each cost-account manager must "psychologically own" the result, they must generate bottom-up estimates in which they believe, while "noticing" the top-down guidance that you have provided (in the form of the prediction for that cost account from the proposal version of the predictions). You must therefore *negotiate* with each of them, clarifying/reducing scope, so that they end up as close as they can to your top-down guidance ... while still *believing* in the baseline predictions to which they commit.

Some cost-account managers will meet your top-down guidance, some will not. You must try to make it all balance at the project level. But do not "force" them to agree to numbers in which they do not believe! That *always* leads to later failure.

So, you and the team get to do the schedule and cost predictions *all over again* – and in far more detail – right after you win (in order to create this baseline version).

It is often the case (and *always* the case if your contract is with the US Government) that people from outside your project team will actually come in and *audit* your new baseline, to see if you and the team did it right; that is, that your baseline is both *procedurally correct* and *faithfully represents the requirements for the project* as defined in the contract.

Passing this audit is often *both a company and a contractual requirement.* Your financial staff will know how to pass this audit. Not passing the audit the first time is a sure way to annoy your customer and your management. My advice is to ask for help from corporate personnel that have done this before!

Note that when you pass the audit, you likely will "pass with findings." This means that there are some things that the auditors think are incorrect or non-compliant in your baseline, but the auditors also believe that these can easily be fixed. In this case, you fix those items, write a memo describing what you did for each, and submit that memo to the auditors. If they are then satisfied, they will send you a letter saying that you have now resolved all of the issues that arose out of audit, and your baseline is approved. This counts as passing the audit the first time.

Even if your contract is not with the US Government, you probably still have to conduct and pass this audit. This is because many non-US-Government contracting agencies have adopted similar audit requirements, and also because many companies now require such an audit, even if the contracting authority does not.

In Chapters 10 and 11, I described a process by which we update the schedule and cost predictions every month, and write variance analyses in which we compare the set of recent schedule and cost predictions from the previous several months (so that we have a *time series*) to our *baseline.* In Chapters 10 and 11, I did *not* describe the process which created the baseline; you may have thought that the baseline was the version of the schedule and cost predictions that was created during the proposal. In the discussion above, however, I have now made it clear that the proposal-time version of the schedule and cost predictions is *not* the baseline, but instead *the baseline is that version created immediately after you win the competition and sign the contract.* You are usually allowed three or four months after the award of your contract to establish the schedule and cost baseline, and to pass the audit; this deadline will likely be specified in your contract, or in your corporate policy manuals.

Of course, you cannot execute the process described in Chapters 10 and 11 – of doing monthly updates to the baseline schedule and cost predictions – until *after you have established the baseline.* So, you don't actually start the processes of Chapters 10 and 11 until *after you complete the baseline.*

You do not, however, need to wait until you *pass the audit* to start the processes of Chapters 10 and 11. Those monthly updates can start as soon as you have a *baseline,* even while you are preparing for and conducting the audit, and correcting any deficiencies found during the audit.

12.1.3 Preparing to Operate at a Large Scale

One of the biggest challenges of running a big engineering project is operating *at a large scale.* You have people who must be recruited, trained, and *aligned* (Chapter 13). You have tasks and deliverable products. You have a complicated activity network that represents the schedule. You need processes and plans. You have stakeholders that have opinions, and demand attention. You need an organizational structure that makes it clear what each person and each subteam is responsible for. You need a way to monitor progress, and to notice if someone is falling behind. You need a way to help those people who are having trouble or falling behind. You need to know on what you need to spend your own time, and what you must delegate. All of this requires planning and thought, and ends up being documented in project procedures, processes, policies, and directives, and implemented via tools and training. The project start-up procedures described in this section are intended to establish the infrastructure that will get you ready to operate at such large scale.

A special aspect for you, as project manager, on preparing to operate at large scale is the question of *what tasks you should spend your own time on.* We answered a portion of that question in Chapter 11: you participate in the *periodic management rhythm.*

But a large portion of your time is not so determinate. People will want and/or need to see you, both because of *planned,* but also because of *unplanned,* occurrences. Your daily calendar will be scheduled in 30-minute meetings, for the entire 10-hour work day. Most of these people will be asking for a *decision.* You must learn when to decide on the spot, and when to say "let me think about this for a day or two," so that you can gather some additional data, consult with others, and just think about it. You cannot dither, but you must make good decisions. In addition, it can be valuable to instill (through your example) in your team members the notion of thinking carefully before deciding.

What you will learn is that often, your only decision variable, your only degree of freedom in making such a decision, is "who can I assign to go and fix this problem?" So, you need to build up a list of people you consider trustworthy ... and be constantly on the look-out for new people you can add to the list. This is a big part of the social benefit of the periodic management rhythm process that we described in Chapter 11; you get to see members of your staff in action, and can figure out who is ready to become one of your set of problem-solvers. You should make it clear through actions that successful problem-solvers are *rewarded,* both financially, through recognition, and through promotions and choice assignments.

12.1.4 Summary for Starting a Project

- Starting a project is fun, interesting, stressful, and a lot of work.
- You have great freedom to design the project's operations to be the way you want.
- All projects require plans, infrastructure, and processes; on big projects, the scale of these items is very significant.
- As the project manager, you have to see to the creation of all of these items.
- A special task is the creation of the schedule and cost baselines, and getting through the corresponding auditing process.
- You will need a start-up checklist, and a set of people whose full-time job is to get all of those start-up tasks completed.
- Your company likely has outlines and example artifacts of the plans that you will need to create.
- Projects usually operate at a large scale, and the start-up process is where the foundations for the capacity to operate at a large scale are developed.

12.2 Systems and Projects With Large Amounts of Software

Most systems today have large amounts of software in them. Cars, airplanes, elevators, telephones, washing machines, and almost everything else contain staggering amounts of software. A modern car is reported to contain 200 000 000 lines of software code. Your laptop computer has even more. Your phone – with all of the useful applications that you have loaded onto it – likely has several tens of millions of lines of software code, all in the palm of your hand.

12.2.1 The Benefits

There are good reasons for this: software turns out to be a mechanism by which we can implement desirable functionality into our systems. It also allows us to improve our systems *after we deploy them*, by loading new and (presumably) improved software onto them. We can even *add functions to our system* that we did not envision when we built the system. Think of all those useful applications on your phone: many of them did not exist when the phone was designed and manufactured.

12.2.2 The Problems

What comes, however, to the mind of many engineering project managers when they think about software is *software as a source of problems*. When an engineering project is late or over budget, it is all too frequently the case that *software* is the portion of the work-breakdown structure where these problems arise. It is not unheard of for projects that entail a large amount of software development to end up taking three times as long as originally planned – and be five times over budget. Some projects without software can have larger schedule and cost over-runs too, but projects with large amounts of software seem especially at risk for this problem.

We might be tempted to say that this must be a problem with the original estimation of schedule and cost for the software. No one apparently even *tried* to create a methodology for estimating the schedule and cost of software until the 1970s, when Dr. Barry Boehm of TRW (now at the University of Southern California[2]) started working on creating systematic methods for estimating the magnitude of a software development effort, both in terms of time (schedule) and cost.

Dr. Boehm likes to tell a story about how he came to work on this problem. In the early 1970s, TRW was getting ready to submit a proposal to a customer to build a satellite. At the final review of the cost before submitting the proposal, the experts from each segment – structure, power generation and distribution, sensors, and so forth – stood up and presented the estimation methodology for their segment of the project. The software team stood up and said "this is our estimate," but had no rationale for how they arrived at the numbers, or why that might be a credible estimate or not. Dr. Walquist, the executive reviewing the bid, pulled Dr. Boehm aside after the meeting and said

2 This is the same Dr. Boehm whom I cited in Chapter 2 for his invention of the spiral model of development. He called his original estimating model the *constructive cost model*, often abbreviated as COCOMO.

"I don't want ever to have to make a bid that includes software again until we have a credible method to estimate the schedule and cost of that software. Go and create such a method."

Certainly, having no systematic method for estimating the time duration and cost of a software project is not a happy situation for a company that must make a bid for an engineering project that includes large amounts of software. Thanks to Dr. Boehm (and to the others who have since created similar estimation models for software), we now have estimation models, but like all other models and estimates, they are *imperfect*. The desire to win the bid, of course, always creates a temptation to be optimistic in these estimates, which probably reduces their credibility even further.

Having read Chapter 2, you can provide a more sophisticated, additional answer about why software projects are often late and over budget: with so many lines of code, the opportunity for *unplanned dynamic behavior* is very large. Having read Chapter 9, you can then improve this answer by pointing out that the entries on the project's *risk register* about what could go wrong with the software development are often superficial. And having read Chapter 8, you can conclude by pointing out that the *measurements and analysis* that we use to monitor the technical performance of the software are subject to all of the errors and misinterpretations that we discussed in that chapter.

12.2.2.1 Scale

Many of these problems derive from *scale*. Those 200 000 000 lines of software code in your car are in many ways like having 200 000 000 separate parts. In contrast, I read somewhere that the Apollo moon rocket at take-off – the booster, the cabin for the crew, the module that landed on the moon, the equipment that kept the three astronauts alive and guided their return to Earth, and everything else – was comprised of just a few hundred thousand parts. This means that each new BMW on the road today has something like 1000 times as many parts as did the Apollo moon rocket.

Humans are imperfect, so there remain *latent defects* in all of our creations. In a creation that contains 200 000 000 parts, there will be a large number of such latent defects. A latent defect rate considered typical for software is one such latent defect per 1000 lines of software code. For a human creation, one error per 1000 steps sounds pretty good. But this means that there are likely to be *200 000 latent defects* from the software in that car which has 200 000 000 lines of software code! Of course, the mechanical and electrical portions of the car will have latent defects too.

Technical Note

Just as discussed in Chapters 7 and 10, the *distributions* of phenomena *need not be symmetric*. There are, in fact, plenty of systems with a latent defect rate *far worse* than one per 1000 lines of software code; some that are *much* worse.

Latent defects are those errors that remain in the product after we deliver the product to the customer. Of course, through the integration and test activities described in Chapter 3, we endeavor to find and fix all such defects before we deliver the system. In general, if a system has more lines of software code, it will contain more defects upon completion of the coding, and it will take more time and money to try to find and remove some reasonable portion of these before we deliver the system. But there will still be defects that remain after delivery; on average, at that one latent defect per 1000 lines of code rate that I cited above. So, systems that have more lines of software likely have more latent defects, just due to the linear increase in the number of lines of software code they contain. But it gets worse: in addition, there are data which indicate that systems containing more lines of software code usually *have higher defect rates*, both at the end of software coding and upon delivery (e.g. *latent defects*). That is, *scale makes this problem worse*, and therefore, as our systems contain ever larger numbers of lines of software code, the average latent defect *rate* likely increases. The effort to remove a reasonable number of these before delivery therefore increases as a percentage of the total effort. Scale, all by itself, *increases the cost per line of software code.*

There is another effect that makes software "parts" more prone to problems than hardware parts, in addition to their sheer quantity: they are likely to be more densely interconnected. In principle, any line of software code can call to any other line of software code; there is no inherent limit on such interconnections. This is part of what makes predicting *unplanned dynamic behavior* in software so difficult. In contrast, the potential interactions between hardware components are far more limited than those for software. For example, electronic devices are connected by wires only to a small subset of the total number of electronic devices in a chip or system; mechanical parts are connected only to adjacent mechanical parts.[3] Because of the explicitness of the interactions, and the far smaller number of potential interactions, it is far easier to envision and correct for sources on unplanned dynamic behavior in *hardware* components than it is in *software* components.

So, the sheer scale of the software that we put into our engineered systems today creates problems that contribute to software development projects taking longer and costing more than planned.

12.2.3 Lessons Learned for the Project Manager About Software

Your team will want to be optimistic in their estimation of schedule and cost for the software; they want to win the bid. You must drive them to *realism*, by grounding the estimates in past rates of software productivity. You do this with the top-down parametric estimation techniques discussed in Chapter 7, under the heading "Injecting realism into our estimates."

In today's systems, *software* is where most of the adverse *unplanned dynamic behavior* (discussed in Chapter 2) comes from. This is, in my experience, where *most of the schedule and cost problems for software originate*; optimistic and unrealistic original estimates are only the second most important contributing factor. This is why I focus so much on the goodness of the design, and provide you with tools and methods both to create better software designs (Chapter 2) and to monitor the risk (Chapter 9) and progress (Chapters 10 and 11) of the software development effort within the project.

Software gets behind schedule and over cost to a shocking degree; often not just a few percent, but actual entire multiples (2×, 3×, 5×, 10×, and more). The two main causes are those cited above: (i) adverse unplanned dynamic behavior and (ii) unrealistically optimistic initial schedule and cost estimates.

3 There *is* a possibility of secondary interactions: unintended electromagnetic coupling, vibration, etc.

Software has the ability to *multiply its adverse effect on schedule and cost* too. If your software is late, you may have to keep *other* project personnel around until the software team finishes. If you are building an airplane, if the software is years late, you have to keep the hardware designers, the engine designers, the testers, the human factors personnel, the safety personnel, the project finance personnel, and many others around, so that they can participate in integration, testing, and training after the software is finally delivered. Therefore, not only do you have to pay for the extra costs of the software, you also incur additional extra costs in order to keep your access to all of these other personnel. *This multiplicative aspect* – sometimes called the *marching army effect* – is often overlooked.

It is very tempting to your team, and to the customer, to expect that you can reuse existing software from other projects on your project. Of course you can, but experience repeatedly shows that this reuse *seldom works nearly as well as expected.* This reused software can require subtle adaptations in order to be suitable for your project, and it is easy to underestimate dramatically the complexity, difficulty, and therefore the time and expense of making these modifications. This is a place where looking at your company's historic archive of project activities will be helpful: you will discover that many previous projects suffered significant schedule and cost over-runs as a result of problems with their planned level of software reuse. You can use these data to lessen your project's predictions of credible software reuse to more realistic levels.

It is also very tempting to buy software products on the market that seem to be able to perform some significant aspect of your project's mission. There is a significant risk in this, however: there is often a *major discrepancy between the capabilities and levels of performance claimed for these products, and those actually provided.* By all means try to make use of purchased software components in your system design, but *never depend on the claims of the manufacturer.* Always get the software and *experiment with it yourself **before*** committing to it in your design. Benchmark its capabilities and performance yourself; do ***not*** depend on data from the manufacturer. Try out the interfaces, data-importing, and data-exporting capabilities. Set up a test bed that mirrors your operational conditions, and measure achieved performance rates under those conditions. For software, seemingly small differences in operating conditions can result in a 10× (or more) difference in performance and capacity!

To some extent, this is true for anything that you buy: what the manufacturer promises, and what the product actually delivers, may be different. But the magnitude of the discrepancies between promises and delivered capabilities seems generally *to be larger for software than for other purchased components.*

Another, related problem is that when you *buy* software, *you cannot control* when problems get fixed, which ones get fixed first, and which ones get deferred. Neither can you control when new features get added, and in what order. Vendors who build software are often serving gigantic markets, and are often unwilling to adapt their plans to the needs of a single customer, even for a single customer as large as the entire US Government. My experience is that the large software vendors will *not adapt their plans at all,* in response to requests from even a large customer. Beware! They may *say* that they will be responsive to your needs, but unless such commitments are *spelled out explicitly in your purchase order* with that vendor, they have the ability to ignore such promises.

There is another, often under-utilized source of software: something called *open-source software.* Open-source software is, as its name implies, software where the *source code* is made available to customers. You may not realize this, but when you buy software from a commercial vendor, almost all of the time you get *only executable software code,* not the source code. Not having the source code of the software that you buy means that you have *no ability* to diagnose problems, to correct those problems, or to add new features to that software; *only the manufacturer* can do that. It may even be illegal[4] for you to try to do so. With open-source software, since you are provided with the source code, you do have the ability to diagnose problems, correct those problems, and add new features (and the ability to decide *when* to do these things). You can also analyze that software. For example, you can analyze it for security vulnerabilities. When you buy commercial software, you have little to no ability to analyze that software; again, it may be illegal even for you to try to do so.

Note that when you use open-source software, you may be required to contribute any improvements that you make back into the open-source library. This will allow other organizations, *even your competitors,* to be able to benefit from those improvements! Open-source software is not for everyone.

Open-source software is often free, or priced very low, compared to commercial software. As I said, it may not be the answer for every software development need, but in my experience, open-source software is often overlooked as a resource for your project. Just to give you the perspective from some actual data, at one point in time, my former company used on average about *100 open-source packages* (some large, most small) in *every* software-intensive system that we delivered.

Software is *intellectual property,* and therefore comes with intellectual property and license provisions, whether it is commercial software or open-source software. Beware: the specifics of these provisions vary fantastically from case to case. Never make assumptions about what the license terms for software are likely to require. Always have your legal staff review these provisions before you purchase commercial software, or before you decide to adopt open-source software.

Since you are likely to be using *both* commercial and open-source software on your project, you need to have someone keep careful track of these intellectual property and license provisions, keep appropriate records, and make sure that your project's planned use complies with the intellectual property provisions and license terms.

Just because the customer asks you to do something, that does not make it either possible or legal. I have seen government contracts that required the delivery of source code for all software, even purchased software. As noted above, when you purchase software, you almost never get the source code, so of course you cannot sign a contract that obliges you to deliver that source code. Similarly, I have seen contracts that required a particular piece of open-source software be used, even though the license terms for that open-source software prohibited such use.[5] You cannot sign a contract that requires you to violate the terms of a license agreement; even the government (at least, not in the United States) cannot authorize your company to do that.

For that matter, your company probably creates intellectual property and licensing provisions on the software that *your* project develops and delivers to your customers.

4 Because it may be a violation of copyright or other intellectual property conditions that you agreed to at the time of purchase.

5 For example, some open-source software contains language in their licenses that prohibits their use in military systems, or by for-profit companies.

These need to be in compliance with the terms of your contract too. Do not overlook the potential for intellectual property and license provisions to be a potential *competitive differentiator* in a competition to win the contract for an engineering project; at times, customers react very positively to offers that provide them with generous intellectual property and software licensing terms.

Patents are treated slightly differently. If your contract is directly with the US Government, you may have a clause in your contract that provides for a *patent indemnity*. This type of clause basically says that if, in the course of the project, you infringe a US patent owned by some other entity, it will be the US government, rather than your company, that will be liable for any damages. This is not a "get-out-of-jail-free" card; your company may still be sued by the patent holder, and you may incur costs to defend that suit, but in the end, if you lose, the government will pay the damages (but perhaps not all of the costs). Some software – but not nearly all software – is covered by patents. Much software is not covered by patents, but instead covered by *trade-mark* law, and by its licensing provisions. In any case, this is a complicated area, and you should have your legal staff (and specialized patent counsel – not all lawyers are knowledgeable about patent law) look into these matters and advise you.

12.3 The Agile Software Development Methodology

People will ask if your new engineering project will be using something called *agile methods*. In this section, I will define this term, help you understand when it might be appropriate for your project, and describe the principal hallmarks of the agile method.

As discussed in the previous section, since many of today's systems that you will be building through your engineering projects have a large amount of software, you, as the manager of an engineering project, need to understand a fair amount about software. *Agile* is a particular methodology for performing software development. It has become somewhat fashionable, and as a result, the label *agile* is attached to different types of methodologies. So, be warned that your company or your customer may use the term slightly differently than I do herein.

Developing large, complex, software-intensive systems is hard. Not surprisingly, problems are frequently encountered. As we discussed in the previous section, the ones that attract the most attention are: (i) software often takes far longer to develop than was originally planned or promised; (ii) software often costs far more to develop than was originally planned or promised; and (iii) when delivered, the software does not perform as promised. We have already discussed all of these.

Another problematic aspect of system development is ensuring that the values, needs, and desires of the eventual users are addressed adequately. We already discussed this aspect, in Chapter 4 and other places as well.

An additional, interesting, and important problem with software (and one that sometimes is the cause of schedule and cost problems) is that it can be hard to make *small* changes to a *large* piece of software; one segment of the software sometimes seems interconnected to another, and by the time you run down all of these little interconnections and the unexpected interactions (a form of *unplanned dynamic behavior* – as discussed in Chapter 2), the process of introducing what you had hoped would be a small change ends up taking longer (and costing a lot more) than you expected. If you have a very large, highly integrated piece of software responsible for operating a large portion of your business or enterprise, this can be a particular problem. If you find a problem with one aspect of (for example) an inventory control system, you do not want to wait 24 months until the next scheduled major upgrade in order to fix it; you want to create a small change (often called a *patch*) and distribute it *now*.

This is the special virtue of *agile* methods: they are designed to create software systems that are *agile*, in the sense that these systems can be readily adapted and improved frequently in small ways.

Side Note

There are different senses in which the term *agile methods* (or *agile software development*) is sometimes used: development methods that look for steps to simplify or eliminate, so that software development can be completed in less time. There is some indication that this may work, but *only for the very smallest software projects*. There are (unfortunately) *many* examples which show that it *does not work* to apply this form of *agile* (which might better be called *streamlined*) to the large, complex projects upon which this book focuses.

So, what projects are appropriate for agile development methods? Systems that need to be readily adapted and improved frequently in small ways. Here is my opinion:

If you have a project that:

Would benefit from *frequent, small updates*, ***and***
The project is *large*, ***and***

The resulting *product will be in service for many years*

then *agile* may be a good fit. *Agile* will allow you to structure your product and your methods so that you can provide those frequent updates over the long anticipated use life of the product. I suggest that you *only* do this if your project is large, because the costs of implementing all of the infrastructure that it takes to do agile properly are not small, and in most cases it is only worth expending these funds if those costs can be amortized over a large development cost and long use life.

Many consumer-oriented products and services *do* fit this profile well, and many of the large, consumer-oriented software platforms that we are all familiar with in our personal life in fact use agile methods, so that they are able to create updates and patches to the software on a frequent basis. I read somewhere that Amazon puts out a patch to some portion of its software every 17 seconds.

If, in contrast, your product has stringent licensing, certification, or credentialing requirements – an airplane, for example, which requires an update to its airworthiness certificate with every modification (which would most definitely include any modifications to its software), or systems with stringent safety requirements (such as medical diagnostic systems, the control rooms of complex industrial facilities, and most military systems) – you are *unlikely* to want to follow a model of frequent small updates to the product. This is because each update requires extensive testing and some type of partial recertification; this is very expensive, and trying to do it frequently can itself become a source of errors. Therefore, agile methods are usually *not* a good choice for systems of this sort. Instead, systems like this are usually updated at far *longer* time intervals, so that each update can be tested to the degree of thoroughness required, and all of the licensing, certification, and credentialing requirements addressed carefully and in detail.

Often systems like these latter ones require extensive training of their users too; and most of the time, when you release a software update to a system of this sort, you must provide and perform some type of actual training on the revised software. It would probably not be practical to require a weekly training session for a safety-critical system, mandated because you were putting out patches for that system every week. This is another motivation for the longer time interval between updates to such systems.

This is quite different from the case for large-scale consumer systems of the sort that we said are likely to be suitable for agile: these systems almost never involve training of the users – consumers just won't pay attention. Since these systems usually do not have significant safety issues, this is not a problem; consumers seem to prefer just to figure out how to use the systems (and any changes to the operation of these systems entailed by one of their frequent patches) by themselves. They make mistakes, but since these systems are not safety-critical, this is only an inconvenience, and not an actual hazard. As discussed in the section about the *user experience* at the end of Chapter 4, you need to know which category of user your system is going to have (e.g. formally trained experts, or untrained members of the general public), and you must design the system appropriately.

One thing that I like very much about the agile approach is that it places *explicit* focus on the *delivery of value to the users* who will be using the software; this is exactly congruent to what was discussed in earlier chapters about understanding your customer and how they determine value, and making sure that the system you are designing will in fact deliver value to those users and other stakeholders. The steps that I described in Chapters 2 and 4 give you tangible methods and tools for accomplishing this important goal. That is to say, the project management approach advocated in this book *already incorporates this important aspect of agile methods*; we can accomplish this aspect *without* calling our project *agile.*

What are some of the characteristics of agile development methods? Here are a few:

- **Significant focus on the structure of interactions** between different portions of the software, and significant focus on simplifying interfaces and service calls. These are accomplished through techniques called *encapsulation* and *controlling visibility.*
- **Frequent but small releases of fixes and capabilities**. These can be viewed as small cycles around the spiral method of Chapter 2. Agile practitioners embrace the key aspect of the *spiral model*: each iteration is *actually used by real customers*, so feedback can be provided to the development team, and that feedback incorporated relatively quickly into another near-term iteration. Some agile practitioners even go so far as to advocate a regular interval for such iterations, believing that the social benefit to the customer of having a predictable timeline for improvements, and the social benefit to the development team of having continuous deliveries (which feel to the team like progress and success) and constant feedback from actual users, make the increased tempo and expense of such frequent and regular iterations worthwhile.
- **Automation of development-team tasks**. Agile projects usually develop automated test scripts and data sets for each small package of software, and deliver these test artifacts into the testing process at the same time that they deliver the source code for

the software itself. This enables the project to automate much of the testing of these small, individual packages of software. In fact, agile projects will often run such automated tests on some regular, frequent interval: once per week, or even every single night. This allows changes that may have accidently "broken" a portion of the software to be spotted quickly. Many non-agile projects do this as well. Such automated tests are usually confined to the smallest units of software; they seldom reach into the integration activity that we described in Chapter 3. This limits the type of error that such tests can detect; they are unfortunately *unlikely* to detect problems that derive from the unplanned dynamic behavior that we discussed in Chapter 2.

- **Plans can be adjusted**. Like all other large engineering projects, projects using agile methods require a large amount of up-front planning, just like we have discussed in previous chapters. But agile projects also usually explicitly embrace the idea of *continuously replanning* and optimizing the approach to the work. I discussed how actually to accomplish this continuous replanning in Chapters 9 to 11. Again, we can do this without calling our project *agile*.
- **Small, cross-functional teams**. There is a natural tendency for us to organize our projects around individual technical disciplines: we form teams entirely of software coders, for example. This provides focus, but just as we have discussed that it is important for the designers to understand the customers, it is similarly important for the different technical disciplines to understand *each other*. Software designers, coders, integrators, testers, trainers all benefit through regular interactions. Agile development methodology codifies this by recommending that teams be formed which include *all* of the skills necessary to successfully deliver each piece of software, including analysis, design, coding, testing, writing, user interface design, planning, and management. This approach has potential virtues, but has often proved difficult to scale to large projects. This is an aspect that I believe must be assessed on a case-by-case basis for your project.

None of this is *easy* or *inexpensive*. This is often a source of confusion: people sometimes expect *agile* to mean *inexpensive* and *quick*. But that is not what agile software development means; *agile* software development means *readily adaptable* through many small patches, and thereby *adjustable to many small changes in the requirements* and able to accommodate *new insights* about the customers and users.

12.4 Ending Your Project

Way back in Chapter 1, we defined a project as a *temporary* activity, intended to create a specific item or items. Therefore, every project eventually *ends*. Let's talk for a moment about the end of a project.

There are many reasons why a project will end. We may have planned to finish the project in a particular manner and at a particular time, but the project might also end *prematurely*. Figure 12.4 depicts the ways that a project can end.

All of the possibilities described in Figure 12.4 will be defined in a section of the contract called the *termination provisions*.

The reason a contract between two commercial companies might contain a clause that permits the prime contractor to cancel the contract upon a change of control is that

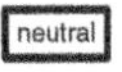

- We reached the contract's expiration date
- We finished the contract's requirements
  - That is, we delivered all of the contractually specified deliverables, and they have been accepted by the customer

- For cause

  - We are being "fired" because the customer believes that they can prove that we violated a mandatory contractual requirement, (e.g. missed a mandatory delivery date, are no longer likely to achieve the required capability and performance, or some other mandatory provision of the contract)
- For convenience

neutral

  - We did nothing wrong, but the customer wants to stop. Perhaps they lost funding, or no longer need our product
- Upon change of control

neutral

  - Our company has been bought by another company. Contracts between commercial companies often contain a clause that such a change of control allows the prime contractor the opportunity to cancel the contract

- The above should all be governed by the *termination provisions* within the contract
- Internal projects might also end because of a *change of organization*

Figure 12.4 The ways that your project can end.

the company that purchased your company might be a *competitor* to the prime contractor. The prime contractor might prefer not to do business with a competitor, or the government might not allow it (because of potential restraint-of-trade issues).

Internal projects might also end because of a *change of organization.*

Figure 12.5 depicts a list of steps that are usually required in order to end an engineering project. It is a long list! The main point that I want you to take away from this list is that ending a project takes time, money, and people. You therefore *need to incorporate these items* into the project plan and the project budget, in the proposal.

12.5 Your Role in All of This

We discussed four separate topics in this chapter: starting a project, projects with large amounts of software, the agile software development methodology, and ending a project. I grouped them together into a single chapter because they interact.

Your engineering project likely has large amounts of *software.* When you start planning your project, you will have to decide whether you will use an *agile development methodology*, or perhaps just selected aspects from that methodology. I have given you some guidelines to help you make that choice, and showed you how methods that we taught you earlier in this book provide your project some of the same benefits while still allowing your project to operate at large scale. Agile is *not suitable* for every project, but for those projects that meet the characteristics identified, it can be very helpful, especially at allowing your product to be adapted through small changes at frequent intervals.

All projects, but most especially big projects – whether agile or not – require *plans, infrastructure,* and *processes.* As the project manager, you have to see to the creation of them all.

The *project start-up process* is an important and non-trivial activity, which requires planning, time, and resources. A special task within the project start-up process is the

- Build a project close-out checklist, validate it with the buying customer, management, and internal specialty organizations (e.g. finance, sub contracting, human relations, etc.)
- Verify and document project deliverables, and their formal acceptance by the customer
- Collect all payments due from the customer, and any refunds due from suppliers
- If you did well, ask the customer for reference letters (for future proposals, which often ask about your company's past performance)
- Collect all final deliverables from suppliers
- Collect all company property loaned to suppliers, including data
- Finalize all activities, collect and archive materials / data from project personnel
- Complete final project performance appraisals for all project personnel
- Arrange for appropriate "thank you" to project personnel
- Arrange for new assignments for all project personnel
- Return all customer-furnished property and information, and obtain receipts for same
- Dispose / transfer all company equipment and property
- Decide what data should be archived and retained (and for how long), and arrange suitable disposal for the rest
- Finalize and document any intellectual property issues (terminal licenses, ensure that privacy markings are in place, ensure that patent applications are filed and that supporting materials are archived with the law department, and so forth)
- Arrange for specialized disposal of materials that require it (batteries, nuclear materials, biological materials, personnel records, classified materials, company-proprietary materials, etc.)
- Lessons-learned report: successes, failures,what could we do better in the future
- If your company maintains a library of project artifacts, provide those to the library, together with the necessary software and instructions
- If your company maintains a library of previous-project actuals, provide those to the library, together with the necessary software and instructions
- Arrange for and support any required audits, by the customer, by the government, by the company

  If project terminated before completion, document reasons
- Close purchase orders, sub contracts, and all other external obligations
- Verify compliance with applicable laws and regulations (e.g. advanced notice to close a facility)
- Close the financial accounts and ensure that bookkeeping is up-to-date

Figure 12.5 Steps to end an engineering project.

creation of the *schedule and cost baselines,* and getting through the corresponding auditing process. The most important engineering projects operate at large scale, and the project start-up process is where the foundations for the capacity to operate at a large scale are created and institutionalized.

Engineering projects usually contain large amounts of software and software development, because software has proven to enable certain desirable characteristics to be incorporated into the end product. Unfortunately, getting that software completed *on time* and *within budget* frequently proves beyond the capacity of a particular project manager and the team. We discussed methods that can allow you and your project to do it better.

Since a project is by definition a temporary activity, it must eventually end. In many ways, ending a project is similar to launching a project: there is a long list of activities that need to be planned and executed, and you must make sure that the money to perform these activities is in your schedule and your budget.

12.6 Next

At many places in this book, we have brought up the question of dealing with *people*, of *motivating* and *aligning* the people on your team, of *resolving conflict* in a productive and polite manner, and many other aspects of the *social responsibilities of the project manager*. In the next chapter, we tackle this aspect of the project manager's job systematically, discussing many aspects of the people-oriented and social aspects of the job.

12.7 This Week's Facilitated Lab Session

In this week's facilitated lab session, you will continue working in your teams. Today's topic is team exercises about *project start-up*.

Refer back to the same project that your team selected earlier in the course. Using the guidance provided in this chapter, create a project start-up check list of no more than eight items for your selected project. Conduct a group discussion of findings and insights.

This will be turned in as a section in your team project, at the end of the course.

13

The Social Aspects of Engineering Project Management

It is a cute cliché that engineers lack social and interpersonal skills. In actual fact, being a good engineering project manager is a highly social activity, and you will find that exercising effective interpersonal skills is an important part of your success as an engineering project manager. In this chapter, I teach you what you need to know: aligning and building an effective team, motivating and inspiring people, managing conflict, and other topics. The good news is that we engineers can actually learn to do this aspect of the job just fine ... despite the clichés.

I will then discuss how you can get ahead in your career. I conclude with a couple of special topics: how to deal with the special problems presented by those projects whose work is geographically distributed across more than one work site, and those projects that include teams located in multiple countries.

You will spend *most* of your time as the manager of an engineering project dealing with *people*: employees, subcontractors, vendors, buying customers, paying customers, supervisors, other stakeholders, your corporate law department, your corporate human relations department, and so forth.

Since you will be spending so much of your time dealing with people, the social aspects of engineering project management – motivating people, aligning your team, resolving conflict, managing expectations, and so forth – are *hugely important* to your success, and to the success of your project.

If I had written this book so that the page count dealing with these social and people-oriented aspects of managing engineering projects reflected their importance to your success, this topic would have made up more than half of the book. But I will instead content myself with this one chapter (which, at least, will be a long chapter). Of course, I have already discussed many individual topics within these social aspects of project management; in this chapter, we deal with this topic systematically.

It is my view that this chapter and Chapter 2 – about requirements and design – are likely to be the most important and most useful chapters of the book for aspiring managers of engineering projects.

I also hope that you will find the material in this chapter useful to you outside of work; we deal with people in almost every aspect of our life. This is a chapter about life.

Engineering Project Management, First Edition. Neil G. Siegel.

Companion website: www.wiley.com/go/siegel/engineering_project_management

13.1 Dealing With People, Becoming a Leader

As an engineer, you certainly expect to deal with *technical* issues: requirements, design, implementation, testing, and so forth. But when we become a *manager* – most likely, starting as the manager of a small segment within a larger project, and working our way over time up to more responsible positions – we also deal with *management* issues. This includes items such as the following:

- Cost estimation
- Change control
- Scheduling
- Management and direction of resources
- Contract negotiations
- Customer understanding and satisfaction
- Selecting and motivating people
- Listening to all of the stakeholders
- Conflict resolution
- Balancing the need for good decisions against the need for prompt decisions.

But in our role as an engineer, and most especially in our role as a manager, we will also spend a lot of our time dealing with *people*. That is, there are important social aspects to the role of an engineering project manager. This chapter, therefore, is about these social aspects, and *how to deal with people.*

When you are the manager of an engineering project, you lead and integrate all aspects of the project: technical, financial, contractual, safety, quality, and everything else that we have discussed so far in this book.

You also lead the interaction with all of the stakeholders:

- Your team
- Your corporation's management
- Your corporation's stockholders or owners,[1] and the investment community
- Your corporate specialist staff functions (e.g. human relations, finance, law, contracts, quality, etc.)
- Your buying customer
- Your users
- Other interested and affected parties (including, perhaps, the general public).

We will talk about how to do this in multiple stages, starting with what I call *alignment.*

13.2 Alignment

As the project manager, it is part of *your* job to synchronize and *align* all of these people. By this, we mean that you take responsibility for bringing all of the project's stakeholders to agreement about the items listed in Figure 13.1.

1 In the United States, a corporation owns itself. The stockholders are *not* the owners; they have acquired certain rights (e.g. a right to a share in the dividends declared by the board of directors) in exchange for their investment. The origins of this nuance have to do with efforts more than a hundred years ago to make it legal to tax corporate income. If you work for some other form of legal entity (e.g. partnership, sole proprietor, etc.), there are people who are the *owners.*

By the term *alignment*, in the context of an engineering project, we mean that all of the project's stakeholders:

- Are motivated by the purposes of the project, are willing to take actions and make an effort so that the project can succeed, and at times, will accept less than they want (e.g. will compromise with other stakeholders)
- Understand and agree with the *goals* for your project, and understand and agree with the *limitations* (e.g. schedule, cost, capability, etc.) within which you are going to attempt to realize those goals
- Understand and agree with your *approach* (methods, tools, locations, facilities, key personnel, sequencing of steps, the design, etc.)
- Are willing to work together, and to reach reasonable compromises in order to resolve difficulties
- Desire to keep the project sold, and see it through to a successful conclusion

Figure 13.1 The definition of *alignment* within an engineering project.

Back in Chapter 1, I introduced in graphic form the concept of alignment (Figures 1.10 and 1.11). Figure 13.1 defines the actual criteria of successful alignment. As you can see by looking at the figure, this can be a complex and difficult task. How do you go about accomplishing it?

All of the items on this list interact. I usually start with *motivation*: Why should someone want to pay for this project? Why should someone want to work on this project? Positive answers to questions like these will likely derive from your ability to explain the *purposes* of the project (e.g. "to improve the safety and decrease the cost of air travel, through the creation of a new and improved air traffic control system"), and to create a sense that you have a *feasible* approach to achieve that purpose, and that there are aspects of that approach which are *better* than competing approaches (this last item is a part of that *positive competitive differentiation* that we discussed in Chapter 5). We have already (Chapter 4) discussed the need for you to understand your customers and other stakeholders, and how they determine value; this knowledge will allow you to express the above in their coordinate system of value, so that they will react positively.

You must similarly provide motivation to your employees. I find that the best and most enduring motivation usually comes from engaging the employees in the value proposition of the system: Why is this system going to benefit society? If you are building a military system, your employees are participating in the defense of your country and its allies. If you are building a health-care system, your employees are participating in improving the health and lives of many fellow citizens. Additional motivation comes from pride in good work; you can make it clear that you and the company expect – and will reward – employees who do good work. Further motivation comes from participation in the team, the sense that they are a part of something larger than themselves. Usually, this sense of being part of something larger than themselves is stronger for the tangible engineering project on which they are working every day than for the less specific role of the company as a whole; that is, the people working on a good project actually self-identify as a member of the project team more strongly than they self-identify as an employee of the company.

No approach to a complicated engineering system is perfect in the value system of all the stakeholders, however, so you must go further: you must engage their interest enough so that they will at times be willing to compromise, so that you can in fact create a consensus among all of the stakeholders. This particular stakeholder will pay a bit more than they wanted to, this particular stakeholder will wait a bit longer than they wanted to, this other stakeholder will accept slightly less technical capability than they wanted to, and so forth. This allows you to *balance* all of the differing wants and needs of the stakeholders with an approach that is feasible. People will in general only be willing to compromise in this fashion if their interest is truly *engaged*; that is, if they really want what you have proposed, if they really believe that your approach is better than all of the others, if they perceive you as an effective and fair leader, and so forth. *You* must go out and create this consensus. You must also keep *re-creating* it every day; as things change, so will the perspective of your stakeholders, employees, supervisors, and the others involved, and your consensus will slip away if you are not constantly working to rebuild it.

You must also remember that there is *always* competition for your project, even *after* you have won the competition and signed the contract:

- Person 1 will come along after you have started and propose to replace your contract with a new contract, using their new-fangled design.
- Person 2 will come along and propose to replace your contract with a new contract that does the work in the district of some particular Congressman.
- Person 3 will come along and propose to replace your contract with a new contract that does the work using only small, disadvantaged businesses.
- Person 4 will come along and propose to stop your contract because they (perhaps incorrectly) see some safety or environmental or equity issue in your approach. Even if they are completely wrong, they might be able to stop your contract.
- Person 5 will come along and propose to replace your contract with *nothing*, because they prefer that the money be allocated to some completely different purpose.

You must, therefore, be on the lookout every day for these types of threats to your contract, and continuously be reselling your contract, your project, its benefits, and the feasibility and benefits of your particular approach. You may discover that there are additional stakeholders out there too, that must be added to your list.

Some of these threats to your contract may come from your own stakeholders; just because they are a stakeholder does not imply that they are a supporter!

Note that the process described above for getting your team aligned is in some ways similar to what we discussed in Chapter 4, about how we would go about trying to understand our customers. In Chapter 4, we said that you must strive to understand the value system of your customers, so that you can marshal the right approach and the right arguments as you try to win the competition, and as you operate the project through to a conclusion. Similarly, you must strive to understand the value system of your *employees*: what do they value? Then you must use that insight to create consensus around the value of the project (this creates motivation), and around the approach (this creates alignment). And later in this chapter we will show that you do the same thing with your supervisory chain.

Also notice that you need to take the time to work out how – in the context of your specific project – you want to go about obtaining this alignment. This is why you need a *team-building plan* on your project start-up list (Figure 12.1).

Achieving alignment clearly involves getting people to listen to us, to give our opinions a fair hearing, and, at times, to actually follow our suggestions. So, next, we will talk about how you can accomplish this.

13.3 The *Sine Qua Non* of Leadership

Sine qua non is a Latin phrase that means something akin to "essential condition." Therefore, the phrase "the *sine qua non* of leadership" means "the condition essential to being an effective leader." My concept of the *sine qua non* of leadership is presented in Figure 13.2.

The key word in Figure 13.2 is *persuade*. Even if you are in a position of authority, you will be more effective, more often, if you operate by persuasion than just by giving orders.

I believe that this applies in all settings. As a defense contractor, I frequently had the opportunity to see senior military leaders in action. The good ones use persuasion too, even though they are operating within a structure that technically requires blind obedience. But these leaders know that they do not get the best efforts from their people when they operate by requiring blind obedience; they know that they get more out of their team when those people have been *persuaded* to do the desired task.

I am not the first person to notice this, of course. Here's a quote along these lines that predates my birth by more than 125 years:

> I like to convince people, rather than to stand on mere authority.
>
> *Arthur Wellesley, The 1st Duke of Wellington*[2]

Think about the setting for this quote. The 1st Duke of Wellington probably qualifies as *the* outstanding military leader of documented history. He enjoyed enormous prestige due to his military successes, including the defeat of Napoleon. He was a nobleman of the highest rank within a society that placed great prestige and respect in such positions. He had a huge popular following. He was the Prime Minister of Great Britain for many years. We could easily imagine that, in such a position, he might feel that he could just give orders. But he didn't; he preferred to *persuade* people instead.

If the 1st Duke of Wellington believed that it was more effective to work through persuasion than merely to give orders, how much more so for the rest of us!

In the United States today, I believe that in fact we can really *only* get people to do what we want through persuasion. In our culture, just giving orders is not very effective, at least after the first time. People have *options*; they are not serfs or slaves, bound to a job or to an owner. They can go to work for someone else. This is especially true for engineers, who are often in short supply and high demand.

The condition essential to being an effective leader:

You must *persuade* people to want to follow you

Figure 13.2 Siegel's *sine qua non* of leadership.

2 Quoted from John Wilson Croker, *The Crocker Papers*, 1826, vol. 1, p. 346.

In some other cultures[3] one might argue that things work differently, and just giving orders *can* be effective. There is a factor that the sociologists call the *power distance*, which reflects the relative variation in social standing and authority between layers of a hierarchy. This might be between a worker and a supervisor, between a student and a teacher, or between a child and a parent. What the sociologists say is that the *power distance* in most cultures outside of the United States is greater – often far greater – than it is in the United States. That is, the difference in social standing between adjacent layers of a hierarchy is often far greater than it would be in a similar situation in the United States. This leads some people to conclude that they can be effective in these cultures just by giving orders. I believe this to be wrong. First of all, in many of these cultures, the power distance is decreasing rapidly, and approaches to managing people that worked yesterday may not work tomorrow. Furthermore, even in cultures where the power distance remains significant in the general culture, within the special subset of those cultures that encompasses their engineers and scientists, my observation is that the power distance between adjacent layers of a hierarchy is likely to be much smaller than the same difference in the general portion of that same culture. And even in those cases where the culture retains the traditional power distance, operating by persuasion still works, and is in my opinion more effective.

The essence of the desirability of persuasion (in addition to its being politer, which is not a trivial consideration) is that persuasion helps to create *motivation*, and all of the data indicates that *motivated employees accomplish more work* than unmotivated ones. This difference is not small either: studies in the field of software development indicate that motivated employees are likely to perform three times as much work per unit of time as unmotivated ones. Three times as much work! As a manager, enabling that increase in work efficiency is very much worth doing. If persuading your employees is part of what it takes to capture that increase in work, you should do so. Now let's talk more about motivating your team.

13.4 Motivating Your Team

We all respond to multiple, different motivations, all at the same time. We have the most basic needs: food, water, shelter. Striving to satisfy these forms a first layer of motivation. Interestingly, once we are in a situation where these needs are met, obtaining more of them is only a minor additional motivation; once we have eaten a big meal, we are not strongly motivated by opportunities for food for at least a few hours.

Many motivations share that characteristic: once they are *met*, they *cease* to be major motivations. If we want to motivate our employees every single day, we therefore need a *different* type of motivation, one that is enduring and not readily satiated. Fortunately, there are many of these types of motivation: the quest for things such as respect, honor, the opportunity for accomplishments, recognition, and a sense of purpose do not appear to plateau in the same way that the quest for basic needs does. We can use this insight to design our approach to motivating our employees. Every employee will respond to these items a little differently, but by using a set of such motivators, we can reach most of our employees.

3 I have worked, and led, project teams in several countries outside of the United States.

Interestingly, *compensation* appears to form a middle example; it does not plateau quite as rapidly as the quest for the basic needs of food, water, and shelter, but the beneficial effects of more compensation become diluted quite markedly once compensation passes some adequate level. My conclusion, therefore, is that you *cannot buy* motivation from your employees; instead, the items cited above – respect, honor, the opportunity for accomplishments, recognition, and a sense of purpose – are better ways to motivate your employees.

Let's talk now about how you can use these techniques.

- Be an *inspiration.* Create a compelling vision: why this project is important, how (once it is complete) it will improve the world, and how each person can specifically contribute to that success.
- *Remove obstacles.* Make sure that conditions exist that allow each person to succeed within their role, and to feel that they can and are succeeding. This usually involves a combination of *tangible* items (e.g. facilities, information, equipment, the mini-contract that we discussed in Chapter 6, etc.) and *intangible* support (acting in front of others in a fashion that displays confidence in this person to do their assigned tasks, helping them privately to find ways to solve problems – but don't jump in to solve the problems for them, it is disempowering to do their job for them – and providing emotional support).
- *Align* the team. Drive the team to a real consensus about how we will perform and complete the project. Do all of the items listed in Figure 13.1.
- *Recognize the problems.* Recognize and articulate what aspects of your engineering project remain as hard and unsolved, and drive the team to a real bottom-up consensus about how to solve those problems, or at least how to approach creating a solution.
- Provide the opportunity for *personal growth.* Assign each person with work that is challenging, but not impossible or unreasonable. Create mechanisms for many types of personal growth for all team members: special assignments, opportunities to present in front of customers and corporate leadership, in-house training, mentoring, continuing education, and so forth.
- Be a *role model.* Always be polite, display good work habits, be calm and rational, listen before talking and before deciding, accept responsibilities, make apologies, and accept blame. In other words, be what you want your team members to emulate. Act promptly to curtail unprofessional or disruptive behavior by others; make sure that an offender is aware of exactly what behavior was inappropriate, and that they understand such behavior is not to be repeated, and that apologies and/or amends are to be made.
- Value *personal growth.* Encourage and reward people who acquire additional skills, especially technical people who learn how to *translate their technical skills into business success.*
- Provide *empowerment.* Delegate actual authority, always making sure that you have provided the necessary enablers for the assigned person to succeed. Provide them with someone to teach them how to do those assigned responsibilities, if they need that. Provide a mentor, from whom they can informally obtain guidance. I also try always to let many people participate in the creation of objectives and plans; feeling that they have had a chance to influence these items is very empowering to your people.
- Insist on *responsibility.* Meet your commitments, and expect other people to meet theirs.

Enumerate – If you want people to know it, write it down! No one can remember everything.

Stipulate – Work your way to a consensus and to a decision. Then make it clear *what the decision was*; lots of things get said, and lots of things get written down. Make it clear which is the actual decision, the actual policy, the actual procedure.

Disseminate – When you make a decision, lots of people need to know. Everyone on the team needs to be able to find the right guidance document when they need it. Therefore, you need *dissemination methods*; you need index and search facilities; you need some training for your team members about what are all of the key decisions made for your project, and about the process for making and disseminating decisions.

Figure 13.3 Siegel's trio for effective communication.

- Value reasoned *risk-taking*. Show that you will seek out and succeed at the hard assignments, and that you will reward others who do the same.
- Be a *communicator*. Become an effective communicator – especially in *listening* and *writing*. Effective communication always includes three steps, which I call "Siegel's trio for effective communication": *enumerate, stipulate,* and *disseminate* (Figure 13.3). Effective communication to the team always includes written materials (e.g. plans, schedules, etc. – these are a part of achieving *alignment*), but also includes you spending time with your team, face-to-face, both in *structured settings* (e.g. all-hands meetings, staff meetings, the meetings involved in the periodic management rhythm, etc.) and in *unstructured settings* (walking the hallways, engaging in informal conversations, encouraging unsolicited one-on-one conversations through what a former colleague called the *open-door philosophy*[4] – e.g. any person on my staff could, if my door was open, just walk in and talk, or alternatively, could get onto my calendar by talking to my secretary).
- Model and enforce *ethics*. Talk about and be seen always considering the ethics of each decision, and always stay far from the boundary of questionable behavior, even if the company risks losing business as a result. Ask for advice from your experts: the law department, your contract manager, your business manager, your human relations manager. Take prompt action if ethical violations occur. If serious violations occur, you need to be seen making serious responses. We will say more about ethics in engineering in Chapter 15.
- Deal with the important aspects of *diversity*. Every person has differing norms, differing values, differing styles, and so forth. Differences that are neither illegal nor disruptive should not only be allowed, but should actually be cultivated. This is because *diversity of opinion and diversity of thought process* is important to successful engineering projects and successful engineering: we need people with a range of skills, experience, points of view, opinions, and so forth. This variety of ideas and thought processes is important in order for us to solve the complex problems that our project will face. We cannot, however, look inside the heads of our employees and measure

4 Thank you, George Petteys.

these things directly. We have to use the attributes that we can see – sex,[5] ethnicity, educational background, and so forth – as surrogates for those attributes in which we really do want diversity. We make the assumption that people who vary in these attributes that we can see potentially have the variation in experience, point of view, and so on that we need and want, and use these surrogates as a way to build the necessary diversity into our team. In those instances where we have actually worked with people in the past, and thereby know something about their work styles, their strengths and weaknesses, we should use that knowledge too in order to create the right kind of diversity in our teams. It is also, of course, the case that we aspire to our work places and ourselves being better than was the norm in the "bad old days," when women and minorities in the United States routinely suffered overt discrimination in our field, and in society in general.[6] My mother (Figure 1.16) became an engineer in 1952, and she personally lived through years of overt discrimination against women in the field of engineering. My father told me of seeing signs in storefronts in Brooklyn, New York, where he grew up, saying "Help wanted. No Jews or Irish need apply."

- Create the *right reward system*. Establish and operate rewards that encourage people to see success in terms of the accomplishments of the *team*, rather than in terms of their own personal accomplishments. Act in a fashion that causes them to believe that the team's success is truly a viable path to their own individual success.
- Tell your employees *how to get promoted*. Everyone is interested in promotion. My recommendation is that you be transparent about what will get your employees recognized and promoted. Figure 13.4 presents the list that I used with my employees.

Career growth is facilitated when one shows capability on some combination of these axes:

- Technical excellence
- Deep understanding of a customer domain
- Leadership (inspiring, motivating, consensus-forming, challenging, team-building, accepting responsibilities, meeting commitments, etc.)
- Ability to work in a team – seeing team success as the path to individual success; multiply your effectiveness
- Ability to translate technical skills into business success

Figure 13.4 Tell your employees what will get them recognized and promoted. This is the list that I used with my employees.

5 My wife, who has a PhD in linguistics, suggests the use of the word "sex" for the biological attribute of people and animals, and the word "gender" for attributes of *words*, like the difference in French between the words *ami* (a male friend) and *amie* (a female friend). Out of respect for her erudition, I try to follow this usage.

6 Having lived and worked outside of the United States, I can assure you that discrimination is rampant and overt in most of the rest of the world, although the groups that are being discriminated against will vary from place to place. Nor is the struggle over within the United States. Although our progress in this matter, even over my working career, has been staggeringly good, things still can and should be better, and we ought neither to be smug nor to sit on our laurels. Good managers, as social leaders, must embrace improvement in this area, as in many others.

It is my experience that the quest for motivation through items such as these does *not* get satiated, in contrast to the quest for motivation through the basic needs for food, water, shelter, and even compensation. These items are tools for motivation that never run out.

One last thought about motivation: there are many actions that can be *demotivators* for your team. Doing the opposite of the items on the above list is certainly likely to become a demotivator for your team members. There are others; I will mention only one:

> A leader never complains, but always sympathizes with those who do.
>
> *paraphrased from William F. Buckley, Saving the Queen (Doubleday, 1976)*

We all have a tendency, at times, to want to complain. I recommend that you *always* suppress this tendency while you are at work. No one really wants to be led by a constant complainer; it is not inspiring behavior. Furthermore, complaining does not create confidence that you have the ability to figure out how to solve problems!

At the same time, showing *empathy* for your employees when there are matters which are causing them discomfort is a motivating characteristic. Sympathizing with the complainers can be done without committing that you agree with their position, or that you are agreeing to fix the causes. You can just be a good listener and a soft shoulder.

13.5 Recognizing and Resolving Conflict

> People are a problem.
>
> *from The Hitch-Hiker's Guide to the Galaxy, by Douglas Adams*

There is an old joke: "Two people, three opinions." When more than one person is involved in an activity, there will be *conflict*. As the manager of the project, we must learn how to recognize and resolve conflict.

There are valid and productive reasons for conflict (Figure 13.5); the mere existence of conflict is actually not a sure sign that "people are a problem." The conflict only signifies that there are multiple opinions, multiple candidate approaches. Remember that when we talked about systems engineering trade studies (Chapter 2), we went so far as

- Reasons for conflict that *may* be *productive*:
  - Individuals have different goals
  - Individuals have different concepts and preferences for method, sequence, and priority
  - Individuals have different preferences for various technical decisions and options
- Reasons for conflict that are *unproductive*:
  - Perception of needing to assert themselves in order to gain their own aspirations
  - No one taught them how to be effective in a group, or what is expected of them when they are a member of a group
  - Dysfunctional behavior: immaturity, rudeness, selfishness, dishonesty

Figure 13.5 Reasons for conflict: both productive and unproductive.

to state that creating multiple candidate approaches was a *virtue*, and even a *necessity*. The problem is *not* the existence of multiple opinions – that is, the problem is not the conflict itself. Instead, the problem is the *poor methods* that many people – and, unfortunately, many project managers – use to *recognize* and *resolve* conflict. Let's talk about how to recognize and resolve conflict appropriately, so that we can gain benefit from the diversity of opinions, without letting things get personal and/or disruptive.

Notice that in Figure 13.5, I have divided the list into two parts. The first I have called reasons for conflict that may be *productive*; the second reasons for conflict that are *unproductive*. In the previous paragraph, I described situations where there might be valid productive reasons for conflict: differences in goals; different concepts about methods; and different preferences for technical decisions. Of course, now that you have read the section above about achieving alignment, you know that some of this should have been talked out and decided early in the project, captured in plans and specifications. By thereby being a part of the project's baseline, we have already discussed and reached agreement on these matters.

Of course, on something as complicated as an engineering project, you cannot realistically hope to sort out everything in advance. Each day, you will discover new matters that either require a rethinking of previous decisions, or are items that have not been previously decided. That is to say, achieving alignment is a *continuous* process that gets worked on every day until the project is completed.

Figure 13.5 also contains examples of unproductive reasons for conflict. These are the stereotypical "bad behaviors" to which we often attribute *all* conflict. Try to remember that this is *not true*; some conflict is valid and productive! Furthermore, just because some of the behaviors that I have called *unproductive* are present does not mean that there are not valid and productive issues that are *masked underneath* that unproductive behavior. We ought not to decide an issue solely on the basis of bad behavior, tempting though that may be. The person behaving badly might be right, even if they are expressing it inappropriately.

This distinction between productive and unproductive conflict makes clear another motivation for striving to achieve the alignment that we discussed above: if the team is aligned, they share a sense of the value of the mission; they have enthusiasm for the project; they share a definition of what the project is to produce; they largely agree on methods, tools, and sequencing; they have a shared definition of what constitutes success; they understand how they can contribute (and the role to be played by others); they understand how the project's success may translate into success for them personally; and they have some sense of a shared destiny – we all succeed or fail together.

With all of that, the team members can develop some level of *trust* in one another. And the presence of that *trust* – in combination with some *guidance*, and some *methodology*, both of which I will talk about in a moment – ought to allow them to *avoid or minimize unproductive conflict.*

We can say this in a negative form too. Here are some reasons that teams fail:

- Poorly developed or *unclear goals*
- Poorly defined project *team roles* and interdependencies
- Lack of project team *motivation*
- Poor *communication*

- Poor *leadership*
- Perception of *unfairness, unequal treatment,* and *inconsistency* in how people are treated
- Bad news is perceived as unwelcome
- *Turnover* among project team members
- *Dysfunctional* behavior.

The perception of *unfair* or *unequal* treatment can raise hackles among your team very quickly. To avoid this takes effort. I find that it helps if I think in advance about how I will respond to various situations. I essentially have practice conversations and scenarios in my head. This allows me to develop model responses in advance, and I find that this helps me a lot in being consistent and fair.

The ancient Greek dramatist Sophocles, in his play *Antigone,*[7] wrote "στέργει γάρ οὐδεὶς ἄγγελον κακῶν ἐπῶν," which can be translated as "no one loves the messenger who brings bad news." No one *likes* to receive bad news, but you must not respond by being rude or disrespectful to the person who brings you that news. In fact, you should thank them!

In the book *The Lives of the Noble Greeks and Romans,* by Plutarch,[8] he writes that an arriving messenger who gave a particular commanding general (Tigranes) bad news (in this case, the pending arrival of an opposing army) had his head cut off. Plutarch goes on to say that, naturally enough, no one else would thereafter dare to provide this commander with any further information, and that he suffered on the battlefield from this lack of intelligence. If you act in a fashion that discourages people from bringing you bad news (even if your reaction is less extreme than that cited by Plutarch), people will stop bringing information to you, and your ability to be effective as a project manager will suffer.

Here are some examples of the last item on the list, what I call *dysfunctional behavior*:

- *Rudeness.* A few immature and/or disruptive people really can set you back; this can be infectious behavior!
- Not meeting commitments
- *Selfishness.* "Me first" rather than "team first"
- Weak ethics
- Dishonesty.

In my experience, there is *not* a lot of overt dishonesty in engineering. But engineers are people, and all of the other dysfunctional behavior occurs amongst our engineering teams as frequently as it occurs in any other group of people ... unless we achieve *alignment,* set expectations for behavior, and use problem-solving techniques that inherently lessen the personal confrontation aspect of the situation.

One item on this list warrants some extra discussion: the question of "me first" rather than "team first" behavior. We engineers spend 16 to 20 years in school, where the value system is "we do our own work, and we are judged almost entirely on our own work." At times, we do work on team projects in school, but the overwhelming majority of our grades derive from our own work. We are even taught in school that it is generally

7 Written around 440 bce.
8 Written around 96 ce.

improper and unethical to collaborate! This value system is embedded very strongly in most of us, because we experienced it for such a long period of time.

Then we graduate from college and arrive at work. In the next section, I will describe what an engineer actually does at work. But I will preview one key aspect of that now: at work, everything worthwhile is accomplished by a *team*, rather than by individuals. Often these are *large* teams. Therefore, suddenly, usually with no preparation, we find ourselves in a social setting where the value system that we used for 20 years is no longer valid! Often, no one even tells us about the necessity to make this transition.

In fact, we must transition:

from

The *school* value system of "*I must do my own work, be evaluated almost exclusively on my own work, and derive my satisfaction only from my own accomplishments*"

to

The *workplace* value system of "*Everything worthwhile is accomplished through the combined efforts of a team, often a large team, and I must learn that I will be evaluated on the basis of my contribution to the team (rather than exclusively on the basis on my own work), and I must learn to derive my satisfaction from the accomplishments of the team (rather than exclusively from my own accomplishments).*"

For many of us, this is a difficult transition, especially the fact that we must derive our satisfaction from the *accomplishments of the team*. In fact, usually no one tells us that we need to make this transition! But now *I* have told you. You need to tell your employees. Then you need to be seen taking actions that *reward the team*, not just individuals.

So, how do we resolve interpersonal conflict in our engineering projects? I use the following approach:

- Focus on the *problem*, not on the *people*. I make a decision by analyzing the problem causing the conflict, not by considering the characteristics or position of the people involved.
- Understand what each party wants and needs, and why. What do they have at stake? Why do they care about this outcome?
- Before starting the decision-making process, create a wide *range of candidate solutions*. This is akin to the large trade space that we discussed in Chapter 2.
- Before starting the decision-making process, create *decision criteria*. Insist on criteria that are both *objective* and *reasonable*.
- Be prepared. Do not "make it up as you go along"; getting facts wrong decreases your credibility and negotiation effectiveness. It is better to stop a meeting and request additional facts and analysis than to proceed in correctable ignorance. Task specific people in advance of meetings with the job of preparing analyses of the problem. What are the premises? What are the limitations and constraints? What are the available analysis methods? What are the data? What are the results of the analysis? What are the magnitudes of the uncertainties? What methods, if any, might be available in order to lessen those uncertainties?

- Make the purpose of the discussion clear: are we just gathering facts, creating options, assessing analyses, or have we arrived at the decision-making stage? If the latter, make it clear who has the authority to make the decision.
- During the decision-making process, keep the discussion about *premises* and *analysis methods*; do ***not*** discuss *positions* or the *desired answers*.
- During the decision-making process, use a structure to moderate the discussion that allows everyone to have a chance to express their views, and to respond to the views of others.
- Create rules that will allow an orderly and calm discussion. Cut off inappropriate discussions (e.g. those that make use of *ad hominem*[9] techniques, those that insist on jumping to a discussion of the desired answer, those that interrupt someone who is already speaking, and so forth). Make it clear that a rule violation has occurred and identify the nature of that violation. Restart the discussion. Do not prevent the offender from participating again, as long as they participate appropriately. Have a person designated as moderator in any meeting, charged with the responsibility to enforce this approach. It need not be you.
- If you have delegated the decision-making authority to someone else, do not take any action that appears to pre-empt that authority.
- As the project manager, you are likely to be the senior person present. You must not pre-empt other viewpoints by speaking strongly in favor, or against, any particular approach too early in the discussion. Because of your senior position, once you make your views known, many people will be reluctant to bring forth opposing views. *Try to speak last.*
- Remain calm and polite. Insist that everyone else does too. Interestingly, you will likely find that remaining calm and polite is an effective negotiating tool, and not just at work, but in all life situations.
- Thank all participants. Make it clear that everyone contributed, even those who did not obtain the decision they desired.
- Keep notes about everything. Archive all artifacts; you may need to recreate the analysis later, and will want the data, the spreadsheets, and so forth. I assign a note-taker role in every meeting; I cannot do it myself and still provide the attention that each participant deserves. Allow the note-taker to stop the discussion and ask for clarifications, so that the notes can be accurate.

The most important and most subtle item on the above list is the one about "keeping the discussion about *premises* and *analysis methods*; do ***not*** discuss *positions* and the *desired answers*." Let's say that you support political candidate X for the US Senate, and I support candidate Y. If we just talk about our desired outcomes (the election of X for you, and the election of Y for me), nothing can be solved, and no new insights gained. You just keep saying that you prefer X, and I keep saying that I prefer Y. We might even be tempted to slip into *ad hominem* arguments, or worse. There is actually nothing being discussed.

---

9 "Ad hominem" is Latin for "to the man," and this term is used to designate an argument that is *not* based on the facts of a situation, but is instead based on attacking the characteristics or ability of the person presenting the argument.

If, on the contrary, we each discuss our *premises* and *analysis methods* (that is, our lines of reasoning that we use to go from our premises to our conclusion), we can have an actual discussion, have the potential to learn something, and might discover an approach that we can both agree on. Your premises might include that social security must be protected at all costs; mine might include that the government should stop running at a financial deficit. We can each explain why we think these are valid premises, and how we reason from those premises to preferred legislative programs, and eventually to our selection of a candidate. One or both of us might learn something, change our mind, or even determine that the election of neither X nor Y is likely to advance our desires. We might decide that we should both vote for Z instead.

That is, a sensible discussion proceeds *forward* from the premises and analysis methods to *reach* the conclusions; it does not *start* with the conclusions. This is a paraphrase of Aristotle, and 2300 years later, it is still good advice.

In the world of engineering projects, this method works well. You may prefer that we use the C++ programming language for our project, and I prefer that we use Java. If we start our discussion with our desired conclusions, we have nothing to discuss, and are likely to annoy each other, and our colleagues too. If, on the other hand, we each start by identifying our *premises*, we can have an actual discussion. For example, you may say that the use of strong typing and inheritance is, in your opinion, the key aspect of the programming for our system, and that language features A, B, and C in C++ are the best-in-class mechanisms for those purposes. I may then say that we have 1400 user forms to create, and finding a way to make that affordable and maintainable is the key aspect for the software on our system that we ought to optimize through the selection of the programming language, and that the X, Y, and Z features of Java are just the right things to achieve this purpose.

In engineering, we usually have more flexibility and opportunity for nuance in the final answer than we do in the voting booth. For example, the software for our system will probably consist of hundreds of modules, and they *need not all be in the same programming language*; perhaps we will decide to build the user forms in Java, and the rest of the software in C++. We might even come to believe that this is a *better* approach than using either language for all of the software. Or, having identified the premises, our analysis might lead us both to realize that we both would actually prefer to use some other programming language in lieu of C++ and Java for *all* of the software. That is, we can create better answers through the discussion than either of us entered with at the beginning of the discussion. This is a true *win–win situation*, but we can get there only if we start with *premises*, work our way through *analysis methods*, and *only then* arrive at conclusions.

Doing this takes practice. I can only urge upon you that it is worth the effort. Not only does it often lead to better decisions, but it can also be very empowering to people, and a great team-building activity. People really feel that they are working together and creating something better through that collaboration than any of them could have created by themselves. As the project manager, that is what we want! Also, just having your employees experience this process, and thereby believing that they were allowed to participate, makes it easier for them to *accept* the decision, even on those occasions when that decision is not the one that they advocated.

Conflict ends with someone – often *you*, since you are the project manager – making a *decision*. People hate ambiguity, and so the tendency for most people is to make decisions *too quickly* and on *too little data*; to throw out alternatives too soon; to close the trade space too quickly. Do not be afraid to ask for more analysis, more options, more time. But you cannot be seen as dithering either, and of course decisions need to be made in order to allow the work to progress. Life is tough as a project manager.

Only judgement and experience can help you determine when it is the right time to make a decision. In a few pages, we will introduce the concept of a *mentor*; he or she can help you with this type of problem.

Good management can lessen the opportunities for conflict. Notice, however, that I did *not* say "eliminate conflict." Basically, through management actions, we get issues and potential issues identified, premises and analysis methods identified, people assigned to gather data and perform analysis, and then (likely through a series of face-to-face meetings), we actually reach a point where we are ready to make a decision. That decision then gets documented (including the rationale for the decision), and all of the supporting materials archived.

Many decisions have "natural" homes for the documented decision. For example, if a decision determines a portion of the design, the decision will be reflected in the corresponding design documents. But it is not always appropriate to capture all of the rationale in the "natural home," so I also follow a practice of capturing all decisions in a project *decision log*.[10] That decision log ought to include all of the data, analysis, models and tools, the various positions advocated (including those not selected), the rationale for each, and the final decision.

Documenting agreements is a vital aspect of project communication. Therefore, the project communication plan (Chapter 12) should include your plans for establishing both the project decision-making process, the decision log for capturing decisions, and the methods that your project will use for communicating decisions (and their rationale).

Not all conflict is about technical matters. Often, it is about people: who is responsible for what pieces of work, who gets to make what decisions, who gets priority access to people or facilities, who gets promoted to an open position, and so forth. Clear statements of scope in the mini-contracts (Chapter 6) and other project documents are the resources you use to lessen these occurrences. When such issues do arise, you must address them promptly; they do not go away by themselves and can get worse over time if left unaddressed. You use the same "Aristotle-based method" to solve these interpersonal conflicts as we described for solving the technical conflicts. Be aware that these types of conflicts have more potential for hurt feelings than do the technical conflicts. All the more reason to be calm, polite, and fact-based, as we described above.

Summary for resolving conflict:

- *Discuss premises and lines of reasoning rather than discussing candidate answers!* (Aristotle)
- Bring issues into discussion (openness; don't bury the problem or shoot the messenger)
- Be willing to listen ("We have 2 ears, but only 1 mouth ...")
- Understand your and other parties' needs/wants (their value system)

---

10 On one project that I worked on early in my career, this was called a *design notebook*, even though many of the decisions had nothing to do with the design. Because of this experience, I used that term for the rest of my working career. But *decision log* is a better name.

- Get the facts (premises) agreed upon before starting the process of trying to reach a decision
- Don't speak for others; that is disempowering. If you state an opinion, use "I." Ask that others do the same
- Ask questions (the "5 Why's" from Chapter 9) to get to the bottom of an issue, paraphrase and restate what others say, so as to make sure that you understand what they actually meant
- Be willing to expand the candidate solution set – this is often a path to resolution of conflict.

13.6 Siegel's Mechanics of Project Management

I have found that certain process steps aid in my mastering the social aspects of project management. These *mechanics of project management* are shown in Figure 13.6.

Let's discuss each of these in turn.

- Take decisions *only* in writing:
  - I recommend that you create a written project policy that nothing said verbally carries any force, and is *not* to be acted upon. Actual decisions that are to be acted upon will be committed to writing with an appropriate level of approval; only then are they to be acted upon, not before.
  - Create a formal record of all written decisions (as noted above, I generally called this the *design notebook*, but it might better be called the *decision log*). It should include both the actual decisions (with approval by the appropriately authorized person), but also opinions, analyses, and other non-binding but potentially informative items.
- Implement *Siegel's trio for effective communication*:
  - This was introduced and discussed in Section 13.4 above.
- Always capture the *rationale and methodology used* – including the premises, the actual data, software tools, spreadsheets, databases, decision criteria, and anything

- Take decisions *only* in writing
- Implement *Siegel's trio for effective communication*
- Always capture the *rationale and methodology used* for reaching decisions, not just the decisions themselves
- Do not use face-to-face meetings simply to convey information – that is done more effectively in writing
- Do not use electronic mail or any similar electronic forum (e.g. social media) for *discussion*. Instead, hold discussions in face-to-face meetings
- Organize meeting agendas around discussion of those items about which there is *not yet a consensus*
- You must have some meetings with *all* of your direct reports; you cannot decide most things just by meeting with the relevant people one-on-one
- Work with, and take advice from, *only* those people who either have *skin in the game*, or are bound to you by genuine affection

Figure 13.6 Siegel's mechanics of project management.

else used to inform the decision, in addition to a description of the decision-making process – for reaching decisions, not just the decisions themselves:
  - This can be vital later on, when new information arises, and the analyses have to be redone.
  - The rationale and methodology used to reach a decision can be documented in separate papers from the decision memorandum, if desired.
  - Both should be right next to each other in the decision log, and the decision memorandum should reference the paper providing the methodology.
- Do not use face-to-face meetings simply to convey information:
  - That is done more effectively in writing, and
  - It also makes for boring meetings.
- Do not use electronic mail or any similar electronic forum (e.g. social media) for *discussion*:
  - Electronic mail and similar electronic forums are fantastic for conveying small pieces of information, such as a phone number, or the location and time for a meeting.
  - **But** electronic forums are weak, and even counterproductive, as a mechanism for discussion. Psychologists and sociologists tell us that something like 70% of all information conveyed between people in a face-to-face meeting is through mechanisms other than the words spoken by the participants. You lose all of that 70% in electronic forums. Furthermore, I find that the use of electronic forums tends to allow people to disconnect from the forms of behavior (described above) that I believe are essential for effective teams.
  - Therefore, discussion is done better at *face-to-face meetings*.
  - I recognize that in our private lives, lots of people do use electronic forums for discussion. I believe that electronic forums do not work well for discussion in our private lives either, but I cannot control what people do in their private lives. I *can* control what they do at work, and need to control them enough to ensure that they conform to what I believe are constructive forms of interpersonal behavior.
- Organize meeting agendas around discussion (not around just conveying information), and (most importantly) discussions of those items about which there is *not yet a consensus*:
  - There is a natural tendency for a group to want to spend their time talking mostly about the things about which they are all already in agreement; that is, after all, comfortable.
  - The tendency is also to avoid spending time talking about things on which they have not reached a consensus ... yet those are exactly the things that need to be talked about.
  - Spending most of your face-to-face meetings talking about the items which you already agree on does not make for very productive meetings; nor does it move us forward toward solving the unsolved problems. I always tried to build meeting agendas to spend most of our time discussing the matters on which we had not yet reached a consensus. That is at times uncomfortable, but productive.
- You must have some meetings with all of your direct reports as a group – you cannot decide most things just by meeting with the relevant people one-on-one:
  - Most of the matters that you will be called upon to decide will involve interactions (and sometimes these interactions are subtle and easy to miss) between different

aspects of the project, and therefore between the areas of responsibility of multiple people. If you decide such matters solely via discussions with the people involved one at a time, you will miss out on the opportunity for these people to react to, and comment upon, the opinions and points of view of the others. My experience is that such multi-participant discussion leads to deeper insights and better decisions.

- Such team discussions – assuming that you keep everything polite and respectful – are very important team-building and consensus-building opportunities. You need not just to reach a decision, you need to reach a decision that all of your direct-report managers are willing to support. Giving them a visible opportunity to participate in the decision process through group discussions is an essential step in forming such a consensus.

- Work with, and take advice from, *only* those people who either have *skin in the game*, or are bound to you by genuine affection:
  - By the phrase *skin in the game*, I mean that they will be materially affected by the outcome of your project, or of a decision on which they are advising. *You and your employees* may lose their jobs if the project fails, and/or suffer reputational damage. Your *using customer* will lose capability (or at least, the improved capability will be significantly delayed) if the project fails. These people have real *skin in the game*; they care about the outcome of your project, and their own personal outcome will vary in both a positive and a negative sense with the outcome of your project. Consultants – even the most distinguished experts – famously do *not* have skin in the game; their outcome (whether they are billing you for their time or serving *gratis*) *will not change* depending on the outcome of your project. For this reason, I suggest that you *never* use consultants, contract labor, job-shops, or the like; they have no *skin in the game*, and their advice may therefore not be aligned with the needs of you, your customer, and your team. I therefore prefer to hire employees (who always have skin in the game), or another company as a subcontractor – this subcontractor (if you write the subcontract to have real responsibilities and real deliverables, not providing bodies) will have skin in the game, both financially and reputationally, just like your own employees. Retirees – even when hired on a consulting basis – may be different. They may be bound to the company or to you by genuine affection, and therefore have *emotional* skin in the game, even if they do not have material financial skin in the game. Retirees also always have a concern about the long-term financial health of their former employer; their pension may be reduced (or at least, interrupted) if the company goes bankrupt; this provides retirees (but not pure consultants) with *skin in the game*.

13.7 Dealing With Special People

13.7.1 Your Management

The social aspects of project management include dealing with your management and your customers, not just your employees. I will talk about that aspect now.

Your management is one of your stakeholders too. Therefore, just like for your customers, you must identify *their* coordinate system of value. What do they need from your project, and from you? What do they want from your project, and from you?

Just as you have multiple customers, you have multiple managers too. You do have a direct supervisor, and that is likely to be one person. But that person has a supervisor too, all the way up to the president or chairman of your company, and the board of directors. There are also the heads of the specialist staff organizations (law, contracts, human relations, engineering, and so forth), who in some ways also have a supervisory role over you and your project. Therefore, the voice of your management is not a single, unified voice. You are in the same position of having to find a suitable balance between competing and conflicting goals, just like you are with your customers.

In some real sense, dealing with your management is easier than dealing with your customers. For example, there will usually be lots of written guidance about what the company (and therefore your management) expect of you and your project. These include items such as:

- Corporate policies and procedures
- Corporate project management handbooks
- Corporate financial goals, and methods of calculating these
- Management incentive goals
- Your contract
- Applicable laws and regulations.

You need to know *all* of these – get help and training from your staff specialists, such as the corporate law department! Do not be afraid (or embarrassed) to ask for help.

When things are going well, dealing with your management is not so hard. But *every* project worth running is complex enough that there will be problems at some point in time; this is the most important time in your relationship with your management.

My advice is do not wait until there is a problem to start building a relationship with all of the people who occupy roles in your management chain. Build a relationship now, discuss your plans, discuss risks (e.g. what might go wrong), incorporate statistical effects into your predictions – and do all these things when everything is still going great in your project.

This helps you understand how they think, and what they value. It also builds a relationship, some trust and credibility, and all of this forms some emotional "credit" that you can draw upon later, when a problem occurs.

When a problem arises, that information does not improve with age. Talk about it right away, even if you are not yet asking for help, and (this is really hard) even if you do not yet know exactly how you are going to solve the problem. You can talk about what you do know, how you will go about preventing the situation from getting worse, how you are going to go about gathering more information, and how you will go about creating a solution. Do not fudge or gloss over the problems, and do not downplay the potential significance of the problems. There is a strong desire in all of us to downplay the significance of problems. Don't do it; you will lose credibility later.

Do not, however, delegate *your* job to your management. Present them with options, show decision criteria, show your current thinking. They will offer advice (especially in the form of recommendations regarding who are the people that they think can help you solve the problem), but do *not* ask them or expect them to solve the problem for you. As the project manager, that is your job.

If you become stuck, say so.

13.7.2 Your Customers

In many ways, dealing with your customers is similar to dealing with your management. We already talked a lot about customers in Chapter 4. We must start by trying to understand what *their* coordinate system of value is: What do they need? What do they want?

Our customers often have lots of *constraints*: rules that they must adhere to which determine how they interact with you, limit how fast they can at times respond, and so forth. You need to understand those constraints, as they will in part constrain your degrees of freedom. Your staff specialists (especially contracts and law) can help you with this. So, can the customer's staff specialists; they have contracts and law staff, too, and they are usually glad to meet with contractor project managers and explain how they operate.

One final piece of advice about dealing with your management and your customers: do not say one thing to your customer, and something else to your management. You will find that they do talk to each other!

13.7.3 The Human Resources Department – An Important Partner

Your project will have a human resources specialist. This is the person who makes offers of employment, deals with the administrative aspects of pay raises and promotions, and other administrative tasks relating to employees.

It is easy to think of this person as an administrator. But in fact, this person can become an important asset to you, a partner. I recommend that you get to know your human resources specialist.

People are obviously an essential element of your success as a project manager; you cannot do the job alone. Your human resources specialist is the person who can help you find and entice the right set of people to come and work on your project.

Hiring people is complicated; the world of business has created entire vocabulary about this subject. Your human resources specialist will ask you to define your personnel requirements:

- How many people with each skill set (e.g. software programmers, etc.) do you require, when, and for how long?
- What level of capability? Novice, right out of college? Seasoned professional? How many of each?
- Which of these people require special professional certifications?
- Which of these people must be US citizens?
- Which of these people require security clearances?

This is not sufficient though. Someone – hopefully not you – has to worry about questions like these:

- Are these people available *inside* the company, or do we have to plan to hire them from *outside*?
- Is it best to hire some of these people on a temporary basis (this is often called *contract labor*), or is it best to hire them as regular, permanent, *employees*?
- What is the going market rate of *compensation* for each of these types of people?

- What types of *employee benefits* (e.g. medical coverage, sick days, vacation, pension, educational reimbursement, etc.) are needed in order to attract the right talent to your team?

Your human resources specialist will tend to all of these matters.

In order to hire a person (whether as a regular employee, or as contract labor), someone will need to write a *requisition* (sometimes called a *job description*) defining all of the above characteristics. This job description is *posted* in various places where potential applicants might see it; this could be an internal company web page, a company web page that is available to the general public, or other arrangements. Specific potential applicants might even be solicited to apply for a particular position.

Then some time is allowed for the applications to come in; a deadline for applications might have been specified in the job posting. Once the deadline has passed, the applications can be reviewed.

In most companies, the process of selecting which candidates to hire or select for a position is quite structured:

- Usually, you will need to *interview* prospective team members. You do not make your selection solely on the basis of the written applications.
- There may be a selection process before the interviews; perhaps your company received so many applications that it is not feasible to interview all of the applicants. More frequently, applicants whose written qualifications appear markedly weaker or markedly less suitable are eliminated from consideration before the interviews.
- Learn the rules of your company regarding the interview process – failing to follow those rules can land your company in court. Your human resources specialist knows these rules.
- You should have some written selection criteria, which are prepared in advance of conducting the interviews. The questions that you ask of candidates during the interview should relate to these criteria.
- It is a good idea to develop *written* interview questions that you ask each applicant:
  - Use the *same* questions for all candidates, and
  - Avoid questions or discussion about irrelevant subjects – ethnicity, age, marriage and children, personal matters, religious beliefs, etc.
- For more senior positions, your company will probably require that more than one person participate in the interviews. This might be a series of individual interviews, a group interview, or a mixture of both.
  - This group of people might include the human resources specialist, the relevant functional manager (e.g. head of the law or engineering department), and some of your existing senior technical staff.
- During the interviews, make notes that will allow you to score the applicants against the established selection criteria.
- After the interviews are complete, the group of people who conducted the interviews will usually get together and formally score the applicants. The scoring is not necessarily the sole criteria for selection, especially if the scores are very close to each other; scoring is an imperfect process. The purpose of the scoring is to keep the discussion of the merits of the candidates to the identified characteristics advertised and needed for the position. The results of the scoring and the notes about the discussions are captured for the record.

- The designated person, using the scoring and the discussion, makes a selection.
- Only certain people are authorized to make actual job offers; usually, only the human resources specialist is so authorized. You, as project manager, are usually *not* authorized to make an actual job offer.
- Job offers must usually be made in writing; it is fine to make a courtesy call telling an applicant that he or she has been selected and that a written offer is on the way, but the *terms* of the offer ought to be conveyed only in writing. You must be careful never to appear to make a verbal job offer.
- It is good practice to obtain written acceptance of the offer from the candidate that you selected. A start date can then be selected, and the candidate processed into the company.
- Always get back in a timely fashion to *all* of the applicants (i.e. all of those whom you did not select, and even those whom you elected not to interview). People who were not selected for interview can be provided with written notification that they were not selected. My practice, however, is that if we interviewed a person, they ought to receive a personal phone call to tell them that they were not selected. This is uncomfortable for many people, but it is polite, and makes a good impression on most people; who knows, you may someday want them to apply for another position on your team or at your company.

The process can be faster if all of the candidates are already employees of your company, but many of the above steps are still required, especially for positions with management responsibilities (e.g. cost-account managers).

As you can imagine from the above description, hiring does not occur in a day. In fact, it can often take *months* to develop a requisition, post the job, select the interview candidates, set up interviews, conduct interviews, reach agreement on whom you will extend an offer to, make the offer, receive candidate acceptance, process the candidate into the company, notify unsuccessful candidates, and so forth. I have seen sources which state that, on average, it now takes 90 days to find, interview, and hire a single engineer; engineering is a particularly competitive job market at present, and engineers generally have lots of options. You have to allow a realistic time line for staffing your project, especially at the beginning.

13.8 Your Career as an Engineer

Some of you reading this book may already have substantial work experience. But many of you may not. I will therefore describe what an engineer actually does at work.

I spent many years working for a large aerospace company. For 18 of those years, I was a vice-president and officer of the company. I held many different job positions over that time: working on engineering projects, managing engineering projects, running business units within the company, running proposals to win new business for the company, work assignments in other countries, and many years as the chief technology officer. I had as many as 12 000 employees. It was a big company, and it offered me big jobs and many interesting challenges.

I also at one point left my large company to participate with five other people in the formation of a start-up company. This experience also offered me a lot of challenges, many of which were quite different from those I experienced at the large company. I stayed with that start-up company until a year or so after we went public (and had grown to around 1 500 people). At that point, I felt that I had learned what I was going to learn from working for a smaller company and wanted to work on big engineering projects again. So, I returned to the same large company where I started my career, and stayed there until I retired.

The aerospace industry, of course, hires a large number of engineers and engineering project managers. There are many well-known companies in this industry. I worked for Northrop Grumman,[11] but other companies include Boeing, Lockheed Martin, Raytheon, General Dynamics, and BAE.

Many other industries hire engineers, engineering project managers, and systems engineers, including:

- Shipbuilding
- Telecommunications
- Energy
- Medicine/health-care
- Government
- Construction
- Banking
- Retail (especially those that do their own supply-chain management)
- Entertainment.

Some of these are relative newcomers to the engineering project management business. For example, it is only in the last couple of decades that the entertainment industry (movies, video games, etc.) and health-care (electronic medical records, digitized medical diagnostic equipment, etc.) have transitioned so many of their products to digital and computer form, which naturally required them regularly to undertake engineering projects in order to create these products.

That is good news for you; there is a diverse and growing demand for engineering project managers, at all levels of seniority.

You get to choose not only an industry to work in, but also the size of the company for which you wish to work. I found that my experiences at big and small companies were quite quite different (Figure 13.7).

But what do you actually do at work? In college, we spend much of our time learning technical skills: computer programming, digital logic design, mechanical stress analysis, and so forth. We do some of that at work, of course, but such tasks are always set in the context of the *project,* and as a result, the focus is always on that *context.* Here is my summary of what we engineers actually do at work:

- Just as we discussed in Chapter 4, we strive to understand the *customer* for the project, and how they determine value.
- Just as we described in Chapter 2, we create metrics for measuring the project's design that align with the customer's value system.

11 I started at TRW, which was acquired by Northrop Grumman in 2002.

Big companies	Small companies
Significant resources	Few resources
Significant expertise, in many different fields	Often deep expertise, but usually in only a few fields
Expert practitioners who can become your mentors, and are usually willing to do so	Founders who can become your mentors, if they will take the time
Investment in processes and tools	Limited investment in tools, processes, and training
The opportunity to work on the biggest and most interesting projects!	Usually, only a few opportunities from which to choose, and generally, smaller projects
A lot of freedom, in the sense of a larger variety of opportunities: many different lines of business, many interesting projects within each line of business	A lot of freedom, in the sense of a chance to get involved in many aspects of a project (e.g. marketing, sales, production planning, etc.)
Rules, which sometimes can be perceived as bureaucracy and slower response times (but this is not always actually true)	More freedom, in the sense of fewer rules, fewer specialists (so you get to try more things)
Your experience: mentoring by experts, better training and institutional practices. You start working on a small piece on big projects	Your experience: being in at the "front end," high risk, usually more stress, broader set of experiences, but each is less deep than at a big company

Figure 13.7 The size of your company influences what you experience at work.

- Just as we described in Chapter 2, we create an engineering and management trade space, and traverse it to create a feasible and suitable solution. We measure the effectiveness of that solution using the metrics that resonate with the customer.
- Just as we described in Chapter 9, we use engineering techniques to reduce risk and exploit opportunities.
- Just as we described in Chapters 2, 10, and 11, we use *processes* to do the job right the first time, and to help the team do it at scale.
- All of this is quite creative; we are solving hard problems for which there is no answer already provided at the back of the book. We have to use our understanding of the customer to identify the pressure points of the design (Chapter 2).
- We are part of a large team, and must work in collaboration and harmony with our colleagues, and derive our satisfaction largely from the accomplishments of the team, rather than our own individual accomplishments (this chapter).
- Hard science and engineering drives our analysis, but does not create the candidates that we assess. Our *imagination* creates those candidates.
- Since so many of the factors that we must consider are in tension and competition with each other (Chapter 2), we strive not for a "correct answer," but instead for a *balance*. In school, we aimed for the "correct answer," but that seldom exists at work.
- Since the resulting system or product must be *both* effective and suitable (Chapter 3), we must exercise judgment, and employ a sense of fitness and artistry to make our design selections.

As you can see, I have already discussed many of these items, but here I take the opportunity to bring them all together in a single list.

As engineers, we have the opportunity to work in two principal types of roles: we can work as *engineers* (often called being an *individual technical performer*), or we can work as *managers* (meaning that we supervise other employees). This is not an either/or choice; in fact, if you are willing to move back-and-forth over your career, your company is likely to value that flexibility, and likely to reward you for that flexibility. I did so during my own career (Figure 13.8); my career started in the lower-left-hand corner of the figure and proceeded upwards.

I found this transitioning back-and-forth interesting, and my willingness to do so increased my value to the company, because I could fill more roles, and serve in whatever capacity was needed at the time. It also opened up opportunities for me to serve in roles that I would not even have asked for; I will say more about this when we discuss mentors below.

I believe that what facilitated my career path upwards to senior positions was the *business benefit*: I developed the ability to use my technical skills to bring business into the company. Refer back to Figure 13.4.

As we discussed in Chapter 5, the key business of many companies requires the performance and successful completion of engineering projects. Therefore, much of your career as an engineer and an engineering project manager will be spent working on these types of projects.

As we have already discussed, it is a *team* activity.

For me, this was a great career. It was interesting work. It was important for society; this combination of *interesting* and *important* made for a great life experience. Perhaps it will for you too.

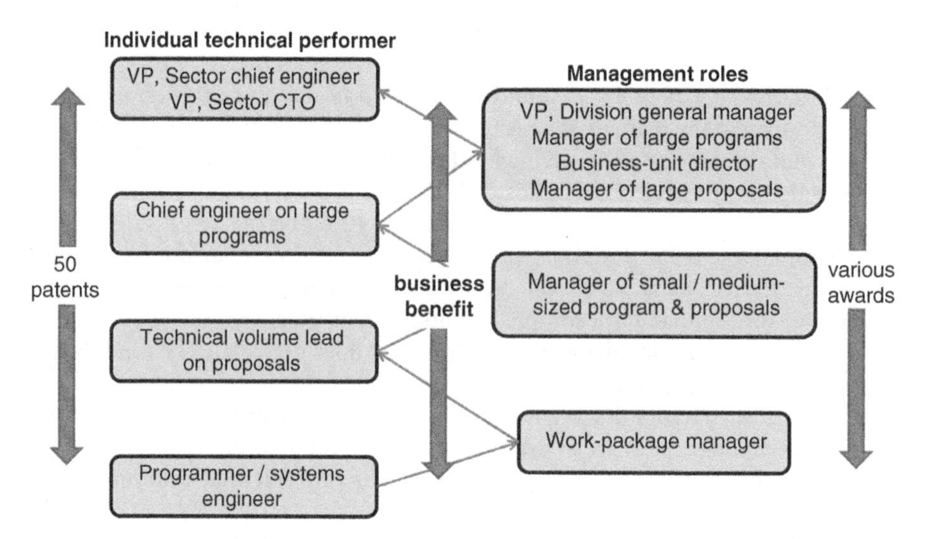

Figure 13.8 The author's career included stints *both* as an individual technical performer and as a manager.

With your training as an engineer, and your eventual experience as an engineering project manager, there are roles that you can play other than engineer and project manager too, such as:

- Testing
- Marketing/sales
- Business development
- Management
- Teaching.

I call these *adjacent roles*. We will say more about them in the section below about *coping with career change*.

13.9 Change on Your Project

Change usually makes people uncomfortable. But change is a fact of life on engineering projects; we just cannot know everything in advance.

Therefore, you must:

- Cope with it yourself, in your role as the project manager
- Help your team and customer cope with it
- Manage the changes, so that you can receive the benefits of the change, and do so without causing too much disruption.

Generally, the bigger the project, the more the amount of change there is that takes place, and the harder it is to cope.

These changes come from two different sources:

- From the customers and the other stakeholders (externally), and
- From within the project team (internally).

Usually, change from these two different sources must be handled somewhat differently, due to the fact that if the change is sourced externally to your team, there will be additional people involved in reviewing and (eventually) approving the change.

What might change? Almost anything: a requirement, an aspect of the design, the schedule, the budget, predictions of technical performance, personnel assignments, the strategy for training, the tools for software development, the legal and regulatory environment, and on and on.

How do we manage and control change? The steps are largely the same, whether the change is sourced internally or externally:

- Noticing that a change is called for
- Discussion and analysis of the impact of the proposed change (see below)
- Approval or non-approval of the change, and of its implementation method
- Configuration control of documents and products
- Validation that the implementation of the change is correct.

One of the hardest aspects of change is that change causes "ripples"; a seemingly innocuous change to one item (e.g. contract, statement of work, specification, schedule, work-breakdown structure, design, piece of software code, etc.) can impose unexpected (and often significant) changes on many other aspects of the project.

Therefore, before accepting and approving *any* change, you must exercise a process to look for any such derivative impact of the proposed change, and estimate its impact to technical performance, schedule, cost, risk, people, customer relations, and so forth.

Because candidate changes will occur almost every day, you likely need a standing process and capability to do this. This is often organized as a standing committee called the *change-control board*. The change-control board has access to a set of tools and people (usually in the systems engineering team within the project) that can perform the required assessments. The change-control board is usually chaired by the project's manager of systems engineering.

Since changes can have such unpredictable and important effects, the question of who can actually *authorize* change is important.

In your buying customer's organization, there is usually only one single person who can authorize changes to your contract, deliverables, terms, schedule, and price ... and it is usually *not* the customer's project manager! In fact, it is usually the customer's *manager of contracts* (often called a *contracts officer*). Do not act on requests for change until after they have been directed *in writing* by the authorized person!

Similarly, you need clear definition of who is empowered *within* your project team to authorize a change to controlled documents, designs, and products. Usually it is only the project manager, acting through the recommendations developed by the project change-control board.

13.10 Coping With Career Change

Your career in engineering is a *journey*, not a *destination*. By this I mean that it ought not to be about achieving some single position or status, but instead ought to be about doing important and interesting work over a period of time, while continuing to develop your skills.

At age 20, when I received my bachelor's degree, I did not know what I wanted to do for a living. No one really expects that you will, either. So, embrace the possibility of mid-career changes.

Over time, you will learn more about yourself, and you will learn more about the possibilities. Entire new fields and opportunities will likely emerge as well.

Here is my recipe for how to cope:

1. Acquire some *foundational* knowledge (e.g. knowledge that does not age quickly).
2. Understand and accept that learning is a lifelong activity.
3. Understand and accept that a lot of what you will use to succeed in a career is acquired post-hire, on the job. It increases your value, and it brightens each day.
4. Over time, strive to learn more about yourself: what you enjoy, what provides enduring stimulation, and so forth.
5. Keep investigating possibilities!

Let's discuss each of these.

13.10.1 Foundational Knowledge

My first full-time job included programming on a computer that had real core memory, discrete transistors, a programming language called FORTRAN, and used punched cards. Imagine if my personal knowledge base was locked to the relevancy of that particular technology base ... by now, I would be unemployable as an engineer.

But most of my college course of study was not about technology. Instead, I took courses in:

- Mathematics
- Logic
- Astronomy
- Physics
- History
- Music
- Russian literature.

These are examples of foundational knowledge. Almost all of these courses are still in the course catalog at my university, and what I learned is still helpful in guiding my thinking and my understanding. In contrast, the computer programming courses that I took have long been obsolete.

You may not be able to fit in as many courses in foundational knowledge as I did; colleges seem to have longer lists of required courses now than they did when I was a student. But you can do some, and also keep up those types of foundational studies as a lifelong learning activity.

You should do this because of the continuous intellectual stimulation and challenge that it will provide. It will also, of course, allow you to keep up with the changes in your field.

13.10.2 Lifelong Learning

Here's a sad case study. I once heard a story about a lady who worked for a phone company for 30 years and (by her own admission) did nothing during that time to improve herself. One day, her job was declared obsolete, and she was laid off. She was devastated; she had no idea what to do. She had no marketable talents. Worse, she had no sense that she could take charge of reclaiming her life.

Don't let this happen to you! Embrace continuous learning. Taking control of your own destiny is not only effective, but empowering and good for your well-being.

I advocate that you include subjects outside of your nominal field of expertise in your lifelong, continuous learning. I have found that such adjacent knowledge is often useful.

There are many mechanisms available to you: additional formal degrees, on-line education, specialty schools, just doing reading on your own, side avocations (e.g. charity, public service).

13.10.3 On-the-Job Learning

You need to understand and accept that a lot of what you will use to succeed in a career is acquired post-hire, *on the job*. Universities are great, but they are not specific enough to teach you everything you need for a specific job. Furthermore, whole fields have been invented since I graduated, and will probably be invented after you graduate too. You might want to take one of those up. On-the-job education is part of how you do that.

This too can take many forms: in-house training courses offered by your company, institutionally supported job rotations, broadening assignments, mentoring programs, asking the person in the next office, seeking out the established experts and asking them to teach you, and many others.

13.10.4 Know and Grow

How can you spot new things that you might want to try? Here's one way that I did it: for some reason, by the time I was five or six years into my career, my only career ambition was to be the chief engineer on a big program. One day, a few years later, I woke up and realized that I *was* the chief engineer on a big program. It was a great job and a great life experience. But I recognized that someday this project would end, and that I had *no idea* about what I would/could/ought to do next.

My solution was to find a *mentor.*[12] A *mentor is* someone who advises you, who helps you mature and grow.

Why do they do it? They want to return something to the institution which has done well for them. Perhaps they see something they like in you; often they see something that makes them think of themselves in you. Or perhaps they are just nice; this happens much more often than you might think.

What my mentor did for me was to see career path possibilities that I never would have considered on my own; *I never would even have aspired to do some of those roles.* He also had ideas about what experiences would help prepare me for those roles, and he saw that improvement in my people skills would help me. Finally, he saw that all of the above would not only help me, but would *increase my value to the company.*

Mentors are usually not your direct boss; they are often at least one level up from your boss in the organization. They actually need not even be in your company or in your institution.

Mentoring is not about technical skills, but instead is about everything else: people, relationships, trust, accepting criticism, courage in your convictions, "when to challenge, when to support." These are often called "soft" skills, but they are important, and not just for work. These are life skills.

If you are an engineer, you may also be mentored in the business aspects of your company: how accounting is accomplished, what financial measures the company uses to measure its progress, and so forth.

The combination of technical and social skills needed to get ahead is complex enough that most of us need mentors. We will talk about "getting ahead" later in this chapter.

In summary, do not expect that you need to get your career choice right the first time and forever. Don't worry about the potential for mid-career change; it might be wonderful.

---

12 I had several important mentors and am very grateful for what they did for me. Let me mention Jack Distaso by name in this context.

Think of change as an opportunity and/or a sign of more knowledge and maturation, rather than as a sign of some sort of "failure."

13.10.5 Summary: How to Cope With Career Change

- Acquire some foundational knowledge
- Understand and accept that learning is a lifelong activity
- Understand and accept that a lot of what you will use to succeed in a career is acquired post-hire, on the job
- Mentors!

13.10.6 Examples of Mid-career Changes I Have Known

Mid-career change can take many forms. Below, I list some examples, so that you can grasp the range of possibilities:

- *Technical to management*
  - The classic career change – a change to *facilitation,* rather than *doing.*
  - For some people, at some times, this is very satisfying.
- *Technical to an adjacency*
  - Such as marketing or teaching.
  - A great case study: my friend who realized that he wanted to talk about astronomy, rather than to *be* an astronomer. He became a teacher.
- *From one technical field to another*
  - My own (so far): music to mathematics to software to systems engineering to teaching.
  - My friend Carol: PhD in biology; couldn't find something she liked in biology, so she went to law school and became an intellectual property attorney for the biotech industry.
- *The long-term dream*
  - *Method 1.* Save up money and learn by being an employee in the field that you want to be in, and then start one's own business (in the same field). We used to call this "apprenticeship."
  - *Method 2.* Do something that pays the bills for a while, and then jump over to pursue your "dream career."
- *The new field*
  - Jump to something that became available only *after* you started working (e.g. a new field, like cyber security or biotech).
  - Via going back to school, on-the-job training, or some mixture.
- *From management to technical*
  - My friend Dave realized that he derived more satisfaction from *doing,* rather than *facilitating,* and therefore transitioned back to engineering from management.

- *Recover from a disappointment*
  - My friend Louise who became a dentist, only to discover that she hated it.
- *Recover from an obsolescence*
  - What the phone-company lady could have done.

13.11 Getting Ahead

13.11.1 Preparing Yourself for Leadership

Every day is a new set of challenges. They come on you fast; you must be prepared largely in advance. How to do that?

As I said in Chapter 8, I always recommend that managers of engineering projects, and systems engineers, do vast amounts of eclectic reading; you never know what bit of remembered reading will trigger a thought or an insight during a crisis on an engineering project. I have certainly benefited many times by making such connections from my reading in many technical fields.

Furthermore, since the *people* and *social* aspects of project management are very important – taking up more than half of your time as the project manager, if the project is of any size or complexity – this reading ought *not* to be confined just to technical readings. I myself read vast amounts of history, old novels, and the classics of Greece, Rome, and Persia. I firmly believe that Rumi, Austen, Bulgakov,[13] and Aristotle have taught me more about how to deal effectively with people, and to identify and manage interpersonal conflict, than all of the many modern management books that I have read.

These books are *old*, I can hear you say. But my view is that if a book is still available hundreds of years after it was written, it is *more likely to have content of value* than a randomly chosen new book. It has *stood the test of time*. I urge you to read old books!

> The person who wants to think will have to practice patience and master fear.
>
> *Alan Jacobs, How to Think*

As a project manager, you must *think*, not just react. You can do some of this *in advance*. As I described above, you can have practice conversations and play out scenarios in your head. But you must also consider the actual circumstances of a situation, which may not conform to the scenarios you expected. Hence the advice I received from one of my mentors: do not be afraid every now and then to say "Let me think about this for a day or two."[14]

13.11.2 Getting Ahead: Understanding Your Boss

The definition for what it means to get ahead is not the same for everybody; "getting ahead" means different things to different people. Actually, it is likely that "getting

13 His masterpiece is a novel called *The Master and Margarita*.

14 Thank you, Dr. Joe Mason.

ahead" will mean different things to *you* at different times in your life and career; what you value highly will change over time.[15]

But here are the most likely possibilities of what it means to get ahead:

- Moving up to higher levels in the organization
- Having subordinates
- Increased immediate financial compensation
- Increased long-term financial compensation
- Increased retirement compensation
- More generous medical insurance
- Increased prestige
- A more impressive job title
- Getting to spend more of your time doing what you want to do
- Not having to spend a lot of time doing things that you don't want to do
- Having work that you find more interesting
- Having work that you find more important.

How do you get there? It starts by *delivering value to your organization*. Some of the ways that you deliver value to the organization are obvious: technical competence, good work habits, honesty, ethical work behavior, collaboration with your colleagues, and so forth (and the items listed in Figure 13.4). If these are all obvious, how do you find things that you can do which will distinguish you from your colleagues, so as to mark you as someone who is ready for advancement within the organization? One way to accomplish this starts by *understanding your boss*. Let's role-play for a moment: a day in the life of a senior executive.

OK, you have made it. You are the vice-president and general manager of a large operating unit within a great company. Profit and loss responsibility. Multiple facilities. 3000 employees. Corner office, car, and so forth.

Here's your typical day:

- Your calendar is booked solid from 8:00 a.m. to 6:00 p.m., including lunch. 30- and 60-minute chunks.
- One person has 30 minutes on your calendar to tell you about a great success.
- 5 teams have 30-minute chunks to seek your advice/approval for an element of future strategy (proposals, expansions, etc.).
- 10 people /teams have 30-minute chunks on your calendar each to describe a problem, and seek your advice/approval for a corrective action.

Actually, on a typical day, you have to do all of this while on the road.

That is, most of each day is spent dealing with problems. There is too little time devoted to each topic to really learn the issue, reflect, discuss, and reach a sensible decision. In general, they will have waited so long to get to you that decisions are needed fairly quickly.

15 I created some of the material in this section while team-teaching a course with Dr. Donald Paul of USC, who was formerly the Chief Technology Officer of the Chevron Oil Company. I am grateful to Dr. Paul for the inspiration and mentoring that he provided during this experience.

So, what do you need? You need *people who have proven they can go and solve problems,* so that you can assign one of them to go and work on each of today's problems. And you will need another set of such problem-solvers tomorrow.

Usually, the executive's *only* degree of freedom is whom he/she selects to go and work out the problem.

One proven path to get ahead, therefore, is to *become one of those people: a proven problem-solver.*

How do you do that?

- Be willing to take on the hard assignments ... that is, the assignments that the boss needs accomplished; be the one who will go and fix the @#$%$^ problem!
- Be reliable; always do what you say you will do. If circumstances change what you can feasibly do, always say so promptly.
- Be success-oriented, while always being ethical.
- Focus on success for the institution and for the team, and trust that that will eventually translate into success for you.

A word about these fix-it assignments: many people think that these assignments are "risky" and stressful, and therefore shun them. What I learned was different: these programs are already "at the bottom"; they can only go "up."[16] It is possible to arrange things so that *you can get the credit* if you pull it off, but *not get the blame if it fails.* After all, the project was already in trouble *before* you got engaged. Yes, it is stressful. Be engaged and focused, but keep a sense of proportion.

What am I saying is this: Understanding how to get ahead starts by understanding what is *valued by your institution,* and your boss is a proxy for that institution. Take the time to learn what the boss and the organization need and want.

Understanding the boss also involves understanding the institution. Big and small companies, for example, are likely to be different in some important ways, and therefore what constitutes effective behavior and success will differ (Figure 13.9).

<table>
<tr><th>Big companies</th><th>Small companies</th></tr>
<tr><td>Operate effectively, but within the parameters of the enterprise</td><td>Be flexible and adaptable</td></tr>
<tr><td>Preserve the brand and reputation of the enterprise</td><td>Create the brand and reputation of the enterprise</td></tr>
<tr><td>Deliver on the mission of the enterprise</td><td>Create the mission of the enterprise</td></tr>
<tr><th colspan="2">At both big and small companies</th></tr>
<tr><td colspan="2">Deliver on your personal commitments</td></tr>
<tr><td colspan="2">Be innovative</td></tr>
<tr><td colspan="2">Act with integrity</td></tr>
</table>

Figure 13.9 What constitutes effective behavior and success will vary with the context.

16 Another big "thank you" to Dr. Joe Mason.

13.11.3 Enablers

I already told you about the key *enablers* who will help you get ahead back in Figure 13.4. I repeat these here:

- Technical excellence
- Deep understanding of a customer domain
- Leadership (inspiring, motivating, consensus-forming, challenging, team-building, accepting responsibilities, meeting commitments, etc.)
- Ability to work in a team – seeing team success as the path to individual success; multiply your effectiveness
- Ability to translate technical skills into business success.

These are the principal attributes that will get you noticed, appreciated, and, over time, rewarded and promoted. Remember what I said about the boss: your boss needs a lot of *effective, reliable,* and *skilled* people to go and solve all of those problems. If you show the potential to be one of those people, your boss will be motivated to give you a chance.

The above is not a comprehensive list, of course. Here are a few other good attributes to demonstrate:

- Constantly looking for opportunities to learn, grow, and share
- Seeking out hard assignments
- Becoming an effective communicator – especially in listening and writing
- Modeling a positive attitude.

13.11.4 Leadership vs. Management

Management – the act of supervising other employees – may not be to everyone's taste. You can get ahead in your career without taking on management roles. But *everyone* can be a leader, and everyone can derive satisfaction from providing some form of *leadership.* By *leadership,* I mean inspiring, providing motivation, driving consensus, and so forth. You can provide *leadership* every day, or whenever you are so motivated; your position in the hierarchy is not relevant.

Participation in management is optional – many people have success and satisfaction without it. And as I said before, you can go in and out of management over the course of your career – it is not a one-way journey. If you are willing to do management roles at times, it increases your value to the company.

13.11.5 Disablers and Pitfalls: How to *Fail* at Getting Ahead

Here is a list of attributes that will slow down or prevent you from getting ahead. The list is self-explanatory:

- Not being able to do the work
- Poor work ethic
- Excessive focus on yourself
- Rudeness
- Temper

- Not being able to derive satisfaction from the success of the team (e.g. only viewing success in terms of yourself)
- Failing to meet your commitments
- Any sort of ethical lapse, including sexual affairs with other employees
- Failing to ask for help when you need it
- Failing to bring forward problems that you see, hiding bad news
- Continuous complaining
- Lack of engagement with other members of the team, and with the customer's mission
- Not trying continually to improve yourself
- Conflicts of interest – for example, taking on consulting work outside of the company without obtaining the company's permission in writing first. This would include even activities that you might believe are acceptable, such as acting as an expert witness.

Employers like employees to show some reasonable level of ambition, but not a consuming level of ambition. That is likely to prevent you from being an effective team player.

Employers also like employees to have a reasonable work–life balance – we are all in this for the long run. The company wants you to take vacation. The company wants you to have a home life.

When I first started my career, employers were sometimes wary of people who had held a lot of different jobs. I believe that this is an evolving area; some employers are now more tolerant of people who change jobs every two or three years than they were when I started my career.

13.11.6 Summary: Getting Ahead

Your working career will consume a lot of your life. Therefore, *satisfaction* is at least as important as *success*. Just as in systems engineering and project management, you must strive to achieve *balance* between satisfaction and success.

To me, such balance derives from having work that is *important* and *interesting*, in combination with never acting in a fashion that can cause you to have ethical regrets. Then you can evolve other factors (such as success and compensation) to suit your character.

The definition of "getting ahead" is not static. Your definition will be different from that of your colleagues, and your own definition will evolve over your lifetime.

Understanding how to get ahead starts by understanding what is valued by your institution; and your boss is a proxy for that institution. The key enablers and disablers are all discoverable.

"Getting ahead" varies with context. For example, it will be different for large companies than for small companies.

Strive to display *leadership*, which is separate from being a part of management. Not everyone needs to be a manager in order to be a success.

You must find your own path. Mentors can help you understand the possibilities. Be prepared to learn, discover, and adapt along the way.

13.12 Two Special Topics

13.12.1 Special Topic 1: Projects Whose Work is Geographically Distributed Across More Than One Work Site

Much of this chapter is about how to get along with the other people with whom you work. I now address two special aspects of such "getting along": geographic distribution of work and collaboration across countries and cultures.

It is common practice today for engineering project development to be split across multiple geographically distributed work locations, often in multiple times zones and/or countries.

This is driven by many factors: wanting to make use of particular company facilities, needing to make use of other companies that possess specialized skills, a requirement from the customer that some of the work be performed at specific customer locations, and so forth.

What you need to know as a project manager is that the geographic distribution of work creates *additional costs* and *additional risks.*

Some of these additional costs result because people must travel, and at times, certain facilities need to be replicated at multiple sites. More subtly, and in my view the main issue resulting from geographically-distributed work teams is the fact that having the team geographically separated *always decreases the efficiency of the team*; coordination is harder and less effective; so are communications. Creating options, analyzing them, and reaching consensus are all harder too. These effects usually cause the schedule to require significantly more time than if all of the work was being performed in a single work location. This, of course, also increases the cost of that work.

Because of the decreased efficiency of the team, the risks that might arise from poor coordination and communication are increased.

You must plan your geographic distribution so that it will be feasible; you do this by determining which tasks require only relatively weak coordination and which require extensive coordination. Then, locate tasks that require extensive coordination with each other at the same site.

You must account for the *additional schedule, cost, and risk* in your proposal and project baseline. Your company probably has performed geographically distributed work in the past; be sure that the past projects you use for calibrating your schedule and cost predictions are those with similar geographic distribution.

People will tell you that you do not need to worry about this, that modern information technology infrastructure and tools (email, speaker phones, video conferencing, shared file repositories, and so forth) are so good that the geographic distribution does not matter. *Do not believe this!* There is a body of research on this subject,[17] and all of the careful, reputable studies show that *there is a significant degradation of team efficiency* with geographic distribution of the participants. My own research indicates that, whereas modern information technology infrastructure and tools are almost adequate to perform geographically distributed *review* of project and engineering artifacts, they are grossly inadequate to support the geographically distributed *creation* of these artifacts. Be warned!

17 For example, that by Dr. Ann Majchrzak of USC.

You will need to have significant travel budget to allow team members to meet face-to-face, and at times be co-located for extended periods. But even this is less than perfect; research has shown that many people do not like to be away from home for long periods of time on business travel, and that what you therefore get are a set of people who are *willing* to travel, but who may or may not be the people who the ones who are the *best qualified* to do the work.

Geographically distributed projects, therefore, *always* incur inefficiencies. Your job as project manager is to: (i) be aware; (ii) partition the work in a feasible fashion; (iii) get some appropriate compensating means and resources into the proposal and the baseline; (iv) work hard to get the right people actually to travel; and (v) constantly monitor this as a gigantic risk to your project. It needs to be on your risk register.

13.12.2 Special Topic 2: Projects That Include Teams Located in Multiple Countries

Many projects are not only geographically distributed, but some of those geographic sites are located in multiple countries. Obviously, all of the issues that we just talked about pertain to these projects, but there are some additional factors that apply to these projects which are distributed across multiple countries. These additional factors pertain to *law* and *culture*.

The factors of law are items such as these:

- What law and regulations are applicable under what circumstances, and in what locations?
- There are almost always restrictions about moving information, equipment, and people to and from certain countries.
- How are disputes that might involve parties in multiple countries to be resolved?
- What expenses are to be paid in what currencies? There are risks involved in the fluctuation of currency exchange rates too.
- Are there limits or procedures that must be followed to move funds from one country to another?
- Are there risks created by the particular employment law and regulations in particular countries? For example, in some countries, your company may have an obligation to pay employees for quite a while after a project completes. Such costs must be accounted for in your proposal.

The factors of culture are items such as these:

- Work ethics and work habits vary significantly from culture to culture. For example, in some countries, everyone goes home exactly at 5:00 p.m.; in others, if there is a vital task that needs doing, they will stay and get the task done. Do you have data which allows you to calibrate for this in your proposal? Have you trained your team to know these differences? Do you have a strategy for how you will operate in the presence of these differences?
- Skill levels vary significantly from culture to culture; people with nominally the same job title in different countries will know very different things, and have far different levels of skills. Do you have data which allows you to calibrate for this in your proposal? The same thing applies to education; a bachelor's degree in computer science may mean very different things in different countries.

- We already alluded to *power distance*. This varies hugely from culture to culture. Other factors of culture will also vary, and therefore the methods that are effective in dealing with your team members will vary from culture to culture. Do you and your management team have the requisite understanding of each culture that will participate in your team?
- Cultures vary significantly in their willingness to bring forward bad news; in some cultures, this is just not done. Do you know the culture of every country participating in your team? Do you have mechanisms in place to compensate for these cultural variations?

Your responsibility as project manager is to get these issues out on the table and considered; you may not be the expert on all of them, so you will need assistance. You must allow time and budget for training, team-building sessions, and many other things. Having your project split across multiple countries will *always* significantly increase the schedule, the cost, and the risk for your project; you must account for that in your proposal and in your management reserve.

13.13 Summary: Social Aspects of Engineering Project Management

- The project manager is the leader of a set of people, and the interface to an additional set of people. You interact with all of those people, and facilitate their interaction with each other. People are at the center of the project manager's universe. Therefore, there are important *social* aspects to the role of a project manager.
- You must *persuade* people to want to follow you; I provided a set of specific actions and attributes that will help you.
- Since people are involved, there will be conflict. You must learn to handle and manage that conflict, and to channel it to productive discussion. *Discuss premises and lines of reasoning, rather than discussing candidate answers.*
- Management and customers are people too, and you must deal with them. Build credibility in advance of a problem.
- There are useful and important "mechanics of management." "Take decisions only in writing" is number 1. Work with, and take advice from, *only* those people who either have *skin in the game*, or are bound to you by genuine affection.
- Your working career will consume a lot of your life. Therefore, *satisfaction* is at least as important as *success*. Just as in systems engineering and project management, you must strive to achieve *balance* between satisfaction and success.
- To me, such balance derives from having work that is *important* and *interesting*, in combination with never acting in a fashion that can cause you to have ethical regrets. Then you can evolve other factors (such as success and compensation) so as to suit your character.
- The definition of "getting ahead" is not static. Your definition will be different from that of your colleagues, and your own definition will evolve over your lifetime.
- Understanding how to get ahead starts by understanding what is valued by your institution; and your boss is a proxy for that institution. The key enablers and disablers are all discoverable.

13.14 Next

Many management books tell you that you need to focus your management efforts on schedule, cost, and technical capability. In the real world, that is insufficient. Factors such as reliability, safety, low defect rates, and environmental friendliness increasingly play a role in product and company success. We can group all of these types of factors under the title *quality*.

13.15 This Week's Facilitated Lab Session

Today, you will continue working in your teams.

Today's topic is team exercises about the social aspects of the role.

Assignment:

- Refer back to the project that your team selected previously.
- As a team:
  - *Boss exercises*. List and describe your boss's coordinate system of value. Create measures that reflect his/her value system.
  - *Staffing profile*. Create a staffing profile for your project. Discuss where the profile is easy to achieve, and where it is difficult to achieve. Discuss what you might have to do to the activity network and other project artifacts to make the staffing profile feasible (e.g. will you have to slow down the beginning of the project and extend the schedule a bit, to reflect the reality of how fast you can bring people onto your project?)
  - *People exercises*. Assign everyone on your team to a project role (e.g. project manager, software development manager, test manager, human relations manager, and so forth). Using the materials on charts 7 and following from lecture 12, discuss how you will build effective interpersonal interactions within your team, and with your customer(s), boss, and other stakeholders.
- The above will be turned in as a section in your team project report.

14

Achieving Quality

Many engineering textbooks teach that what you need to focus your project management efforts upon are schedule, cost, and technical capability. In the real world, that is insufficient. Factors such as reliability, safety, low latent-defect rates, and "environmental friendliness" increasingly play a role in product and company success. In this chapter, I teach you the basics of this aspect of your role as an engineering project manager, which we group together under the title of *quality*.

14.1 Defining the Term *Quality*

In this chapter, we will discuss *quality*. By the term *quality* in this context, I mean:

- Products and services that are effective and suitable (Chapter 3).
- Products and services that meet the specifications and the contractual terms.
- Products and services that are easy to use, *or* effective for trained users (remember the difference? Chapter 4), whichever is appropriate for the product or service that your project is creating.
- Products and services that work the first time and every time – consistency.
- The number of latent defects is at or below a defined level.
- Products and services that are reliable and long-lasting.
- Products and services that are safe to use, safe to manufacture, and environmentally responsible.
- Products and services that look and feel good (the user experience, Chapter 4).

There could be other intangible factors that people consider attributes of quality as well, and combinations of factors, such as those embodied in familiar terms such as *good value* and *value for money*.

14.2 One Motivation for Quality: A Good Reputation

I like to approach the motivation for achieving quality from a personal point of view: my *reputation*. I want to have a good reputation.

Good companies (and, in fact, good organizations of all types) similarly value *their* reputations. To some, reputation may seem an old-fashioned sort of word, and so the

Engineering Project Management, First Edition. Neil G. Siegel.

Companion website: www.wiley.com/go/siegel/engineering_project_management

term *brand* is sometimes used instead, especially in consumer-oriented businesses. But it is the same concept. For example, I would suggest that The Walt Disney Company strives to be *family-oriented*, and to ensure that their rides are *safe*: these are the characteristics that they want associated with their company.

Sensible people desire to have a good *personal* reputation too. Just as The Walt Disney Company gets to select what characteristics they want to be known for, *you* get to select the characteristics for which you want to be known. Perhaps you would like to be known as smart, hard-working, and team-oriented. Or perhaps there are other attributes and characteristics that you prefer.

Interestingly, I do not believe that one can work *directly* on achieving a good reputation. Instead, one *accomplishes* things, and does so in an ethical and appropriate fashion (or otherwise) ... and *those accomplishments and methods* determine your *reputation*, your personal *brand*. That is, reputation is *derivative*.

My suggestion is that both you and your organization ought to focus on *achieving high quality*. It feels good, and it can contribute to a good reputation.

14.2.1 Quality Control and Audits

The standard corporate view of *quality control* consists of items such as the following: checklists, work procedures/processes, training employees how to perform specific tasks, and auditing/compliance checking. In my view, these are *necessary* but not *sufficient*; we also need good design and good designers; reduction of *unjustified variation* in design (I will define and explain this term shortly, in the section below called "6-sigma"), manufacturing, and all other activities; products and services that *prevent* errors, or at least prevent errors from leading to significant problems; and a project mentality of striving for excellence.

Checklists, work procedures/processes, and training employees how to perform specific tasks all clearly contribute to getting the work done properly.

The next item on the standard list was *audits/compliance checking*. Why perform audits and check for compliance? We do so because this helps us to identify problems *earlier than we might otherwise notice them*, and (as discussed earlier) it generally takes less time, and costs less money, to fix a problem found earlier in a project rather than later. Audits and compliance checking therefore can improve project performance; they can identify mistakes, remedy them, and minimize their future recurrence (by pointing out which work procedures/processes need improvement, and/or additional training for the team).

Like so many other project activities, we do audits throughout the life of the project. Note that, in addition to any audits *you* might want to have undertaken, your contract and your company policies may have their own list of audits that *must* be undertaken. For example, in Chapter 12, we discussed how you will create a baseline schedule and cost prediction for your project, and how a team will come in and see that you have followed the designated procedures for doing so. This review may or may not be called an "audit," but that is what it is.

14.3 Quality Initiatives

How do we go about addressing the *other* items on my list of steps to achieve quality: good design and good designers; reduction of unjustified variation in design, manufacturing, and all other activities; products and services that *prevent* errors; and a project mentality of striving for excellence? Most organizations try to accomplish these items through *quality initiatives*.

These quality initiatives are based on the reasonable premise that quality does not happen by accident. The organizations therefore initiate activities intended to achieve one or more of these items. I will describe three specific quality initiatives that are used widely within the engineering industry: 6-sigma, ISO-9000, and the capability maturity model.

You will also hear terms such as *lean* and *agile*. These too are part of the quest to find the right processes and achieve quality on a particular project. We discussed *agile* in Chapter 13.

14.3.1 6-Sigma

6-Sigma is a methodology to improve quality. It strives to do this by finding sources of *unjustified variation* and reducing them.

What is unjustified variation? We know (from Chapter 8) that there is always some *variation* in any process or activity, due to unavoidable differences and random chance. For example, every batch of aluminum will vary slightly from all other batches; we can call these *unavoidable* sources of variation *justified variation*. The way we achieve improved quality (and achieve our desired low defect rates) is to reduce possibilities for variations from *unnecessary* sources: human mistakes, machine failures, incorrect instructions, and so forth. If we can truly decrease and control the rate of these *unjustified variations*, then we have a chance to build our parts and our systems with the high quality (and the low latent defect rates) to which we aspire, limiting the variability in our products to the range of *justified* variation.

The 6-sigma methodology was originally focused on *manufacturing*, but is now applied to other work processes, including design, business, billing, customer interactions, and many other situations. Anything that can be decomposed into a sequence of defined steps can be improved through the 6-sigma methodology.

The term "6-sigma" comes from statistics. The Greek letter "sigma" is used in statistics as shorthand for the expression "one standard deviation from the mean of a distribution."[1] So "6-sigma" simply signifies performance six standard deviations from the mean. If a measurement is six standard deviations from the mean, it could be *either* six standard deviations *better* than the mean, *or* six standard deviations *worse* than the mean. For the purpose of the 6-sigma process improvement methodology, of course, we are interested in obtaining results that are on the *better* side of the distribution.

1 In Chapter 8, you will recall that we discussed a number of common statistical mistakes. One additional common statistical mistake is applying concepts like "mean" and "standard deviation" to sets of data whose underlying distribution fails to meet the criteria for which these concepts are defined. You *ought not* to calculate the standard deviation of a data set that is *not* Gaussian (the "normal" bell curve), or is not some specific similar distribution for which the term "standard deviation" is *defined* and *valid*. If you make the calculation for a distribution for which the concept of standard deviation is not defined, you will of course derive a numeric answer, but this numeric answer does not signify anything meaningful about the data. Making such a calculation can only mislead you into making unjustified conclusions about the data.

We can state the 6-sigma level of quality in terms of a *defect rate*: if the underlying distribution is Gaussian, then being 6-sigma better than the mean translates into *3.4 defects per million instances.*[2]

Technical Note

I grew up as a mathematician. In the artificial (but interesting!) world of mathematics, lots of phenomena display what is called a *Gaussian distribution*. This is also called the *normal distribution*, or the *bell curve* (because of the shape of the distribution when plotted on a graph: a hump in the middle, tailing off at a very fast rate, and doing so *symmetrically* to the left and to the right). Unfortunately, in the real world of engineering projects, most of the items that we are interested in are *not* Gaussian – often, not even *approximately* Gaussian – in distribution. This existence of non-Gaussian distributions in engineering is easy to illustrate: we already discussed that many of the items in which we are interested are *not* symmetrical, that is, there are more examples of the "bad" side of the curve than of the "good" side of the curve. We discussed, for example, the fact that there are many more ways for an airplane flight to be late than for that flight to be early, and the magnitude of the worse-than-average outcomes can be far larger than the magnitude of the better-than-average outcomes. Another key way in which many of the items in which we are interested as managers of engineering projects show that they do not follow a Gaussian distribution is that the occurrence rate of adverse effects is far higher than that predicted by a Gaussian distribution. Therefore, do *not* assume that the items that we track in project management obey a Gaussian distribution! This means that terms like "standard deviation," and even "mean" and "average," do not necessarily have any meaning.

There is something called the *Central Limit Theorem* that causes people unreasonably to assume that randomly selected phenomena are approximately Gaussian; this theorem says that many distributions, if extrapolated far enough, converge to a Gaussian distribution. The Central Limit Theorem *is* valid in the artificial and beautiful world of abstract mathematics, but the conditions and assumptions that make the Central Limit Theorem valid in mathematics are *not* met by our messy, but real, world of engineering project management, and therefore, in our world, the Central Limit Theorem seldom actually applies.

See Appendix 14.A for examples of distributions of engineering design parameters from a real project.

The objective of the 6-sigma methodology is to reduce defect rates, through the mechanism of reducing unjustified variation.

Let me give you an example of how this works. I will make it a manufacturing example, because it is very easy to visualize what we are talking about. In a conventional automobile engine, the pistons go up and down; this up-and-down motion is converted to rotational motion by a device called a *crankshaft*. To allow the crankshaft to turn, there are *bearings*, which are devices that attempt to minimize the friction of the rotation, while also holding the crankshaft in its proper location. Each such bearing consists

2 Herein, I will ignore a nuance called the *1.5-sigma shift*. If you are interested in the details, you can find this explained in many books and on-line sources about 6-sigma and statistical process control.

of two circular parts, one inside the other. There is a small space – literally, a few thousandths of an inch – between the two circular parts, so that one can rotate inside the other. This small gap is filled with oil.

Clearly, very close tolerances are required in making these bearings: if the inside piece is just a little bit too large, it will not rotate freely inside the outer piece and this will destroy the bearings through friction. Similarly, if the outer piece is just a little bit too small, you will have the same problem. Also, if the inside piece is too small, or the outside piece is too large, the bearing will have too much free play; this is almost as bad as having too little – the additional space allows too much vibration, which can quickly destroy the bearings.

There is a *range of acceptable deviation* from the exact size desired for each part; but *deviations outside of that range* will cause the parts to fail very quickly. It is clearly desirable to be able to manufacture these parts so that the occurrence rate of defects – which are those deviations that are *outside* the acceptable range of deviation – is very low.

The exact defect rate that you require will depend on your system and its situation; despite the name "6-sigma" for the process improvement methodology, the defect rate that you decide to aspire toward can be anything, and is not necessarily the 3.4 defects per million implied by the title "6-sigma."

You will therefore design a *manufacturing process* (the selection of your materials, the selection of your tools and equipment, the training of your people in a set of steps, and so forth), that will allow you to achieve your desired *defect rate*. You achieve this desired defect rate by selecting processes, materials, and tools that will reduce the errors which lead to *unjustified variation* that could result from the manufacturing process. If we can truly decrease and control the rate of these *unjustified variations*, then we can usually build our parts with the low latent defect rate to which we aspire.

Such *unjustified variation* is a very significant factor in society. Here is an example that caught my attention: an acquaintance of mine is a physician who used to run a large health-care and hospital company; he once told me that he believed *unjustified variation* (he used that exact term) in health-care was the source of about *40% of all expenditures on health-care*, and that such expenditure represented *waste* – it does not deliver value. Health-care is already the single largest expenditure in the United States (something like 18% of our *gross national product*, the total value of all goods and services produced by the US economy). Here are the actual numbers. In 2016, the gross national product of the United States was around $18.5 trillion dollars ($18 500 000 000 000[3]), and health-care was almost 18% of that amount. This means that in the United States in 2016, we spent around $3.3 trillion on health care[4] alone. If my friend is correct, then we waste more than *$1.3 trillion* each year in the United States on *unjustified variation in health-care alone.*[5]

3 World Bank: https://data.worldbank.org/indicator/NY.GDP.MKTP.CD?end=2016&start=1960.

4 US Center for Medicare Services: https://www.cms.gov/Research-Statistics-Data-and-Systems/Statistics-Trends-and-Reports/NationalHealthExpendData/NationalHealthAccountsHistorical.html.

5 Just to help you grasp this number: according to the Congressional Budget Office, the *entire* US defense budget in 2016 was around $600 billion dollars (https://www.cbo.gov/publication/52408). Therefore, if my friend's assessment is correct, the amount *wasted* in health-care alone in the United States is twice the size of the *entire* US defense budget. Or consider this: my car cost about $30 000. The amount we waste in health-care costs each year in the United States would buy more than 43 000 000 such cars. This is more than twice as many cars as were actually sold in the United States in 2016 (according to https://www.statista.com/statistics/199983/us-vehicle-sales-since-1951, 17 500 000 cars were sold in the United States in 2016).

When talking about health-care, wasted money is only a small part of the picture, of course. This unjustified variation in health-care also results in less than optimal health-care outcomes, and in fact hundreds of thousands of unnecessarily early deaths. Decreasing unjustified variation is *important*.

Notice that to achieve the manufacturing benefits we discussed in our crankshaft bearing example, the 6-sigma process improvement methodology must be applied first during the *design* process; you cannot wait until you get to manufacturing to start thinking about how to improve quality. *Design* is at the heart of engineering project management, so the 6-sigma process improvement methodology is highly relevant to us, as managers of engineering projects. I like to think of 6-sigma as a process that helps me *design for manufacturability* and *design for reliability*. Since I am talking about quality in this chapter, let me also phrase it this way: 6-sigma is a process that helps me *design for quality*.

We can use the 6-sigma process improvement methodology to improve processes other than manufacturing; we can improve the processes that we use to design software, to control configurations, and even for routine administrative tasks, such as placing orders for parts. The 6-sigma methodology has proven to be useful in a wide variety of engineering project and administrative situations.

If you end up working at a large company, most likely they have already adopted 6-sigma or something similar. This means that they will have training materials and training courses, and you will have colleagues who understand process improvement. The company will also allow, and indeed expect, you to spend time looking for potential process improvements.

It does not take a lot of time for an engineer to learn the basics of the 6-sigma methodology. There are two levels of certification:

- Green belt – this training course is usually 80 hours
- Black belt – this training course is usually 160 additional hours.

These courses involve only modest training in statistics. You will find that the courses teach you to use a computer spreadsheet, or purpose-built applications (e.g. there is a piece of software called Minitab which is used for this purpose) to do a lot of the detailed statistical calculations; you only need to learn some basic statistics in order to know how to set up the calculations correctly.

14.3.1.1 Defect Rates

Defect rates are measured on a *part-by-part basis*. You need not aspire to the same defect rate for each part; some parts are safety-critical, others are not. Some parts factor more significantly than others into the overall system reliability rate than do others.

What then is the actual defect rate for each part to which we ought to aspire? This is determined by two main factors. The first is the reliability you need to achieve in your completed system or product. This is usually specified in your contract, expressed in the form of a *mean time between failure* or some similar measure. Systems that I built often had requirements to achieve a mean time between failure of 1000 to 5000 hours, although the actual measured mean time between failure of my delivered products was usually much better: in the range of 10 000 to 40 000 hours. I took pride in designing systems that would be reliable.

The other main factor in determining a defect rate for each part is the number of parts in your system, and how they will interconnect. Because system reliability is proportional to the *multiplicative product* of the reliability rate of each individual part, today's use of large amounts of software in our systems and products has therefore dramatically changed our quest for reliability. As we multiply the reliability rate of each part, because the reliability of any part is less than 1, more parts (and each line of software acts something like a part[6]) implies that today's software-intensive systems tend to be *far less reliable* than the systems they replaced[7]! This is not good, and at times is a very significant problem. Today's systems may be more capable, but are often of low quality, as manifest by this low reliability.

Since modern systems have a fantastic number of parts (remember the example we cited before: a modern car might have 200 000 000 lines of software code, in addition to all of its mechanical and electrical parts), the reliability of each individual part must be very high indeed in order to achieve a reasonable rate of reliability at the overall system level. For software – mostly, I believe, due to the *unplanned dynamic behavior* we discussed in Chapter 2 – achieving high reliability has been a challenge that in most systems and products remains unmet. *Systems that contain lots of software are seldom very reliable.*

So, improvement, via methods such as 6-sigma, remains very important.

If you have 10 000 completely independent mission-critical parts in a device, and each part achieves a 6-sigma level of defect rate, on average only one complete unit out of 30 would be defect-free. This is obviously not likely to be acceptable. We therefore are likely to need to achieve defect rates for our parts that are *much higher than 6-sigma*. I have seen real examples of 15-sigma parts. Software, however, is usually nowhere nearly as good as a 6-sigma rate; you will recall that earlier I reported an average latent defect rate for software of 1 per 1000 lines of code.[8]

This has also driven many manufacturers to buy *integrated assemblies*, rather than individual parts (or to buy large *pieces* of software, rather than code all of the software themselves); they impose a quality-level requirement on the provider of the integrated assembly. This is because, when they buy such integrated subassemblies, each large subassembly is – from the point of view of the final assembly process – *one single part*. Therefore, the manufacturer's final assembly process includes many fewer "parts," and their challenge in achieving a desired quality level is eased, because they have passed some significant portion of the problem to their specialist suppliers. Since those specialist suppliers presumably are real experts in their field, this process of buying

---

6 Note that software crashes, even those that can be fixed by rebooting a computer, usually count as a *system failure*.

7 My favorite example: the old Western Electric dial telephone (https://beatriceco.com/bti/porticus/bell/telephones-500.html) – and the analog phone systems to which they were connected – could go *decades* without a significant failure. The Western Electric phone instruments were very rugged and reliable, and the phone switch centers had very reliable and long-lasting back-up power systems, which could operate the switch center for a long period of time if the mains electric power went out. Today's cell phones, although very convenient, drop calls all the time, the mobile handsets seldom last more than a year or two before failing, and the cell-phone base stations have at most 10 hours of back-up battery power. So, not only are the mobile handsets less reliable than their Western Electric predecessors, the modern cellular/digital phone network to which they are connected is also far less reliable than its analog predecessors.

8 Since 6-sigma is 3.4 defects per million parts, a 6-sigma latent rate would be around 1 latent defect per 300 000 lines of software code, not 1 per 1000 lines.

subassemblies rather than parts can actually work; those expert suppliers can make their subassembly very reliable. This, more than any other single factor, is what I believe is motivating the modern tendency to subcontract out ("outsource") more work: a company cannot be expert at *everything*, and many companies have realized that they can build better products and achieve higher quality by using vendors who are true experts in each domain, whether materials or labor.

Despite the use of a statistical term in the title of the 6-sigma methodology, the statistics are *not* actually the core of the methodology; the core of the methodology is the *systematic analysis of how a task is currently performed, identifying through that analysis all of the places where unjustified variation can creep into the outcome, and then creating improved process steps that seek to lessen such unjustified variation.* Learning how to perform such assessments forms most of the time allocation in a 6-sigma training class. The statistics *are* used to measure whether you are making progress or not, and to assess whether you are reaching your goals. For example, to assess whether the system or product will likely meet its contractual reliability requirement.

14.3.1.2 Justified Variation

Notice, however, that we talk herein only about *unjustified* variation as being a problem; we are not asserting that *all* types of variation are a problem. Some variation is *essential*, even *beneficial*. For example, it is the random variation in the gene pattern passed on by parents to their offspring that enables evolution; this variation derives both from the randomness in the mixing of the genes from the two parents, and also from the random mutation of genes. In our engineered systems, we too can design so as actually to *benefit from variation*. Here are three examples:

- The technique called *spread spectrum* uses the random variation in the background radio-frequency environment to hide a transmitted radio signal, so as to lessen the chances that an unauthorized observer would notice the transmission, would be able to locate the position of the transmitter, or would be able to listen in on the transmission.
- The technique called *carrier-sense multiple access* uses a deliberately introduced random variation in transmission timing to allow a larger number of simultaneous transmissions to take place on a shared communication resource (such as a local area network) while lessening the instances of multiple transmitters accidently jamming each other's transmissions.
- Many *routing algorithms* use the variation in network connectivity that occurs over time in a network to create more reliable and more effective communication paths from one participant to another than any fixed (e.g. non-varying) routing algorithm would be able to achieve.

These are examples of *justified* variation, circumstances where the designer has made variation an asset, an actual *benefit*. We want and need such variation; this is why we distinguish between *all* variation and *unjustified* variation, and seek only to reduce *unjustified variation*. That is, unjustified variation is that which occurs by accident, which serves no useful purpose, and in fact often degrades the performance of our systems, products, and processes.

14.3.1.3 Defects in Assembly

Here is another consideration: errors arise not just from defective parts (whether hardware or software), but also from defective *assemblies*. That is:

- Parts inserted upside-down
- Fasteners that are over-tightened
- The wrong part placed in a particular location
- Wrong assembly sequence
- Missing parts (e.g. washers, spacers, etc.)
- And so forth ...

You must therefore *design the assembly processes* too. There can be a significant difference in the defect rate induced by different assembly methods; you can also design your parts so as to minimize the opportunity for assembly errors. You must design the assembly process to match the skills of your assembly personnel, and must account for assembly error in your system reliability calculations.

As you can see, improving our processes through the reduction of unjustified variation is a very important technique for the manager of an engineering project to understand. The 6-sigma methodology is one of the best-known methodologies for accomplishing this worthwhile goal.

14.3.2 ISO-9000

ISO-9000 (its implementing document is called ISO-9001, so you may see this method called either *ISO-9000* or *ISO-9001*) is one of the most common quality management initiative standards. ISO stands for *International Standards Organization.*

ISO-9000 says that an organization ought to have requirements for a quality management system, so as to be more effective at meeting customer needs. The standard does not specify what the objectives relating to "quality" or "meeting customer needs" should be, but instead requires organizations to define these objectives themselves, and *continually improve* their processes in order to reach them. In some sense, ISO-9000 is not a quality standard, but instead is a statement that an organization ought to have a quality standard.

As noted above, ISO-9000 calls for the idea of *continual improvement*; an organization may define a standard, train their people, and start implementing it, but this is not enough. They must also learn over time from experience, and use that learning to refine and improve their quality standards.

ISO-9000 also provides guidance for how such quality standards ought to be created: you start by identifying the *needs of your customers* and potential customers with regard to quality. From this insight, you can then define a set of quality objectives for your organization that will meet the needs of these customers, and determine who within your organization must do what steps in order to create and implement the quality standards.

An organization can be certified by the International Standards Organization as having a correct implementation of the ISO-9000 standard. Many customers respect this certification, and some require it.

14.3.3 Capability Maturity Model

If you think about it, you will realize that not all of an organization's work processes are likely to be equally rigorous or equally effective. Each organization will select or create a set of processes that they believe are well matched for their business environment;

Software Engineering Institute
Capability Maturity Model level

Level	Name
Level 1	Initial
Level 2	Managed
Level 3	Defined
Level 4	Quantitatively Managed
Level 5	Optimizing

Increasing organizational maturity

Figure 14.1 Capability maturity levels, as defined by the Software Engineering Institute.

since this business environment will vary from company to company, and even from organization to organization within a single company, some organizations will have strong work processes, and others will have weaker ones. More realistically, each organization will probably have some process areas where they are strong, and some where they are weak. How can we understand which is which?

This is what the *capability maturity model* is all about: *assessing the strength of organizational work processes.* It does this by assessing the processes used by an organization, and rating this organization's process strength (which they term "maturity," hence the title) on a scale of 1–5.

The standard defines assessment criteria for each of the five levels (Figure 14.1); one interesting aspect of these assessments is that if your organization meets all of the criteria for a given rating except one – one single failure – your organization receives a lower rating. To earn a rating at a given level requires that you meet *every single* criterion for that level. The result is that focus is placed on those aspects of your organization's processes that are the *weakest,* because these are the items that are preventing you from achieving a rating that in all other respects your organization merits.

Is this a perfect rating system? Far from it. But it provides a framework for assessment, and (since some customers require a rating at a particular level in order for a company to be eligible to receive a contract) it provides *motivation* for self-assessment and improvement.

14.4 Processes for Engineering and for Project Management

Engineering is a complicated enough activity that in order for it to be performed in a consistent and effective manner, written guidance needs to be provided. We call such written guidance an *engineering work process*, or sometimes just an *engineering process*. Most organizations that perform engineering as part of their core business have such written engineering processes that they use for guidance to their employees. Not only do they help each employee understand the expectations of the employer, but they provide a shared vocabulary and a shared approach that all members of the engineering team can use to make their collaboration more effective.

In actual fact, most organizations have work processes for many *non-engineering activities* too: how to pay a bill to a vendor or subcontractor, how to hire an employee, how to conduct a meeting, and so forth. These *processes* provide the same benefits to the non-engineering activities as the engineering processes do for engineering: they allow the employees to understand the organization's expectations for how work should be performed, provide a shared vocabulary and a shared approach that enables effective collaboration, and so forth.

A process, in order to be effective, must consist of more than just the written guidance itself. For example, the organization usually has a library of example artifacts and templates. If a particular process provides instructions about how to create a requirements specification (Chapter 2), then the organization also ought to provide an editable *template* – a sort of detailed outline, with instructions for the authors embedded in that outline – of a requirements specification. The organization also ought to provide a few examples of completed specification that it considers examples of good work.

If the process is sufficiently complicated, then the organization also ought to provide *training* to the employees about how to perform the process.

If – as is usually the case when a process has been established – the organization requires or recommends that the work be performed in accordance with the process, then the organization needs to establish a mechanism to check up on *compliance*, and to enforce correction of errors, omissions, or failure to follow the process. This might take the form of periodic *audits*, with lists of *findings* concerning deficiencies issued to employees and their supervisors.

Finally, no process is perfect. So, the organization needs to gather lessons learned from the use of each work process, and periodically to review each work process and make improvements to it. At times, this might even entail *retiring* a process, and replacing it in its entirety.

A good process embodies all of the following elements:

- It describes a series of actions or operations that must be performed.
- It identifies which of those actions must be performed sequentially, and which can be performed in parallel; that is, it identifies the *dependencies* among the steps of the process.
- It identifies who is to perform each step; some steps might be performable by any employee, but others might be limited to employees with specified qualifications, certifications, management level, and so forth.
- It identifies what are the required *inputs* to the process.
- It identifies what the process will produce as *outputs*.
- It characterizes how much effort it will likely take to complete the process; this is useful, as it makes the employees aware of the anticipated level of detail involved in performing the process in the manner expected by the organization.
- It identifies what tools they need to, or ought to, employ.
- It describes how decisions are made, and how agreements are reached.

14.5 Procurement and Subcontracting

On a typical project that I worked on at my former company, *half of our total project budget went to outside vendors*, in the form of:

- Procurement via *purchase orders* (buying *commercial, off-the-shelf* parts, devices, software, and services), and

- Purchase via *subcontracts* (buying *custom-built* parts, devices, software, and services).

With that much being spent *outside* of the company, you probably cannot achieve high quality without getting your *outside vendors also to focus on quality*. Such guidance must be in writing, most likely as *mandatory terms* in their *subcontract* or *purchase order*.

As we discussed before, when something is manufactured, or when a person performs a task, there is *variation*. For example, the diameter of the inner and outer crankshaft bearings that we considered earlier in this chapter will vary from item to item.

If we – and our subcontractors and vendors – can achieve *lower levels of variation* (which is achieved by reducing *unjustified variation*), this will usually result in *higher quality*.

However, it almost always costs *more money* to build parts with lower levels of variation. Therefore, it costs more money to *buy* parts with lower levels of variation. So, if you want to achieve a certain level of quality (e.g. 6-sigma), you *cannot just buy the cheapest part*. You must instead buy (at the very least) the *cheapest part that achieves your desired level of quality or variation*.

14.5.1 Vendor Partnerships

My experience indicates that, in actual practice, obtaining quality parts and services from outside companies seems to work best through *long-term partnerships*, rather than buying each batch from today's cheapest-and-compliant source.

Let's explore this concept of a partnership with a vendor for a moment. It is in fact possible these days to hold an on-line auction each and every day among a set of vendors for the parts that you will need tomorrow, and to buy those parts each day from whichever vendor offers the best price on that particular day. Experience shows that this does not actually work very well. Working in this fashion increases the likelihood that the vendor will misunderstand what parts you want to buy, and will not understand all of your requirements for those parts. It also increases the likelihood that something will go wrong in the shipment process, and cause you to run out of parts in the middle of a manufacturing run. It also makes it very hard, if you encounter a problem, to trace the exact root cause of the problem; you may not even be able to determine which vendor provided the screws that were not properly heat-treated.

A partnership works differently: you commit to a vendor that you will use them as the source, likely the *exclusive* source, of a set of parts for a period of time, as long as they continue to meet your needs and requirements. In the automobile industry, for example, the vehicle assembly companies (Toyota, General Motors, Ford, etc.) usually commit to a vendor for at least an entire year. That way, every 2017 Toyota Camry has wheel lug-nuts made by the same vendor, and if there is a problem, there is no ambiguity about who is responsible.

The bi-lateral commitment allows you, as the prime contractor, to:

- Tell the vendor what you need, and why.
- Help them achieve the desired level of quality, that is, the level of latent defects low enough that your completed product achieves your quality goals (mean time between failure, etc.).
- Help them establish processes that keep their manufacturing at that desired level of quality.

- Establish optimized delivery and resupply arrangements. Having large batches of parts on hand at your assembly plant lessens the likelihood that you will accidently run out of parts, but also costs extra money, both to buy those parts and to store them. Shipping parts in larger batches, however, probably costs less per part than shipping in smaller batches. There is both art and science involved in the question of how many parts to hold where, and how often to resupply them.
- Have appropriate non-disclosure agreements in place with your vendors, so that they can work side-by-side with you on the final assembly line to solve problems, and to create improvements in manufacturing procedures. They can also work with you to analyze problems reported after sales, so as to plan future improvements to those parts.

In summary, such a *long-term partnership* with your vendors causes you to give up some of your leverage on price, but in return allows you to improve quality, lessen defects visible to your customers, lessen the likelihood of supply shortages on your manufacturing line, and many other important benefits.

14.6 The Effects of Quality

Good processes and reduction of variation is not an academic or incidental issue; it can be at the very heart of business success or non-success.

Here's an example. In the last 50 years, the percentage of cars sold in the United States that are made by brands owned by General Motors has *decreased* radically: from nearly 50% to just about 17%.[9] During that same time, the combined sales of car brands owned by Toyota and Honda has *increased* radically: from about nothing to about 23%.

I believe that the *biggest* driver of this change has been the *reliability* of Toyota's and Honda's cars, and the relatively low reliability of General Motor's cars. That is, this dramatic change in the market share of car sales in the United States – a gigantic decrease on the part of General Motors brands, and a gigantic increase on the part of Toyota and Honda brands – is due almost entirely to *quality.*

Toyota and Honda each *explicitly* attribute the reliability of their cars to the effectiveness of their *6-sigma processes.*

Note that most of the Toyotas and Hondas sold in the United States are now assembled in the United States, by US citizens, with a large US-made parts content. The *total US content* of their cars is *not very different* from that of a General Motors car. It is not, therefore, some purported difference between Japanese and American assembly-line labor or parts manufacturing that is driving Toyota's and Honda's higher reliability levels – it is instead the inherent reliability that comes from their 6-sigma design and manufacturing processes. Note also that the 6-sigma methodology was invented in the United States; the higher quality of Toyota and Honda cars is not the result of some quality initiative invented in Japan. It is instead the result of a deliberate decision on the part of these companies to make *quality a part of their reputation,* just as I recommended at the beginning of this chapter, and to stay focused on that quality year after

9 https://www.statista.com/statistics/239607/vehicle-sales-market-share-of-general-motors-in-the-united-states. The information about Toyota (and its subsidiary Lexus) and Honda (and its subsidiary Acura) also comes from www.statista.com.

year. General Motors has never done this; it seems to me that they have preferred instead to focus their reputation on styling, high horsepower, and factors other than quality. As the sales figures show, this has not served them well.

Quality can truly make a gigantic difference to the bottom line.

14.7 The Bill of Materials

Most engineering projects buy goods from outside vendors. These might be simple items, such as screws and nuts; they might be commercially available items, such as power supplies and paint; or they might be custom-created items, things that are invented or adapted just for your project. The cost for each item can literally range from a few cents (a screw) to tens of millions of dollars (a radar, or a jet engine).

You cannot go on a shopping spree without a shopping list! The shopping list for an engineering project is called the *bill of materials*. Not surprisingly, a bill of materials for an engineering project is more complicated than writing "I need 2 quarts of milk" onto a piece of paper.

First of all, you need to specify what you need. Even for something as simple as a screw, this can be complicated: diameter of shaft, length of shaft, pitch of threads. Do the threads go all the way up to the head, or only part of the way? Are the threads right-handed or left-handed? Is that screw made of steel, brass, stainless steel, or some other material? Or do you not care? Is the size in English or metric units? Does the head accommodate a Phillips driver, a Yankee drive, a hex driver, or something else? Does the screw need to be painted or anodized?

All of that just for a screw! Obviously, if you are buying something more complicated, you have even more parameters to specify.

Sometimes, you will accept an item from any manufacturer who can meet the requirement. This is likely the case for that screw, or if you are buying some sort of commodity, such as blocks of aluminum.

In other cases, you will actually only want a product from one or more vendors whom you have reason to believe are best positioned to provide that part. Perhaps there are many companies that build a particular type of an encryption device, but only three of those companies have received some certification that your customers require or desire. In such a case, you would limit the potential sources to these three companies.

Every item on the list needs a description of both mandatory and optional desirable features, a quantity, a needed delivery date (which might be multiple dates by which you need a portion of the total quantity), and so forth.

14.8 Your Role in All of This

You can choose to make quality an aspect of *your* personal reputation. Your company can also choose to make quality an aspect of *their* reputation, just as Toyota and Honda have done.

Quality is not achieved by accident; it takes a deliberate, carefully planned and executed, long-term effort. 6-Sigma is an example of a methodology that you can use to improve quality. As the project manager, you have the authority to undertake such a quality improvement effort. It will, however, require time, money, and people, and therefore such a decision has to be coordinated and approved by your customer and/or your company. Sometimes, your customer will be willing to pay for this effort as part of the contract. More often, however, while your customer will applaud the intent, they will believe that improving quality is a core mission of your company, and therefore believe that the quality initiative should be paid for by the company, and not charged to the customer's contract.

14.9 Next

I have provided you with a set of tools for managing engineering projects. In the next (and final) chapter, I discuss the use of these tools in the real world, and in particular the special matter of ethics in engineering.

14.10 This Week's Facilitated Lab Session

There is no facilitated lab session this week.

14.A Appendix: What Distributions Actually Look Like in the Real World of Engineering Projects

Earlier in this chapter, I cautioned you that few phenomena encountered in the real world of engineering projects display a Gaussian distribution. This is in contrast to the ideal world of theoretical mathematics, where many phenomena are either Gaussian, or converge in the limit to Gaussian.

Here is a real example, from a project on which I actually worked many years ago.

Figure 14.2 depicts the port-to-port time of a key element of a system. "Port-to-port time" means the time that it takes a single instance of this element to execute, from data arriving in the input queue until the final outputs are created. This element of the system is a computer that was controlling the physical movement of a device; the system has a requirement that 90% of the time, a certain accuracy requirement will be met. We determined through analysis that if this particular element of the processing can complete in about 90 milliseconds, and every other element completes in their own allocated amount of time, we would meet this 90%-of-the-time accuracy requirement.

There are two curves depicted in the graph. Let us consider the one that has the *higher* values at the right-hand edge of the graph. The graph depicts the results of running this element about 9000 times, with a realistic range of inputs and conditions. The Y-axis on the graph represents the number of instances out of those 9000 that completed in the corresponding time indicated on the X-axis. For example, you can see that just over 1200 samples completed in around 95 milliseconds.

The graph looks vaguely Gaussian. But there are big, important exceptions:

- The graph is not symmetric! There are more instances to the right of the "hump" in the graph than to the left of the hump.

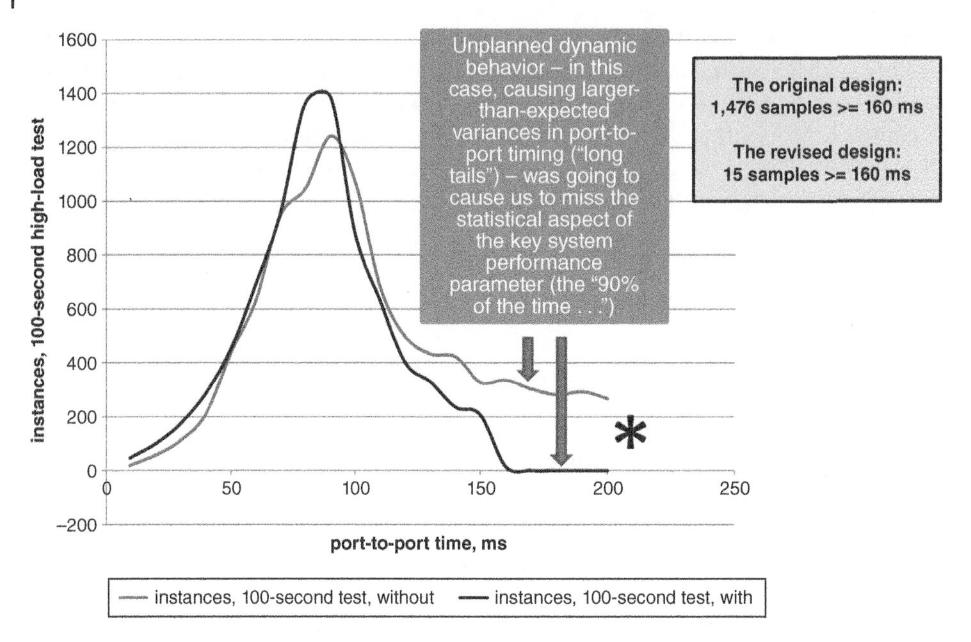

Figure 14.2 A real measurement of an engineering project technical performance measurement, showing that the distribution is not Gaussian.

- The graph has what statisticians call a "fat tail" on the right. That is, on the right side, you can see that the number of samples does not approach zero as quickly as it does on the left. You can see that there are, in fact, hundreds of samples that took between 160 and 200 milliseconds to complete; that is, just about twice as long as we wanted each instance to take. There is a little box in the upper-right-hand portion of the graph which states that in fact, there were 1476 samples that took 160 milliseconds or longer to complete, out of our 9000 or so samples.

This is interesting: the behavior appears to be approximately Gaussian. If you used Gaussian models to predict performance, you would have predicted that this version of the system should work great. In fact, this version of the system actually *failed*; having 1476 out of 9000 samples (more than 15%) taking around twice as long as desired turned out to result in our *not* being able to meet the 90%-of-the-time accuracy requirement. It turned out that *every* sample where we took more than about 160 milliseconds failed to meet the accuracy requirement, and since more than 15% of the samples took at least that long to process, the accuracy achieved was actually just over 80% (some of the faster samples also failed to meet the accuracy requirement, based on causes of failure other than the mere passage of time).

You could fairly characterize those 1476 samples that took so long to process as *unjustified variation*. In order to make the system work, we had to redesign this portion of the system so as to reduce significantly the number of times that it took more than 160 milliseconds to process an instance. You can think of this redesign effort as a 6-sigma process improvement effort to reduce the unjustified variation. You can see a plot of the performance of this revised design on the graph too; it is the curve that has the lower set of values at the right-hand edge of the curve. You can see in the little box that, using this

Figure 14.3 A bi-modal distribution.

improved design, only 15 samples (out of around 9000) took more than 160 milliseconds, a marked improvement compared to the 1476 samples that took more than 160 milliseconds in the original design. Having removed this unjustified variation, the system now met its 90%-of-the-time accuracy requirement.

The key lessons to be learned from this example are these:

- In the real world, phenomena seldom display a Gaussian distribution.
- Even phenomena that *appear* to display a Gaussian distribution are usually *not* in actual fact Gaussian, and the deviation of the real distribution from a Gaussian one – although seeming minor when viewed on a graph – can be *decisive* and *fatal* for your system and your project. A seemingly minor deviation from the Gaussian distribution caused a gigantic degradation in the performance of this particular system.
- The particular aspects of non-Gaussian distributions that will harm your systems are often *asymmetry* and *fat tails*.

I used this particular example because the curve looked so close to Gaussian, and I wanted to stress that in this case *looks were deceiving*.

In many real-world examples, the curve of behavior over a set of samples would not look Gaussian at all. For example, a very common mode of behavior for real systems is *bi-modal*: there are two values around which the samples clump (Figure 14.3).

The "clumps" need not be the same height, width, or shape, nor do they need to be disjoint; they could overlap so that the value does not return to zero in between the clumps.

Clearly, to talk about averages, means, medians, and standard deviations for such a bi-modal distribution has no meaning, and the use of these terms and concepts would completely mislead your analysis. For example, the *average* of all the sample depicted in Figure 14.3 would be at a point in between the two clumps, at a value where in fact there was not a single actual sample!

More generally, there can also be *multi-modal behaviors*, where the samples clump around N values, not just two values. In this case, the graph would have N clumps, each of which can be of a different height, width, and shape, some of which may overlap with the others, but some may not.

15

Applying Our Ideas in the Real World, Ethics in Engineering

In many aspects of engineering and engineering project management, there are factors in tension with each other. In fact, we have seen many examples of this throughout this book: making the airplane lighter achieves better fuel efficiency, but makes the plane more expensive; hiring more people for your project might get it finished sooner, but probably increases the total cost at completion; and so forth. Every engineering management process that we have discussed is subject to similar tension when you go out into the real world and try to apply it. In this chapter, I help you understand how to use these techniques within the limitations of actual people, companies, and customers. I also discuss the sort of knowledge that you will need to acquire on a continuing basis as you move through your engineering project management career.

I conclude with a discussion on the important subject of ethics in engineering.

15.1 Applying Our Ideas in the Real World

I found my time as the manager of large, complex engineering projects to be a great life experience. I hope that you will too. I found it to be *interesting* and *important*, a very nice combination of attributes. It was also stressful; I suspect that anything that is both interesting and important is going to be, at times, stressful.

It can also consume your life for long periods of time; accepting the position of manager of a large, complex engineering project is a big commitment. The typical time for which I held each such assignment was 30 months. By the time that 30 months had passed, I had made my contribution, and it was best for someone else with a different perspective – and a different set of strengths and weaknesses – to come in and run the next portion of the project. I am a big-picture person, and was fairly effective at getting the design right, selecting good people, winning the competition, and getting projects started and off to a good start, but others were probably better than I was as managers during the final stages of execution: implementation, integration, testing, and deployment.

As the manager of an engineering project, you will spend most of your time dealing with *people*; you will learn to be a leader, in all of the aspects discussed in Chapter 13. The skills that you learn for dealing with people will help in the rest of your life too, including knowing when to let others do the leading.

Engineering Project Management, First Edition. Neil G. Siegel.

Companion website: www.wiley.com/go/siegel/engineering_project_management

You learn about the vital importance of *understanding the value systems of other people*. As I discussed in Chapter 4, understanding the value system(s) of your customers and other stakeholders is necessary in order to be a good engineering project manager, but understanding the value systems of other people – your friends, your boss, your spouse – is also a useful skill in the rest of your life.

You learn how to identify and solve *technical* problems, but also how to identify and solve *social* problems.

You do not yourself perform most of the project management processes described in this book; instead, you *establish guidelines and principles for how they are to be done* (e.g. requiring the use of three-point estimates for schedule and cost, requiring the use of Monte Carlo simulations for schedule and cost, and so forth), you *ensure that they are actually performed*, and then you *monitor* the efficacy of that performance. For example, you don't yourself prepare the process-behavior chart and time-series analyses of the operational and technical performance measures, but you mandate that all quantifiable data used for decision-making on the entire project be analyzed as a time series via the process-behavior chart methodology.

In engineering project management, there are a *lot* of artifacts: the work-breakdown structure, the activity network, the statement of work, the requirements specifications, the design documents, etc. We *never* expect to make these artifacts perfect. At any point in time (and even at the end of the project), they are probably incomplete and contain errors. The real point is that they represent a *commitment and a process by which fact-based analyses can be investigated and fact-based decisions made*. The discussions regarding what the customer values, the discussions about risk and opportunities, the discussions about design alternatives, the meetings entailed in the periodic management rhythm – this is where the "magic" takes place that leads to successful projects, through the give-and-take among knowledgeable, motivated people. You continuously drive the team to look at the facts, identify their premises, discuss the candidate approaches and their implications, and drive toward decisions. You ought not be afraid to change your plans if the data indicate that is appropriate. You do all of this while also trying to operate in accordance with a long-range vision that represents a combination of the customer's value system, your value system, and your vision for an effective, feasible, and simple design.

Figure 15.1 depicts *Siegel's six focus areas for an engineering project manager*. This sounds difficult, and at times it is; especially trying always to remain calm. But I have found that this combination of good leadership, good engineering, and good personal behavior can position you to solve all of the problems that you will face as an engineering project manager.

In Chapter 13, I provided you with methods for resolving conflict between people. I did this because engineering projects are *full of conflict*: conflicting goals, goals not in line with resources, different stakeholders not agreeing, your customers and your management not agreeing, different members of your team not agreeing. You *cannot* avoid such conflict; many of these conflicts are inherent in the nature of an engineering project, and others arise from the fact that (as we said in Chapter 13) "people are a problem."

As the project manager, you must recognize when conflict is arising, and manage it. This is both really hard, and really important. It will help, while you are working on

1. Understand the customers (and other stakeholders), and how they determine value
2. Acquire, motivate, and align a team of suitable practitioners
3. This opens the door to the possibility of creating a suitable design, which is usually characterized by simplicity. If the design seems to be overly complex, it probably is, and should be radically simplified. One cannot simplify around the edges – one must attack the heart of the problem
4. There are usually overwhelming social obstacles on engineering projects: conflicting desires, conflict methods, conflicting personal aspirations. Your job is to listen to all of these stakeholders and participants, and to lead them to a consensus that achieves a reasonable balance between these conflicting points of view, but still makes execution of the project feasible
5. Always be calm, polite, respectful, and patient
6. Always act with flawless ethics

Figure 15.1 Siegel's six focus areas for an engineering project manager.

resolving conflict, to think about the characteristics that I described for effective leaders, and always try to remain calm, polite, respectful, and patient. Work hard to achieve consensus and resolution, but do not take the conflict personally. Do not expect that you will always be popular; in the role of project manager, it is more important to move the project toward success than to be liked. In the end, if you do your job well, you will be generally respected, which is something different from being popular or being liked. This comes with the job.

Do not expect to satisfy everyone; that is just not always possible. You have an obligation to keep the project moving on a sound and feasible path. If you have to make a unilateral decision that is unpopular with some people, so be it.

Even after you are promoted to be the manager of a large, complex engineering project, you still have things to learn. Every day will provide practice in your people skills; the sort of reading that I recommended in Chapter 13 (in the subsection entitled "Preparing yourself for leadership") is always helpful too.

You also need to make a modest effort to understand the new technical capabilities and approaches that your team is considering and/or using. You are not going to become the chief designer of your system, but you do need to understand the approaches, options, risks, and opportunities involved; you *cannot* be an effective project manager depending solely on the word even of a trusted technical subordinate.

Life will be busy during your tenure as the manager of an engineering project, but never dull.

15.2 Ethics in Engineering

I give pride of place – the very last topic in the book – to the subject of *ethics in engineering*.

In my view, the most important items in this book are the section on *design* within Chapter 2, the material about the *social considerations of engineering project management* which forms Chapter 13, and this section on *ethics in engineering*.

As the manager of an engineering project, I believe that you have *four primary constituencies* that you must satisfy (I presented a longer list in Chapter 11). These four primary constituencies are the following:

- Your customer
- Your management
- Your team, and
- Yourself.

We have talked a lot in this book about those first three constituencies: your customers, your management, and your team. You must not forget that fourth constituency: yourself! You satisfy yourself by doing the best you can, by flawless ethics, by continuous politeness and kindness to those around you, by maintaining a reasonable work–life balance, and by achieving an effective balance of listening and deciding. We have talked already about all of these, except for ethics. To close the book, I will now turn to that subject.

15.2.1 When Does Bad Engineering Become Bad Ethics?[1]

Many engineering projects create products and services that are important to society; many have explicit safety implications; some are distinguished by explicitly supporting national security. Failures and deficiencies that might be considered "routine" in some settings can in these cases directly and predictably cause injuries and lost lives, in addition to harming national security. In such a setting, decisions regarding quality, testing, reliability, and other "engineering" matters can become *ethical* decisions, where balancing cost and delivery schedule, for example, against marginal risks and qualities is not, by itself, a sufficient basis for a decision. When operating in the context of an engineering project with such important societal implications, established engineering processes must, in my view, be *supplemented with additional considerations and decision factors based in ethics.*

Proper analyses – as described in this book – would indicate when major problems are being overlooked in the specification and design of a system. When those analyses are *not* performed with sufficient detail and rigor, and therefore those problems remain overlooked, *that* is when bad engineering can lead to bad engineering ethics. We have reached a state of bad engineering ethics when, by being insufficiently thorough and rigorous, we allow our systems to endanger the safety or the health of our users, or the health and/or security of our society.

15.2.1.1 How Do Engineers Get Into Situations of Ethical Lapse?

No one enters a career in engineering intending to put lives and missions at risk through ethical lapses; at the very least, this is not the path to promotion and positive career recognition. So how could engineers find themselves in ethical quandaries?

As we noted in an earlier chapter, Taleb[2] writes about the tendency of humans to *underestimate the likelihood of low-probability events.* That is, if an event is reasonably rare, humans tend to act as if the probability actually approaches zero. Taleb even cites

1 A longer version of this section on ethics in engineering appears in the book *Next-Generation Ethics*, Cambridge University Press, edited by Ali Abbas.
2 In his book *Fooled by Randomness*, Texere, 2004.

sources that attribute this tendency to the deep operation of our brains, as developed through evolution. If this is true, it takes considerable active effort to overcome such a tendency.

The problem with this tendency is that rare events do occasionally occur! This is why we ought to buy fire insurance for our homes, and collision insurance for our cars.

In particular, in Chapter 9 I introduced two special type of risks, one of which was the *low-likelihood, high-impact event.* Those low-likelihood, high-impact events are potential sources of ethical lapse; if we assume that, since they are low likelihood, we need not take actions to mitigate their impact, we can have serious, adverse consequences. Recall that I therefore recommended that we do *not* use expected-value techniques based on estimating the likelihood of these events as a method to determine whether we ought to undertake efforts to mitigate the impact. Instead, I said that if we have identified a risk that will have a serious impact which might include harm to people or society, we *must mitigate that potential impact,* no matter how low we estimate the likelihood to be. Failing to do this may put you in a situation of ethical lapse.

15.2.1.2 Characteristics of Modern Engineered Systems that Create the Risk of Ethical Lapse

I believe it is the case that modern engineered systems exhibit *specific technical and social characteristics* that can lead to this specific type of ethical quandary. These characteristics are listed in Figure 15.2.

I have already discussed the human tendency to discount the likelihood of low-probability events. I will next discuss the four additional system characteristics that can trigger a transition from engineering risks to ethical risks.

15.2.1.3 Complexity and Scale Introduce Non-linearities

An important characteristic of modern systems is their complexity and scale. Complexity and scale introduce *non-linearities*[3] in system behavior, so that our intuition – which

- The human tendency to discount the likelihood of low-probability events to essentially zero probability
- The complexity and scale of modern systems
- Reliability and availability tend to be under-emphasized, as compared to functionality and capability
- We tend to accept operator-induced and user-induced failures as being outside our design responsibilities
- We ignore – or seriously under-emphasize – the potential for use of the system beyond the uses that were originally envisioned, and also do the same for potential uses beyond the originally specified conditions

Figure 15.2 The technical and social characteristics that lead to bad ethics in engineering.

3 That is, small changes in input can lead to more than a small change in output. At times, small changes in input can lead to gigantic changes in output. This is not exactly how the term *non-linear* is defined in mathematics, but I believe that this constitutes a useful definition within the context of engineering project management.

many people who study the human brain suspect operates by linear or proportional extrapolation – is no longer even approximately valid. This can cause lapses in consideration of failure modes (among many other system characteristics), for example – which can in turn manifest themselves as unsafe operation. Many systems in fact exhibit serious failure modes that derive entirely from scale, complexity, and the resulting errors in their dynamic behavior – that is, scale and complexity can be actual *sources* of failure. This, therefore, can become a path through which we can create an ethical lapse: we understand the scale and complexity of the system we are designing, but fail to account for the failure modes that such scale and complexity introduce themselves, above normal engineering considerations.

15.2.1.4 Reliability and Availability are Under-emphasized

Reliability and availability, when they are under-emphasized in favor of focusing on system capabilities and functionality, can also become a source of lapses in engineering ethics. There is a natural tendency to focus on functionality and the visible features and capabilities of our systems, and therefore to under-emphasize quality characteristics such as reliability and availability, which are in some sense "less visible" to the eye of the intended user than system capabilities and functions. This tendency is reinforced by our contracts and system specifications: it is often the case that as many as 99% of the requirements for a typical system describe *functions* and *capabilities*, whereas only a small portion of the requirements deal with reliability, availability, and other quality factors. But reliability and availability – and other quality factors too – are quite likely to be involved in safety and other important societal considerations, even though they did not require very many words in the specification to define their requirements. As a result, many system development efforts perform only rudimentary analysis during design of these quality parameters; the fault tree might only identify the most obvious types of faults, completely omitting many more subtle faults that have significant negative impact but are harder to see. The result is that the realized system reliability and other quality measures often fall far short of the levels predicted and/or required. Another result of this behavior is that even if the occurrence rate of faults is as predicted, the severity (e.g. the impact of those faults when they occur on system operational effectiveness) can be far higher than predicted. Having system reliability far lower than predicted, and/or the impact of system faults being more severe than predicted, can have serious safety and other consequences ... and therefore, once again, we have created an ethical lapse through incomplete or inadequate engineering.

15.2.1.5 Treating Operator-Induced Failures as Being Outside Our Design Responsibilities

Yet another characteristic of modern systems that can cause ethical lapses is that we tend to accept the idea that operator-induced failures are *outside* our design responsibilities. That is, if the user or operator of our system does something wrong (or even just something unexpected) and a problem results, we tend to say that it was *his or her* responsibility, rather than saying that *we the designers* should have foreseen the possibility of such a mistake, and made the system react in a safe and predictable fashion, even in the presence of such "wrong" inputs.

This is an ethical lapse because it is 100% certain that at some point in the life of a system, the users of our systems will "punch the wrong button," or create an input outside the nominal range, or provide some other "wrong" input or action. A robust design is *precisely* one that protects the system and its users against excessive adverse consequences from such an action.

Early versions of the DOS computer operating system, for example, did not even have a simple "Are you sure?" check when a command to erase a file (or even an entire directory of files) was entered. This clearly constituted negligent design (quickly corrected by Microsoft in later versions of DOS); queries such as "Are your sure" were already standard practice in many competing computer operating systems at the time.[4] "Negligence" evolves into an "ethical problem" when the consequences can harm people. For example, why should the microcontrollers of a centrifuge or a motor generator accept a command to spin faster than the device is designed physically to tolerate, when the consequence of such a command is that the centrifuge or generator may physically come apart, and people could be injured or killed by that action? Why should the computer built into these devices accept commands to operate outside of their known physical limits?

Those examples are *single-point* instances of "wrong" user inputs causing actual physical damage. Much more common, and much more subtle, is the creation of physical damage through a *combination* of commands – none of which may be intrinsically unreasonable – which only in *combination* cause physical damage (and thereby, maybe injuries or death to people). Think of the chain of valves and pumps (these days, all of which are capable of being controlled remotely by computer commands) that operate an oil refinery or a chemical plant. With heavy, hot fluids moving through pipelines, the commands to a succession of valves and pumps must be properly synchronized, else the momentum of the column of moving fluid can burst a pipe wall, or cause other damage. A command to an individual valve or pump may be within the range of operation for that single device, but in the context of the *total operating picture* of the facility, that same command may be a disaster. I assert that proper design and good engineering ethics require that we design our systems with the appropriate *dynamic* checks and balances to prevent command sequences that can combine to cause damage, and do so whether those commands are intentional or accidental. Very few of today's complex engineered systems meet such a standard. I believe that we have a responsibility to protect our systems even against such operator-induced failures, and therefore also against hacker-induced and bad-inside-actor-induced failures as well.

15.2.1.6 Ignoring the Potential That Our Systems Will Be Used in Ways Other Than We Intended

Another characteristic, somewhat related to the example immediately above, is that we often ignore – or seriously under-emphasize – the potential for using our systems in ways that we did not envision, or beyond the specified operating conditions. Many of us have used a screwdriver as a chisel, or vice-versa.

A more important example of this that everyone knows about is the Internet. The Internet was designed to share small bits of textual information between academic and scientific researchers; no other use was envisioned or specified at the time of its

4 Not wishing to appear to pick on Microsoft, I remind you that early versions of Apple's MacIntosh computer allowed the user to drag the icon for the hard drive into the trash can, and thereby inadvertently delete the entire contents of the computer, with no warning of the severity of the impending action.

creation. Certainly, the use of the Internet for safety-critical missions was *not* envisioned.[5] Yet today, countless safety-critical and society-critical missions are operated over the Internet. Yet the Internet has never been upgraded to allow for safe and reliable operation of these safety-critical and society-critical missions. Because of this unintended use of the Internet for safety-critical missions, many individuals, and society as a whole, are exposed to many large-scale safety issues.

15.2.2 Corrective Actions

Having illustrated the problem, I wish to present some ideas about how to avoid the problem.

First and foremost, in my view, is the matter of placing *proper emphasis on good design*. At present, I believe that it is fair to say that many engineering projects place their primary emphasis on *developing good requirements*, rather than on *developing a good design*. Many texts and corporate guidelines about performing systems engineering, for example, are heavily focused on the matter of requirements. Many academic papers that examine the question of problems in the system development process focus on requirements too, citing factors like incomplete requirements, or "requirements creep" (e.g. the problem that the requirements continue to change, even as the design and implementation progress), as the root cause of the large number of engineering projects that have significant problems. It is natural for the customers and eventual users of the system to focus on the requirements too; after all, the requirements are something they can understand, and are also something they have a natural reason for wishing to influence.

There is no doubt that such problems occur (e.g. engineering projects often end up costing significantly more than promised, usually accompanied by taking significantly longer than promised, failing to implement all of the promised capabilities); the data indicates that a large portion of engineering projects are terminated before their completion, due to these factors. But I spent several years of my engineering career as a sort of "designated engineering project fix-it person," and what I found was *not* that engineering projects which were in trouble had bad requirements. Instead, I found that they *all* had *bad designs*.

I have also seen cases of two completed systems that do approximately the same thing, where one runs 100 times faster than the other. Similarly, I have seen cases of two completed systems that do approximately the same thing, where one is 1000 time more reliable than the other. Having for these examples also had the opportunity to examine the root cause for the slower and less reliable performance, I can assert that the systems at the bad end of these examples had bad designs.

This finding has many interesting implications. First of all, having a 100× or 1000× range of outcomes for a critical parameter from an engineering project is shocking; mature engineering disciplines simply do not have such large ranges of outcomes. Consider mid-sized family sedans offered for sale that meet US emissions control requirements; the variation from best to worst, for example, in gas mileage is no more than 25%, not 100× (10 000%) or 1000× (100 000%). Something is going radically wrong

5 Dr. Len Kleinrock of UCLA, one of the originators of the Internet, said this to me explicitly.

inside the designs of the systems that exhibit such bad performance on such an important metric.

I drew upon my personal experience at fixing troubled engineering projects in my own PhD research. In my PhD dissertation, I developed and attempted to validate a hypothesis describing in exactly what way these designs are bad, and how they could have been improved.

Creating a good design for an engineered system is, in my experience, far more difficult than developing the requirements for that same system. We get a lot of "help" as we develop the requirements for an engineering system; after all, our customers and our users understand well *what* they want the new system to do, and such *what* constitutes a major portion of the requirements.

We do not usually have such a resource pool to help us with the design. The design is far subtler than the requirements; interactions between elements of the design are far more likely to have significant impacts on the system and its performance than are interactions between elements of the requirements, and the design is far more technical (and hence opaque to many observers). Those same customers who are able to help us with the requirements can seldom offer us significant help with the design.

Furthermore, most projects do not have reasonable technical metrics for measuring the progress of the design; they tend to use only *management* metrics for measuring progress on the design (e.g. we held these reviews, we produced these documents, etc.).

As a result of all these factors, weak designs are adopted by many engineering projects.

Engineered systems almost always aspire to create some sort of emergent behavior, a sort of "1 + 1 = 3," where useful things happen due to the interaction of formerly separate elements. But what often happens, while creating a design which leads to the desired emergent behaviors, is that the design fails to *prevent* the arising of unplanned emergent behaviors which appear as unintended adverse consequences.

As a result, unintended adverse emergent behavior often creeps into our systems through such incomplete designs. In my judgment and experience, this is the *true root cause* of most failures of engineering project developments (rather than requirements creep), and such incomplete design – that is, a design that does *not* incorporate features explicitly aimed at preventing such unintended emergent behavior (in Chapter 2 and in my other writings, I call this *unplanned dynamic behavior*) – is likely to exhibit the poor characteristics that I have described (low reliability, poor performance, etc.) and therefore, these are the projects most likely to fail and/or be canceled.

So, improving the design and the design process is "step 1," in my view, toward avoiding the problem of bad engineering transitioning into bad ethics.

After the design, the next most important corrective action, in my view, is the risk management process. Most big projects have some type of formal risk management process; the process itself is usually fairly rigorous. What is lacking, in my experience, is *content*. As I discussed in Chapter 9, I have found that the risks contained on the risk register of an engineering project are often completely superficial and general. I have actually seen the statement "The software might be late" as an entry on the risk register in a multi-billion-dollar engineering project. Such a statement is useless as a risk register entry; first of all, it is true on every project that has a material amount of software (as almost every engineering project does these days), therefore it is not specific in any fashion to this particular project. But far worse is the fact that it contains no insight about what the project should be measuring every month in order to determine if in fact the risk is coming to pass, or what steps should be taken to mitigate the impact if and when they determine, through those measurements, that the risk is coming to pass.

Therefore, the next corrective action is generating far more specific and far more measurable entries on the risk register, and then taking the corresponding actions in the rest of the risk management process (e.g. figuring out what to measure, figuring out how to make those measurements, creating mitigation plans if the risk materializes, and so forth).

Of course, doing this takes intense (and expensive) effort, expertise, and a lot of time. It also results in there being many more items on the risk register! These are probably some of the reasons that this task is not performed properly more often.

15.2.3 Conclusions About Ethics in Engineering

Proper analyses would indicate when major problems are being overlooked in the specification and design of a system. When those analyses are *not* performed with sufficient detail and rigor, and therefore those problems remain overlooked, *that* is when bad engineering can lead to bad engineering ethics. We have reached a state of bad engineering ethics when, by being insufficiently thorough and rigorous, we allow our systems to endanger the safety or the health of our users, or the security of our society.

I further believe that it is the case that modern systems exhibit *specific technical and social characteristics* that can lead to this specific type of ethical quandary.

Societal expectations for engineering are very high; whereas a baseball player has only to get a hit 30% of the time to be considered a major success, in contrast, society's expectations for engineered products and systems is near-100% availability and correctness. I believe that, in turn, this expectation grants us license as engineers to insist on proper designs, based on proper analyses, especially in safety-critical and mission-critical situations.

I also wish to plea for practicing engineers to believe that developing a personal reputation for thoroughness, diligence, and good engineering ethics is a boon, not a liability, to your individual career. I cannot "prove" this, but most of my experience over nearly 40 years as a practitioner supports that conclusion. One last little story. When I was the vice-president and general manager of an operating division at a large aerospace company, we elected to bid on a competition for a new type of system for the United States Marine Corps. The system specification had a "hard limit" of 11 000 lb for the complete system, because this was the weight capacity of the specific vehicle chassis that we were required by the Marines to use in building the system. When it came time to submit our bid, my proposal manager pointed out that when they added up all of the components of our proposed design (which we thought was wonderful, and would offer the Marines a lot of operational advantages), we were slightly over that weight limit; as I recall, 11 045 lb. My proposal manager asked what we should do; the implication was for me to choose between fudging the analysis to make it say "10 999" lb, or submitting the proposal as was (that is, over the specified weight limit) and assuming that the Marines would appreciate the honesty, understand that a thousand design decisions remained between the proposal and the fielded system, and that there was plenty of time to solve the weight problem before we were done. I said to submit it as was. And we won – despite being overweight. Many, many years later that proposal manager[6] came to me and told me how much he and the proposal team had appreciated my having taken the ethical approach in that situation.

15.3 Thank You

Thank you for reading my textbook. I hope that you have found it interesting and worthwhile.

6 Dr. Phil Allen.

Index

Engineering Project Management, First Edition. Neil G. Siegel.

Companion website: www.wiley.com/go/siegel/engineering_project_management

d

e